Paleontological Institute of the Russian Academy of Sciences
The H. Rausing Foundation

Valentin A. Krassilov

ANGIOSPERM ORIGINS:
MORPHOLOGICAL AND ECOLOGICAL ASPECTS

PENSOFT
1997

PENSOFT

A SOFIA & MOSCOW-BASED PUBLISHING HOUSE IN ALL FIELDS OF LIFE & ENVIRONMENTAL SCIENCES

Acad. G. Bonchev Str., Bl. 6
1113 Sofia, Bulgaria
Tel./Fax +359-2-7133460
E-mail: pensoft@main.infotel.bg

ISBN 954-642-016-6
First published 1997

Editors: *Dr. S. Golovatch & Dr. L. Penev*
Graphic design: *M. Veleva & D. Velev*
Cover design: *Dr. D. Dobrev*
Photograph on the cover by V. A. Krassilov: preflower of *Eoantha zherikhinii* Krassilov, 1996

Printed by the Printing-office of "Prof. Marin Drinov" Academic Publishing House
Sofia, May 1997

Contents

Introduction

It has become traditional to open a discussion of angiosperm origins by quoting Charles Darwin who described the problem as an abominable mystery. Usually this is followed by an assertion that, although much has been done since Darwin worried himself with this particular problem, the mystery still remains. But what is so mysterious and so abominable about angiosperm origins that, after 140 years of research and profuse publication, there still is no light in the end of the tunnel?

Certainly, no historical event is fully understood. All the origins are subjects of at least some controversy. Yet this is normal, not abominable. What Darwin seems to have had in mind was that the advent of angiosperms disagreed with the theory of gradual evolution by means of natural selection. This might have been the case, so what? Is it the theory or the angiosperms to be blamed? As far as the theory is concerned, the abominable mystery may refer to either (1) the angiosperms having no evolutionary links with any other plant group among which their ancestry could potentially be found or (2) the appearance of angiosperms in the fossil record being sudden rather than gradual. These are the assertions directing relevant research to either a reconciliation of angiosperm origins with the Darwinian theory of evolution or a falsification of the latter.

In practice, however, like any puzzle, the mystery of angiosperm origins defers clarification, for people are psychologically unprepared to part with it. A few would sacrifice the Darwinian theory of evolution to unravelling the actual history of angiosperms. Even fewer would give up a notion of angiosperms as a fairy separate type of plants defined by a set of unique characters. New evidence casting any, if but indirect, light on the origin of angiosperms does not make the professional people happy. They tend to suspect something wrong about the evidence as an excuse for ignoring it.

I have already written at some length about the factual basis of angiosperm origins (Krassilov, 1989). Here I bring in some fresh evidence. However, this book deals also with intellectual aspects of the problem. Fortunately, several comprehensive reviews published in the recent years (e.g. Hughes, 1994) allow me to bypass some traditional themata of the story.

Making of the type

Although at present the designations "flowering plants" and "angiosperms" are used as synonyms, historically their references are somewhat different. In the pre-Linnaean and Linnaean times, all terrestrial plants were considered as flowering, at least potentially. "Angiosperms" and "gymnosperms" first appeared in the writings of "fructists", the XVII century system-makers who relied primarily on fruit characters. Paul Hermann among them has distinguished (in "Horti academici Lugduni Batavi Catalogus....", 1687, and other works) between the angiosperms with a conspicuous pericarp and the gymnosperms without it. The latter included some members of Ranunculaceae, Umbelliferae, Compositae, etc. (see Linnaei, 1751). In the XIX century, the angiospermy versus gymnospermy concept approached the present-day understanding (De Candolle, 1827), whereas "flowering plants" became synonymous to phanerogamous plants as opposed to cryptogamous, or spore, plants, the latter having no flowers. Cones of conifers were considered as a special type of inflorescence not unlike the catkins of alders or spikes of grasses. However, because angiospermous plants are far more diverse than gymnospermous ones, the idea of a true flowering plant became associated with angiospermy, while gymnospermy seemed an aberrant condition associated with aberrant flowers that perhaps were not true flowers after all.

As far as the notions both of flower and angiospermy are concerned, the formal distinctive characters of the angiosperms were essentially ecological, and the angiosperm concept was primarily of a life-form type. Its subsequent transection into the realm of morphology was not substantiated by explicit reformulation of the basic notions and, for many, it passed unnoticed, leaving a lot of residual definition problems. Far ahead of a plausible interpretation of coniferous seed-scale complexes as condensed bracteate shoots (Florin, 1938-1945), plant morphologists set their minds on the problem of gymnospermous flower: whatever gymnosperms might have was not a flower. Thus the flower has become a charismatic feature of angiosperms, a possession that makes them the flowering plants in the modern sense.

The subordinate and subsidiary characters were treated in much the same way. Definition of the flower as an organ consisting of stamens and/or carpels (with the perianth as a dispensable component) was not meant to discriminate between angiosperms and gymnosperms. Flower could be used discriminatingly only by asserting an exclusive right of the angiosperms to stamens and carpels. A distressing thought of gymnosperms also having clusters of stamen-like and carpel-like structures, the latter in some cases actually enclosing the ovules, was resolutely rejected: whatever a gymnosperm might have as its pollen sac-bearing or ovule-bearing organ was neither stamen nor carpel. A total lack of homology was claimed for the vessels of angiosperms and those of gnetaleans or other non-angiosperm plant groups (e.g. Takhtajan, 1966), the double fertilization in angiosperms and the similar phenomena in *Ephedra* and conifers, etc. Creation of the type has thus become completed, with the ensuing abominable mystery of its origin further sustained by seclusive morphological interpretations.

Angiosperms and Magnoliopsida

Presently there is a strong tendency among system-makers to rename traditional higher taxa, such as angiosperms, gymnosperms, conifers, etc. on the basis of typification with arbitrarily chosen genera serving as types not only for their families, which is a conventional practice, but also for orders, classes, divisions, etc. Thus, there would be no Angiophyta, Angiospermae, Dicotyledones, but Magnoliophyta, Magnoliatae, Magnoliopsida (and the same with the Pinophyta, Pinatae, Pinopsida, etc.). Of course it would be difficult to abandon habitual names used, with no appreciable harm, for over 200 years. In fact, the modern system-makers usually use Magnoliopsida and the like only once, as a title, then immediately switching to dicotyledons (e.g. Cronquist, 1981). They are, however, fully prepared to grapple with this kind of ambiguity for the sake of stability and uniformity of the nomenclature.

But is the nomenclature of higher taxa getting more stable with the advent of Magnoliopsida? Suppose *Magnolia* will be found synonymous to an earlier valid name. Another conservation petition would become mandatory (and the conservation list is too long already) lest all the names from the Magnoliaceae upwards should be changed automatically. Suppose, too, that a nonconformist would separate *Magnolia* and its closest allies from the main bulk of angiosperms. He/she would then be obliged to introduce a new name, e.g. Rosopsida, for the latter, with all hierarchical steps appropriately renamed. All this would hardly add to nomenclatural stability.

As regards uniformity, there are two aspects of the problem, one dealing with nomenclature, the other with classification. For some psychologically not quite obvious reasons, those who approve of Magnoliopsida would also elevate the ranks of higher taxa, e.g. allotting the orders the status of classes, turning classes into divisions, etc. While higher taxa are often thought of as arbitrary, their rank might seem a matter of practical convenience alone. If so, then it will be far more convenient to leave them as they have been for centuries rather than to endeavour all the cumbersome shifts and emendations.

Classes as adaptive types

As there are repeated attempts at restructuring the system of higher plant taxa, there should be some reasons other than convenience. These may be uniformity and objectivity of a higher order classification achievable on the basis of some general principles.

Goethe (1790), borrowing from Spinosa (who, in turn, borrowed from Plato), sought to represent, in an essentialist's spirit, the diversity of plants growing on earth as variations of a single theme, the Urpflanze. To him the Urpflanze was all leaves implying that all the diverse plant structures arose as various leaf modifications. These ideas inspired De Candolle (1827) and were used as the foundation of plant systematics, the major divisions of the plant kingdom being conceived of as consisting of species created upon a common Bauplan. They survived the period of unbounded dominance of the existential evolutionary paradigm (e.g. Stebbins, 1974) and have become, perhaps unawares, revived by modern cladistics to which a taxon is valid as far as it is defined by uniquely derived characters of a common ancestor stamped upon all its progeny.

Classification units are categories that are either introduced for convenience or are contrived to reflect the structure of natural diversity. Natural biological classifications convey our current understanding of biological diversity. Although "natural" is often used evaluatively as equivalent to rational or objective, it may signify neither of these. All that "natural" implies in the case of a biological classification, is that biological diversity is structured in one way or another and that our intention is to arrange our material accordingly. There can be several ways of doing so, hence more than one classification is natural, though not necessarily rational or objective.

The units of a natural classification, whatever the basic criteria may be, are intended to approach the natural entities as closely as our current understanding of the latter permits. Guided by this general principle, we can focus then on different types of natural entities. The more easily discernible among them are those bound by ongoing interactions, such as the reproductive interaction in populations or the trophic interactions in biotic communities. With the advent of the evolutionary concept, we have become aware of the entities bound not so by the current as by the historical relations. These are of two major types, the phylogenetic, or genealogical, and the adaptive, or ecological. Inasmuch as phylogeny is directed by adaptation, they can, to a certain degree, coincide, but their obligatorily one to one correspondence is hardly feasible.

Since an adaptive relatedness is usually more obvious than a genealogical one, adaptive types have been recognized by taxonomists more readily than phyla. From the time of Aristotle and Hipocrates, biological classifications have been built primarily upon adaptive types. What Linné called natural classification (in distinction of what he himself worked out as an admittedly artificial classification) was a system of the most clear-cut adaptive types. He recognized their merits as higher taxa, but found it convenient to complement them with a parallel system of conventional taxa based on a few arbitrary characters. Old time system-makers, notably Andrea Caesalpino (in De plantis libri, 1583) and occasionally even modern taxonomists (Hutchinson, 1969) have used arboreal versus herbaceous life-forms as distinctive characters for classes of a natural botanical classification.

Generally, life-forms reflect coenotic roles of botanical species in plant communities - their ecological niches. Most botanical species are distinct life-forms, although some of them are represented by more than one life-form. As for the higher taxa, a one to one correspondence to the life-form categories, or adaptive types, is even more problematic. Nevertheless, grasses, for example, constitute a well-defined taxonomic order as well as an adaptive type. Although the arboreal versus herbaceous habit is no longer used as the sole basis of a classification, either one or the other prevails in a number of supraspecies taxonomic categories. Therefore the characters traditionally used in higher rank classifications are not entirely independent of the life-form.

Traditionally, reproductive characters are given priority over vegetative characters. Because, in plants at least, reproductive adaptations are much more sophisticated than trophic adaptations, the reproductive structures are, as a rule, more diverse morphologically than vegetative structures. Otherwise there will be no obvious advantage of one set of characters over another. Both the foliar and the floral characters can be obviously adaptive or (e.g. in the case of pinnate versus palmate leaf venation or axile versus parietal placentation) apparently non-adaptive. In the latter case they are ascribed to common ancestry rather than a shared adaptation. As for angiosperms as a whole, their typical reproductive traits, flower, angiospermy and even double fertilization, are certainly adaptive which makes them the most inclusive adaptive type.

Classes as phyla

It seemed that evolutionary approach would dispense both with archetypes and adaptive types, but in practice this did not happen. The changes the evolutionists introduced into the previously existing classifications were not so impressive. The adaptive type-based natural units were, with a few exceptions, reinterpreted as phyla, and even many units of artificial classifications survived on the assumption that Linné had intuitively used non-adaptive characters evidencing nothing else but phylogenetic relations. It was due to the latter coincidence that some XIX century taxonomists thought of phylogenetic classification as just another attempt at artificial classification. Only later did we come to think of phylogenetic classifications as the only natural ones.

But what could be expected of the so-called phylogenetic classifications entirely lacking any factual evidence of phylogeny as a historical process? Paleobotanical evidence was meagre and seemed unconvincing. It even seemed to be of no use, for the problem of ancestry was assigned to the area of pure logic which did not tolerate any intrusion of historical accidents. Logically, it seemed quite plausible that characters of no obvious adaptive value might have been passively inherited and could still be traced back to a no longer surviving ancestor. That this imaginary ancestor came to suspiciously resemble the Urpflanze was taken as another evidence of our great predecessor's wonderful intuition.

Only in the mid-1960's, with the development of computer techniques, was the integrity of traditional taxonomic systems put to a serious test. It was reasoned that the more characters are shared by a pair of taxonomic units comparative with other pairs the more genes they share and therefore the closer their phylogenetic relatedness. No criteria for selecting the characters were introduced in the counting procedure - the more the better, with a perspective of biological classifications turning into a branch of applied mathematics. However, the premises of the numerical classification seemed convincing to mathematicians alone, for in modern biology the idea of one to one correspondence of genetic and morphological similarities was no longer acceptable.

A new ideology was badly needed to make any use of numerical techniques. And eventually it was found in a theretofore obscure entomological classification (Hennig, 1950) that became known as cladistics, a name first coined by its opponents but later accepted by its proponents as well. It was proclaimed as a methodology of genuine phylogenetic systematics going to replace a false methodology of traditional evolutionary systematics. It was thus important to underscore the fundamental difference between them.

In the late 1970's, "Systematic Zoology" sounded more like a periodical in philosophy of science rather than of biology. The founders of a new paradigm of biological classification launched an attack on traditionalists using the philosophical argumentation borrowed from the then popular epistemologist Karl Popper. One familiar with Popperian philosophy of science might wonder in which way the latter might relate to biological classification except for the assertion of a falsifiable hypothetico-deductive scheme against the alleged inductiveness of traditional systematics. Actually, falsification is no invention of Popperian philosophy but it rather a trivial element of propositional logic an elementary course of which would greatly benefit natural scientists. Popper (1972 and elsewhere) used falsification as a criterion of scientific universal statements that cannot be verified for all imaginable worlds. Insofar as biological classifications deal with restricted rather than universal statements, the Popperian theory holds them as both falsifiable and verifiable.

Leaving philosophy of science in peace, one finds a real difference in that traditional evolutionary approach seeks to establish a classification on the basis of phylogeny, while cladists deduce phylogeny from classification, to a certain extent emancipating the latter from historical evidence. While the evolutionists look up to paleontology for phylogenetic evidence, the cladists find no use of paleontological data apart from calibrating cladogenesis (see Smith, 1994). Another distinction is that evolutionists rely on passively inherited non-adaptive characters the sharing of which (symplesiomorphies) indicates common ancestry, whereas the cladists prefer the shared derived characters (synapomorphies) as indicating phylogenetic proximity. It has passed unnoticed that, inasmuch as newly derived features are obviously more often adaptive than residual archaic characters, the cladistic classifications are unawares approaching a system of adaptive types.

Natural classification

Despite the still widely held preferences, there are no purely scientific reasons for giving phyla priority over adaptive types as the basis of a natural classification. Such preferences may better be explained by non-scientific reasons. The public interest in genealogies came from the Old Testament tradition and prevailed in stratified societies. Recently, the interest has shifted to adaptational aspects of natural and human histories brought forth by the growing environmental concern.

The structure of a natural system relates to both its functioning and its history. The structural characters that are not ascribable to the functioning can be ascribed to some historical accidents. But the exclusive use of the latter in natural classification is hardly justifiable.

Julian Huxley (1942), who coined the terms "clades" (phyla) and "grades" ("horizontal" taxa sharing an adaptive zone), claimed that both were "facts of nature" deserving special places in a natural classification. A trivial argument in favour of a phylogenetic approach is that all descendents of a single ancestor, however divergent ecologically, share common genes. But actually they might lose most of the common genes in the course of divergence, while acquiring new ones by mutation and horizontal transfers. On the other hand, the not so closely related but adaptively convergent lineages might increase their genetic similarity by parallel mutations and systemic transformations of their genomes. Thus the genetic similarity and the genetic relatedness are different things.

In all higher organisms, morphology is primarily a product of epistatic interactions and is a poor guide to the genomic structures that can be better studied at the molecular level. But even such conservative compounds as cytochromes seem to have preserved not all of the structural evidence of their evolution - the latter might have been less straightforward than it would seem. *Stellaria*, for example, might have received its cytochromes C by a horizontal transfer from a bacterial symbiont (Sullivan, 1996) rather than by a mutation event at the base of the caryophyllid clade.

But even with a more precise guide from structures to histories, translation of the latter into a classification would still be a problem. In the retrospect, not all historical events are equally significant. For a natural classification to tell us something essential about nature, we have to chose between the more important and less important historical facts as conveyed by taxonomic characters.

Selection of diagnostic characters is a crucial problem for any taxonomic methodology. It may primarily depend on our intention to obtain a practically useful classification system and/or to convey some natural phenomena, such as phylogeny, evolution or the standing diversity. Objectivity of a classification is not a thing in itself; it relates to the intentions of the system-makers. The cladistic techniques are commonly considered as more objective than the "intuitive" evolutionary approach. Using parsimony, taxonomic congruence and consensus analyses, one can construe a total evidence phylogeny from a number of fundamental trees incorporating alternative views on the uniquely derived characters versus homoplasies, the latter's contribution being assessed by the retention index (Farris, 1989; Vane-Write & al., 1992). However the use or abuse of the cladistic techniques depends upon what we are striving at. These techniques provide for rationality and reproducibility of a classification, yet this qualities do not make it natural.

Generally, biological classifications use both natural criteria, pertaining to their objects, the classified organisms, and artificial criteria, pertaining to their subjects, the system-makers. A division of the plant kingdom into poisonous and harmless species reflects upon human users rather than plants, hence primarily artificial. Parsimony or consensus may be attractive to human intellectuals, but both are virtually meaningless in respect to phylogenetic processes in nature unless we, like the XVII century typologists, believe in the parsimony of providence.

A modern classification strategy presupposes an analysis of large sets of characters for which various methods of assessment are available in the literature, notably in relation to their evolutionary conservatism, unique derivation or the likelihood of a parallel development. A single character approach, if there is such nowadays, would seem archaically Linnaean, and even Linné himself recognized it as artificial. However, all modern classifications have the Linnaean artificial classification as their starting point, and even the end-products may not be entirely different. Up to now, nobody has attempted to construe an entirely new multicharacter system of higher taxa without preconceived arrangements. What we actually do with large sets of characters appears rather as an evaluation of the pre-existing single-character arrangements.

In other words, we are not making falsifiably parsimonious classifications out of the primaeval chaos. Instead we start with a single (few) character classification trying to assess and perhaps falsify it by adding new characters and/or weighting procedures. Incidentally, the concept of flowering plants has been put to an integrity test by introducing such initially even not suspected of character as the double fertilization. The flowering plants seem, by the data presently available, to come through the test, thereby attesting to the flower as a well chosen single diagnostic feature. Currently, the double fertilization may even seem a more objectively definable single character of the angiosperms than the flower, although both are related to developmental condensation (of sporophytic and gametophytic developments respectively), their consistency in the taxon thereby depending on the consistency of the latter process.

The use of numerous instead of a few characters is, in plants at least, scarcely advantageous in itself, for the amount of ambiguous phylogenetic evidence tends to increase with the number of characters. If we include in the analysis such characters as stipules and foliar nectaries, for example, we have to bear in mind that in some, but not all, angiosperm species foliar nectaries are modified stipules. Otherwise pluses or minuses in the respective columns of the character matrix would be misleading. Therefore, what matters is the quality of phylogenetic evidence rather than numerical aspects.

In a numerical classification, it is important to differentiate between the functionally related and functionally independent characters. If the characters are functionally related, as are, for example, the enclosed ovules and siphonogamic pollen tubes, they tend to occur together in all members of the adaptive type whatever their genetic affinities. For the latter, sharing of the characters taken from different functional syndromes may be more indicative. One can assume that a coinciding phylogenetic and adaptive relatedness would give a taxon maximal integrity, while their discrepancy would result in a more loosely bound taxon.

The taxonomic practice has convincingly showed that a single well-chosen character may surpass dozen characters picked up at random. Angiospermy is one such character. Nominally, it relates to the condition of ovules, but actually it lies at the core of a functional syndrome including stigmas, extraovular germination of pollen grains, recognition substances in the pollen wall, female gametophyte reduction, sperm competition, etc. Stability of angiosperms as a taxon is further warranted by the combination of the angiospermy syndrome with the flowering syndrome as a functional integration of reproductive structures. However, these two syndromes may be functionally related, thus being not entirely convincing as evidence of monophyly. But even if they denote a grade, then this grade is so important as a fact of nature that we can hardly afford to ignore it.

Maybe it is for the abovementioned reasons that the achievements of botanical cladistics are not too spectacular. To its credit, cladistic analysis of angiosperm origins claims the demonstration of a close morphological proximity of gnetaleans to angiosperms (Donoghue & Doyle, 1989). This is a robust result scarcely depending on the technique. Actually the same was repeatedly stated in botanical text-books during the last century at least. Whether to accept or reject gnetaleans as possible angiosperm ancestors is a different matter, and the cladistic analysis has no bearing on it.

Evolving characters

Historical relations are deducible, in the first place, from chronological relations and morphological continuity. In the classifications conveying dynamic processes, the evolving characters showing certain trends in their historic development may seem preferable to static characters. The palaeontological record suggests that all characters stop to evolve sooner or later, thus having a restricted period of utility for classifications of this kind. A general trend in character evolution can be objectively defined by the relative frequencies of advanced to conservative states through time. For instance, through the Paleozoic, the share of the pollen grains with distal apertures increased relative to those with proximal apertures. It does not matter whether the trend was linear or fluctuating, or whether it occurred within a single lineage or in a number of parallel lineages.

In plants, there are at least three sets of characters, i.e. gametophytic, sporophytic vegetative, and sporophytic reproductive, the evolutionary rates of which are but loosely correlated and which, therefore, could be used alternatively in evolutionary classifications.

Gametophytic characters

The higher plants show four major options of the gametophyte to sporophyte relationships: (1) the gametophytes are free-living, while the sporophytes are epigametophytal throughout their life, as in bryophytes; (2) both gametophytes and sporophytes are free-living at some stage of their development, as in homosporous pteridophytes; (3) the female gametophytes are endosporous, nutritionally dependent on the spore, as in heterosporous pteridophytes; and (4) the female gametophytes are endosporangial, nutritionally dependent on the sporophyte, as in seed plants. While the distinctions of (1) from (2) and of (2) - (3) from (4) are traditionally allotted the highest taxonomic rank, those between (2) and (3) ascend no higher than the ordinal rank, although from the point of gametophyte evolution they seem no less important. This situation can be explained by the fact that heterospory, evolving independently in a number of free-sporing plant lineages (thus being a typical grade feature), scarcely affects major sporophytic characters, while the endosporangial development, though perhaps also of polyphyletic origin, involves not only the sporangial, but also the extrasporangial sporophytic tissues in the nutrition and protection of the developing sporophyte and the ensuing embryo. The care of the latter is consequently transferred from the gametophyte to the sporophyte. This results in a radical restructuring of sporophyte morphology. Since the evolutionary strategies (3) and (4) are entirely different, the seed plants might have never had any free-sporing heterosporous pteridophytes in their ancestry (Chaloner & Hemsley, 1991).

A phylogenetic succession of the four stages of gametophyte evolution might have not been linear. Recent discoveries of the massive, fairly persistent gametophytes in early land plants (Remy & Remy, 1980) seem to be in favour of isomorphic alteration as a primary mode, at the same time suggesting a derivation of the modes (1) and (3) from (2), the more so that some early land plant gametophytes, e.g. *Sciadophyton* (Remy & al., 1980), resemble thalloid hepatics. The early gametophytic plants might have been variably vascularized and even the recent *Psilotum* still retains a genetic potential for developing vascular strands in the gametophytes of polyploid varieties (Bierhorst, 1953).

Thus a neotenic gametophyte reduction is traceable from the rather massive gametophytes of early land plants through the ephemeral free-living gametophytes of geologically younger spore plants and the gymnosperm female gametophytes confined to the ovules but still developing archegonia, to the embryo sacs of angiosperms lacking structurally definable sex organs. In the course of these events, the gametophytes became more and more dissimilar to the sporophytes of the same plants. At the genetic level, this phenomenon is related to the gene compensation effect, i.e. the ability to compensate a haploid gene dose by the doubled transcriptional activity (well-known in respect to the sex-linked genes in animals with heteromorphic sex chromosomes). It is conceivable that, in the course of neotenic gametophyte evolution, the compensatory effect was lost in an increasingly larger proportion of gametophytic genes, resulting in a time-series of progressively deviating gametophyte to sporophyte morphologies.

This is perhaps the only character trend sustained throughout the history of higher plants, thus being of crucial importance to evolutionary classification. Traditionally, the gametophytic features are given a very high taxonomic rank, though perhaps in a somewhat oblique way. Thus, early land plants, or protopteridophytes, might have differed from the subsequent spore plants, or pteridophytes, primarily in their relatively more strongly developed gametophytes. Seed plants differ from spore plants primarily in the possession of ovules that are essentially the sporophytic food-storing structures for female gametophytes and the embryos. Furthermore, the double fertilization, a major distinction of angiosperms (but actually occurring in gnetophytes as well, see below), is related to their archegonia-less embryo sacs.

Thus, as far as gametophytic characters are concerned, the boundary between spore plants and seed plants is fairly sharp, perhaps the sharpest of all the higher plant divisions, with virtually no transitional forms. On the other hand, the gymnosperm/angiosperm boundary is comparatively less significant and, therefore, is of a lower taxonomic rank. This situation is further affirmed by sporophytic characters.

Vegetative sporophytic characters

Insofar as the stem and leaf characters are the most conspicuous in the higher plant sporophytes, it is tempting to use them as the first order distinctions. However, this strategy has been successful in the spore plant high-rank classification alone, distinguishing between the typically unbranched macrophyllous ferns, the helically branched microphyllous lycopods, and the branched verticillate microphyllous sphenopsids. These groups are rooted in different early land plants showing respective distinctions in their incipient form. Macrophylls and microphylls might have arisen from branching systems of different - penultimate and ultimate - orders respectively (rather than having had entirely different origins, as postulated by some plant morphologists, e.g. Takhtajan, 1970). Nonetheless, the distinctions in branching and leaf size, as well as a correlation between the two, remained fairly consistent throughout the history of spore plants.

The macrophyllous unbranched or sparsely branched and the microphyllous copiously branched lines occur also in seed plants, i.e. the cycad and conifer lineages in gymnosperms or the monocot and dicot lineages in angiosperms. The distinction is, however, far less consistent than in spore plants, precluding a higher rank taxonomic division on the basis of vegetational characters alone.

In seed plants, the decussate leaf arrangement is usually considered as advanced relative to the helical one. However, phyllotaxis is sometimes verticillate or decussate in seedlings coming up helical in the adult plants. For instance, two closely related genera of taxodiaceous conifers, *Metasequoia* and *Sequoia*, differ primarily in their phyllotaxis being decussate and helical respectively, with the former acquired by *Metasequoia* perhaps as a pedomorphic feature.

The seed plant microphylls are primarily of a phyllodic origin (corresponding to petioles of macrophylls), representing a trend in leaf evolution involving different lineages of both gymnosperms and angiosperms. This trend is, however, poorly, if at all, correlated with other vegetational trends.

It appears that, in seed plants, the genetic constraints of vegetational characters are more flexible, resulting in their lesser taxonomic significance than in spore plants. The demarcation is again more sharp between pteridophytes and gymnosperms than between the latter and angiosperms.

Most monocots are herbs or pachycaul trees with polyarch roots and phyllodic leaves, but there is a broad overlap with the dicots of similar life-forms. Among gymnosperms, a similar distinction separates cycads and their allies from conifers and their allies, the distinctions also being rooted in the different life-forms. As far as vegetational characters are concerned, the dicot/monocot division is of a lower rank than the angiosperm/gymnosperm division and is of the same rank as the cycad/conifer division in gymnosperms.

Reproductive sporophytic characters

As their names imply, gymnosperms and angiosperms differ in sporophytic reproductive features related to their different ovule protection and pollination modes. The distinctions stood fairly well until the discovery of various extinct groups, such as the Paleozoic pteridosperms or the Mesozoic caytonialeans and czekanowskialeans with enclosed ovules and, in the latter group, with stigmatic exostomes (Krassilov, 1969), making the reproductive demarcation of gymnosperms and angiosperms far less distinct and much of the nature of their gametophytic demarcation discussed above.

The most conspicuous structure used in the seed plant classification is aggregation of reproductive units, including both strobili and flowers. In gymnosperms, the ovules are borne (1) in branching systems, or stachyospermic (Lam, 1950), as in early seed plants; (2) on solitary laminar ovuliphores, or phyllospermic, as in pteridosperms; (3) in coaxially aggregated ovuliphores, or strobili, as in cycads; and (4) in coaxial aggregations of strobili, or doubly-strobilated, as in conifers. Planation of ovuliphores as well as strobilation and double strobilation are the trends sustained by parallel lineages of gymnospermous plants, with occasional reverse trends, such as destrobilation in *Cycas, Ginkgo* or *Taxus*.

In angiosperms, the aggregation modes have also been used as an important distinction between the ranalean and amentiferous taxa and is still evident in the separation of magnoliids and hamamelids. Furthermore, the axial versus parietal placentation in angiosperms is analogous to the phyllospermic and stachyospermic modes in gymnosperms, though no phylogenetic continuity is implied.

Embryonic characters

The highest-rank division of angiosperms is based on a cotyledon condition that is of no utility in the systematics of gymnosperms. This situation may reflect a correlation between directional evolution and taxonomic significance. There is no definite cotyledon trend in gymnosperms, whereas in angiosperms a certain derivative asymmetry may exist between the monocotyledonous and dicotyledonous conditions. A single cotyledon may be terminal, as in *Lemna*, or pseudoterminal, as in the Cyperaceae, Potamogetonaceae, Hydrocharitaceae and, among dicotyledons, *Claytonia* (Portulacaceae). Furthermore, monocotyly may result from reduction, as in palms (Guignard, 1984), or from fusion, as in the "dicotyledonous" Nymphaeaceae. The multiple ways to monocotyly suggest a trend, perhaps sustained by parallel lineages of angiosperms.

However, some "monocotyledonous" embryos seem to have no cotyledon homologues whatsoever, while for the graminoid embryo there is no consensual scheme of homology at all. A consistent application of embryonic characters would give the grasses a taxonomic rank at the level of dicotyledons and monocotyledons. But, until a definite trend is discerned, the embryonic characters will be used pragmatically and, therefore, inconsistently at different levels of the taxonomic hierarchy (as, for instance, the supposed "helobial" endosperm development in the Lemnaceae, see Maheshwari, 1958).

Taxonomic rank intuition

The above discussion may suggest that we have to investigate taxonomic intuition before abandoning it. Intuition relates to the sense of ambiguity that appears rather late in the human cognitive development. Until this, cognition is irreversible and straightforward, tending to substitute all irregular forms, such as went, slept, feet, teeth, etc., by the regular, although intuitively ungrammatical, forms, such as goed, sleeped, foots, tooths, etc. (Cairns & Cairns, 1976) These childish disambiguating attempts resemble the ongoing correction of "intuitive" biological classifications on the principle of objectivity and phylogenetic parsimony.

The examples discussed above seem to suggest that the higher taxonomic ranks are not arbitrary but, instead, they reflect our perception of directedness, continuity and consistency. A character acquires a higher taxonomic rank when there is a definite trend in its evolution which is not totally counterbalanced by the reverse trends. Conceivably, the more lineages partake in the trend the more consistent the character and the higher its taxonomic status. Incidentally, phyllotaxis illustrates a character the directional evolution of which was exhausted rather early, at the spore plant/seed plant transition, henceforth of a minor taxonomic significance. In contrast, monocotyly reflects a trend that appeared rather late, at the gymnosperm/angiosperm transition, henceforth acquiring a taxonomic significance the embryonic features never had before.

In addition, the evolution of a high-rank character has to be consequential for other characters inducing them to evolve. Heterospory in free-sporing plants is an example of a character showing a definite trend of only little consequence for other characters and thereby of a rather modest taxonomic rank (as compared with the "consequential" shift to endosporangial gametophyte nutrition in seed plants).

In a few characters, we are able to perceive a hierarchy of gradational to abrupt changes as well as the degrees of consistency. In both gametophytic and sporophytic character sets, the spore plant/seed plant discontinuity is far more prominent than the gymnosperm/angiosperm discontinuity, whereas the cycad/conifer and the monocot/dicot discontinuities are of the same rank and are both subordinate to the gymnosperm/angiosperm one. Such considerations, partly intuitive, underlie the traditional system of higher taxa that is less arbitrary than it may seem. The above arguments are in favour of a conservative system of the higher taxa:

Divisions	**Classes**	**Subclasses**
Pteridophyta	Protopteriae	
	Filicinae	
	Lycopodiae	
	Articulatae	
Spermophyta	Gymnospermae	Protospermidae
		Pteridospermidae
		Cycadidae
		Coniferidae
	Angiospermae	Dicotylidae
		Monocotylidae

Prehistory

Angiosperm prehistory starts with the appearance of the early Devonian seed plants with protected (cupulate) ovules. That these plants were allied with gymnosperms rather than with angiosperms is due primarily to the pre-existing idea of phylogenetic relations between these groups.

As we noticed above, the initial concept of gymnospermy referred to fruits lacking morphologically distinct pericarps, thus having nothing to do with our present-day idea of gymnosperms. The latter became distinctive by their negative characteristics as lacking in such and such typical angiosperm characters. There seemed to be also clear-cut life-form and commercial differences. The conifers produced soft wood and needle leaves. The less familiar cycads also bore cones but differed in the thick, scarcely branched stems crowned by large pinnate leaves. *Ginkgo*, a fairly isolated genus in the modern flora, joined gymnosperms on account of its naked ovules and motile sperms as in cycads. The latter feature testified primitiveness of both these groups.

Thus, on the present time plane, gymnosperms appeared well separated from angiosperms, with three genera of gnetaleans, or chlamydosperms, as the only possible intermediates. Interpretations of the vessels, cupules, nuclear embryo sacs and double fertilization in gnetaleans as homologous or non-homologous to the corresponding structures in angiosperms were subjects of long-lasting controversies. In any case, the extant gnetaleans formed a very thin morphological link at most.

With the addition of numerous fossil groups, the distinctions between gymnosperms and angiosperms became less clear-cut. Pteridosperms and caytonialeans bore their ovules in ovary-like cupules, while bennettites had flower-like bisexual reproductive shoots. If these plants lived today, the classification of Spermophyta might have been very different from what we are used to. Some extinct Paleozoic and Mesozoic seed plants went, as it were, with gymnosperms, though scarcely fitting the original gymnosperm concept. Thus inflated, gymnosperms seem no longer clear-cut from angiosperms.

One can argue that the division of seed plants into gymnosperms and angiosperms reflects the present-day situation and is inadequate in terms of the evolutionary history. Logically, an evolutionary classification would convey changes in the diversity patterns through geological time rather than an accidental pattern of a single time plane. But the prospects of such a classification are still remote.

There have been several attempts at a superordinal gymnosperm classification (Arnold, 1948; Lam, 1950; Pant, 1959; Emberger, 1960; Chamberlain, 1966a; Meyen, 1987), using such characters as pycnoxyly versus manoxyly, microphylly versus macrophylly, phyllospermy versus stachyospermy (ovules borne on leaves or axial organs), radiospermy versus platyspermy (pertaining to the symmetry of ovules), and phanerogamy versus prephanerogamy, i.e. the shedding of seeds or unfertilized ovules. Some of these characters are adaptive (e.g. wood structure), others scarcely seem so (e.g. seed symmetry: in *Ginkgo*, there are both platyspermic and radiospermic ovules). Their phylogenetic significance is still to be learned. But the reason why most of these features have failed as supraordinal distinctions is that they are either inconsistent within the gymnosperm orders (e.g. seed symmetry) or infringe the ordinal boundaries (e. g. microphylly joining conifers with ephedracean gnetaleans). Since angiosperms are defined by the ovuliphore structure (angiospermy) combined with the ovuliphore arrangement (flower), it may be appropriate to apply the same criteria to the non-angiospermous seed plants.

An adaptive type classification of the nature of the angiosperm/gymnosperm division can be based on the ovules being borne open (gymnosperms proper) or in cupules (chlamydosperms). The latter would include, in addition to the extant gnetaleans, also several extinct groups of cupuliferous plants, such as pteridosperms, caytonialeans, bennettites and czekanowskialeans. The ovuliphore arrangement can be described as non-strobilate in pteridosperms, strobilate in cycads and their allies (including a few destrobilate taxa, such as *Cycas* and *Ginkgo*) and doubly-strobilate, as in coniferids. Combining the ovuliphore morphology with ovuliphore arrangements, in the manner of the angiosperm - flowering plant combination that is most closely approached by proangiosperms, we arrive at the following scheme:

Gymnospermidae		**Chlamidospermidae**	
Cycadidae	Coniferidae	Pteridospermidae	Proangiospermidae
Glossopteridales	Cordaitales	Hydrospermales	Caytoniales
Vojnovskyales	Walchiales	Lagenostomales	Bennettitales
Peltaspermales	Coniferales	Trigonocarpales	Czekanowskiales
Nilssoniales		Callistophytales	Gnetales
Cycadales			
Ginkgoales			

We shall return to this scheme after briefly considering the gymnosperm orders. Although general characteristics of the orders can be found elsewhere, it is necessary to put forth the characters and interpretations that will serve for the analysis of phylogenetic relations.

Progymnosperms

Gymnosperm characters appeared in the Devonian aneuropterids and archaeopterids, arboreal plants with a coniferoid stem anatomy, by virtue of which they are assigned to progymnosperms (Beck, 1960). They produced three-dimensional systems of lateral branches, and their ultimate appendages could be webbed and leaf-like, yet not forming leaf blades. Fertile parts of the lateral branch systems bore paired sporangia. Some archaeopterids were homosporous, the other heterosporous, with megasporangia slightly larger than, but otherwise similar to, microsporangia. In these plants, heterospory is considered as a derived character which, however, was scarcely correlated with the advancement of assimilating organs. For example, in *Tanaitis*, a late Middle Devonian archaeopterid, heterospory is associated with non-webbed ultimate branchlets of a very primitive syntelomic aspect (Krassilov & al., 1987).

That ovules appeared in the direct descendants of progymnosperms, is yet to be proved. First seed plants contemporaneous with progymnosperms had ovules with lobed integuments borne in horny cupules that looked as whorled, repeatedly forking branches. In *Hydrosperma*, such cupules contained also microsporangia (Long, 1977a). *Lenlogia*, a Middle Devonian plant from Siberia (Krassilov & Zakharova, 1995), had clustered fertile branches quite similar to the seed-bearing cupules but actually containing paired sporangia (Fig. 1). Its axial organs showed a primitive protostelic structure suggesting that seeds could appear in small plants totally lacking gymnospermous stem anatomy.

Pteridospermidae (pteridosperms)

The concept of pteridosperms, or seed ferns, is due to a fascinating discovery of seeds taphonomically associated with and later found actually attached to the Carboniferous fern-like fronds that were for many years considered as fern foliage. After the initial distrust, the concept became very popular among plant morphologists and became inflated to encompass the Permian and Mesozoic plants with fern-like leaves but having their seeds borne in strobili, like in modern gymnosperms. In the extant cycad *Bowenia*, the leaves are bipinnate and very fern-like. Is it a pteridosperm? Even *Ginkgo* has a fair chance of being included into pteridosperms (Meyen, 1984). Apparently, pteridosperms have to be defined more strictly so as not to engulf all traditional orders of gymnosperms. The initial idea of leaf-like ovuliphores may suffice as a demarcation criterion setting pteridosperms apart form cycads which have ovulate scales that are gathered in more or less compact strobili as well as from ginkgoaleans with the ovules typically terminal on the simple or forked pedicels. This criterion brings peltasperms and other Mesozoic "pteridosperms" closer to cycads than to the typical pteridosperms (Krassilov, 1990).

There are parallel pteridosperm classifications based on stem anatomy and seed characters (Seward, 1917). In the former, the two major orders, Lyginopteridales and Medullosales, are distinguished by the vascular structure being either transitional protostelic-eustelic or polystelic (actually segmented protostelic). Their stratigraphic ranges overlap and they might have grown side by side in Carboniferous forests. However, lyginopterid anatomy and, especially, its calamopityacean and buteoxylonacean

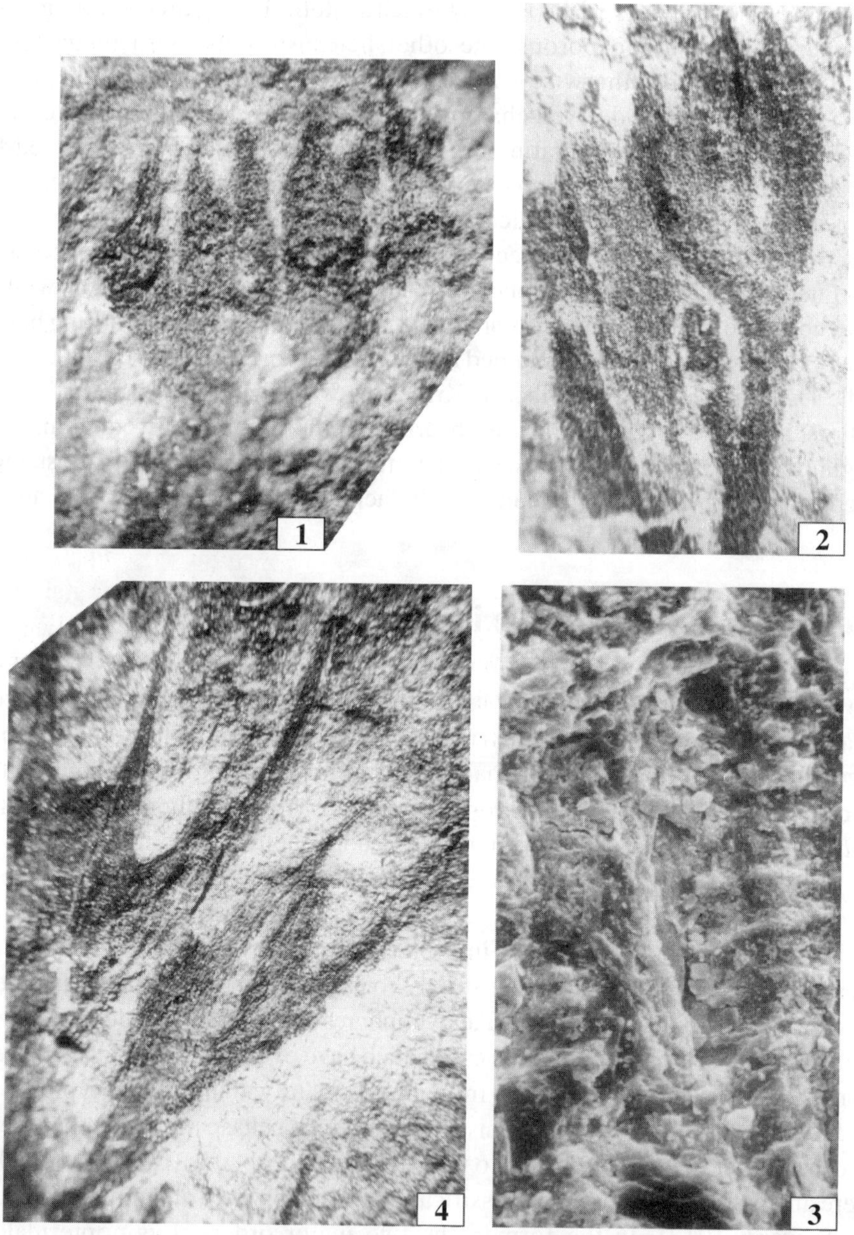

Fig. 1. Devonian precursors of seed plants: (1, 2) *Lenlogia kryshtofovichii* (Radcz.) Krassil. & Zakharova, cupule-like clusters of bisporangiate synangia, from the Late Devonian of Minusinsk Basin, Siberia (Krassilov & Zakharova, 1995), x8, (3) helical tracheids from the protostele of the same plant, SEM, x1 000, (4) *Moresnetia zalesskyi* Stockmans, cupule of an early seed plant from the Devonian of Belgium (courtesy of A.R. Ananiev), x10.

variants appeared earlier and were initially associated with primitive syntelomic assimilating organs and the likewise primitive early seeds. The latter combination deserves the status of a separate order, Hydrospermales, including the dichotomously dissected cupules containing several to many small ovules, triangular in cross-section, with a lobed integument and a distally free nucellus (Long, 1977a; Matten & al., 1980). The nucellar apex was transformed into a special pollen-catching device, lagenostome: a broad pollen chamber was pierced by a central column and extended into a funnel-shaped salpinx. The column, vascularized with spiral tracheids might have functioned as a secretory organ. Occasionally, the megaspore membrane shows a triradiate scar and even the remnants of a tetrahedral tetrad, as in *Lyrosperma* (Pettitt, 1969). In a closely allied but more advanced seed-genus *Gnetopsis*, the ovules were somewhat flattened and were provided with a pappus of apical hairs (Gensel & Skog, 1977).

The advanced lyginopterids had a *Sphenopteris*-type foliage often showing marginal hydathodes, as well as the branching pollen organs of the *Feraxotheca-Crossotheca* type, with the sporangia in erect or pending clusters, proximally fused but separated at maturity and producing spore-like prepollen (Stubblefield & al., 1982). They roughly correspond to the Lagenostomales of a seed-based classification, including the radially symmetrical ovules, solitary in cyathiform cupules that show characteristic secretory glands. The ovules had vascularized integuments displaying a residual lobing but, at least in the advanced seed-genera, forming a conical micropyle over a conspicuous lagenostome. The megagametophyte produced three apical archegonia (Long, 1977c) and a special protuberance, or tent-pole, piercing the bottom of the pollen chamber. In the geologically younger representatives, there was a tendency to smaller and considerably flattened seeds with a reduced lagenostome and the cupule forming an outer micropyle, or exostome.

The Medullosales are linked to the lyginopterids through such morphologically, if not phylogenetically, intermediate forms as *Questora* that had six central and multiple peripheral vascular bundles in the leaf traces, although the leaves were decussate, an allegedly derived feature in pteridosperms (Mapes & Rothwell, 1980). Some medullosaleans might have been lianas with long decurrent leaves. However, the larger Permian species were trees with the stems up to 50 cm thick. In the primitive medullosaleans, the leaf traces were concentric and produced a considerable amount of secondary wood. These "foliar stelae" were much like the numerous additional axial stelae produced at the leaf nodes. In the late species, the leaf traces lost their secondary tissues, and the additional stelae became reduced.

Characteristic of medullosaleans were the pitted tracheids with helical and reticulate thickenings, the phloem fibers, as in cycads, and the sclerenchymous strands in the periderm. The leaves were large, pinnate, with the venation pattern open or reticulate. The typical pollen organs were large cup-shaped or bell-shaped synangia of multiple tubular sporangia arranged in concentric rings (Millay & Taylor, 1979; Taylor & Millay,

1981). Such structures might have been related to pollinivory and insect pollination. Some synangia were pubescent and resembled seeds (Delevoryas, 1964), perhaps encouraging the visits of pollinivorous insects to the latter. The prepollen grains were large, with a monolete or occasionally trilete (in *Potoniea*) leasure, with distal folds and a thick alveolate sexine (*Monoletes:* Taylor, 1978) evolving in the direction of a monosaccus or, rarely, bisaccate.

The ovules assigned to the Trigonocarpales were radially symmetrical, with the integument free from the nucellus and with both integument and nucellus vascularized. Two apical archegonia were observed in the megagametophyte developing from a functional megaspore of a tetrahedral tetrad. In the form of the micropyle and the pollen chamber, with a short nucellar beak, the trigonocarps seem modernized in respect to the lagenostomaleans. Vascularization of the nucellus is commonly considered primitive, but actually the trigonocarpalean nucelli, some of them ribbed as integuments (Smoot & Taylor, 1983/1984), could have evolved by means of fusion of the lagenospermalean or hydrospermalean-type nucelli with their integuments. The trigonocarpalean integuments would then be homologous to the lagenostomalean-type cupules.

The Callistophytales, a minor but phylogenetically important order of Carboniferous pteridosperms, were represented by shrubby plants with creeping spiny stems bearing adventitious roots and 3-4-pinnate leaves (Stidd & Hall, 1970; Rothwell, 1980; Taylor, 1981). They resemble lyginopterids in the vascular structure, sclerenchymatic periderm (sparganum), and glands. Their synangiate pollen organs of radially disposed sporangia seem also derivable from those of lyginopterids. The pollen grains were, however, of a more advanced coniferoid morphology, and the ovules were platyspermic, with lateral wings. Hence, the Callistophytes show a mosaic of typically pteridospermous and coniferoid characters.

The Gigantopteridales were characteristic of the Permian eastern Asiatic, or Cathaysian, palaeofloristic province, but the group extended to the Middle East and North America as well. The 2-3-pinnate leaves showed a tendency to fusion of successively larger units until, in the end of the series, broad entire leaves were formed (Asama, 1962, 1976; Li & al., 1994). Their venation was typically hierarchical areolate, partly retaining the original vein pattern of the constituent pinnae. Both the pollen and the ovule-bearing organs were leaf-like, with linear synangia and submarginal rows of ovules respectively (Li & Yao, 1983). The North American *Spermopteris*, a laminar taeniopteroid ovuliphore with numerous ovules aligned along and slightly protruding beyond the margins (Cridland & Morris, 1960), was obviously related to this Asiatic group. The so-called cycads from the Permian of North America (Mamay, 1976) had similar, though somewhat reduced, ovuliphores. Unlike in cycads, their nucelli were more strongly cutinized than the integuments.

Cycadidae (cycadophytes)

The cycadophytes are usually referred to as the fossil seed plants resembling cycads in the morphology of their leaves and/or reproductive structures. As such, they may include many, if not all, pteridosperm leaf genera recalling of *Bowenia*, as well as the leaf-like ovuliphores resembling *Cycas*. Here they are defined primarily on the basis of strobilate ovuliphores (including a few derived destrobilate forms) that make them different from pteridosperms. The leaf-like ovuliphores of glossopterids are conceived of as formed by fusion of the axillar reproductive shoots with their subtending leaves.

The Glossopteridales constituted a dominant group of the Permian Gondwana Province extending into the Triassic. They are characterized by the entire elliptical leaf blades with reticulate or sometimes open venation. The midrib was formed of separate bundles given off as lateral veins. The anastomoses were of different types, produced as short side branches as well as by splitting and looping of the lateral veins. Atypical leaf morphologies of glossopterid alliance were taeniopteroid, lobed, as in *Pteronilssonia* (Pant & Mehra, 1963), or ribbon-shaped with parallel venation and resembling cordaites, as in *Noeggerathiopsis*.

As for the growth forms, the opinions vary. *Glossopteris* leaves were found attached to *Vertebraria* stems or rhizomes the characteristic articulate aspect of which was due to the aerenchymous wedges divided by xyleme ridges associated with adventitious roots (Schopf, 1982). These structures suggest a mangrove habitat. However, not all glossopterid allies had *Vertebraria* structure, and the leaves were occasionally bunched, as if borne in spur-shoots (Thomas, 1952; Pant, 1968).

The pollen organs were the dichotomously branched androclades, with the sporangial clusters terminal on the ultimate branchlets or strobilate, as in *Kendostrobus* (Surange & Chandra, 1975). The pollen grains were typically saccate striate, with the exine divided into parallel stripes, or taeniae. The ovulate structures were of numerous types, some of them (*Arberia*) gynoclades of the same dichotomous structure as the androclades and bearing ovules on incurved branchlets. The more advanced cupulate structures might have been derived from this basic type (Rigby, 1978). They were cyathiform or discoid, open (or closed at earlier developmental stages), variously lobed or winged, stalked or sessile and axillar (Pant, 1977), or, more frequently, attached to the leaf-like bracts. The ovules were platyspermic with lateral wings or glochidial appendages and with a long micropyle tube that was occasionally funnel-shaped. The nucellus was free, forming a broad pollen chamber. The megaspore membrane was cutinized. The megagametophyte was cellular, with two apical archegonia. The seedlings produced two lobed cotyledons (Pant & Nautiyal, 1987).

In distinction from the typical pteridosperms in which the ovules occurred in fertile regions of the lateral branch systems evolving into leaf-like ovuliphores, the leaf-cupule complexes of glossopterids appear secondary, derived by fusion of the axillar ovulate branches to their subtending bracts. Why such structures were so widespread among glossopterids and in no other group of Permian gymnosperms, remains enigmatic. Functionally, they could have served as dispersal units in the same way as bracteate angiosperm fruits.

The Vojnovskyales were characterized by the radial arrangement of bracts and ovules in the apical zone of scaly short shoots (*Vojnovskya*) or on pedicellate receptacles (*Gaussia*). Short shoots alternated with *Nephropsis*-type rhomboid leaves or cataphylls, the latter showing an open flabellate venation pattern (Neuburg, 1963). Other foliar types associated with the vojnovskyalean ovulate organs were lanceolate or taeniate, with parallel veins and intercostal stomatal grooves (*Rufloria*) and were borne in terminal clusters on short shoots. *Zamiopteris*-type leaves with an indistinct midrib and oblique lateral veins, less commonly with reticulate venation, might have belonged to the same alliance.

All these foliar morphologies occurred also in the *Glossopteris* flora of Gondwanaland localities. Such similarities are currently ascribed to evolutionary parallelisms. In particular, the parallel-veined leaves might share a phyllodic origin. However, parallelisms extend to reproductive structures as well. The radial ovulate structures occurred in both groups, while *Cladostrobus*, a putative vojnovskyalean pollen cone with peltate sporangiophores (Meyen, 1984), resembled *Kendostrobus* of glossopterids (Surange & Chandra, 1975), and the pollen grain morphotypes were very much alike, too. On the other hand, the proclaimed cordaitalean affinities of vojnovskialeans are based on superficial foliar similarities alone, the reproductive structures having had little, if anything, in common with the compound strobili of the cordaites.

The Peltaspermales, another dominant group of the Permian and Triassic seed plants usually linked to pteridosperms, is here assigned to cycadids on the account of their typically strobilate reproductive structures. They are reconstructed as small trees or shrubs with the stem up to 5 cm thick, bearing conspicuous discoid glands and having a rather thin cylinder of secondary wood (Harris, 1964). The leaves were typically 2-4-pinnate with interstitial pinnules, 1-pinnate or simple in the derived forms. The petioles were typically swollen at the base, anatomically resembling the medullosaleans and some transitional forms between the latter and lyginopterids (see above). The rachises were forked or overtopped by the distal pinnae (Kerp, 1986). There was a tendency to segment fusion involving the successively larger units, as in the gigantopterids, and accompanied by a transition from an open to areolate venation (a more detailed account of the latter will be given in the next chapter). The end-members of the morphological series were bilobed or dichopodially lobed flabellate leaves, as in *Furcula* or *Scoresbya* (Harris, 1932). Characteristic of them were thick cuticles showing a radial arrangement of stomatal subsidiary cells that formed an irregular petaloid structure with the stomatal pits encircled by a thick papillate ridge (Fig. 2).

Fig. 2. *Scytophyllum vulgare* (Pryn.) Dobrusk., peltasperm leaves from the Late Triassic of Eastern Urals (Krassilov, 1995) showing a marginal fusion of the pinnae and interstitial pinnules that are marked by a conservative venation pattern: two leaf fragments, x2, and papillate cuticle (bottom), SEM, x400.

The pollen organs were dichotomously branched androclades, with the ultimate peltate branchlets bearing pendent sporangia in rows or radial clusters, free or synangiate. The pollen grains were sulcate asaccate or with a bilobed monosaccus, as in the callistophytes, or bisaccate. Some striate morphotypes, notably *Vittatina*, might have been produced by peltasperms (Meyen, 1984 and elsewhere). The ovulate organs were loose strobili of peltate ovuliphores, with two to many abaxial ovules encircling the stalk. The latter ovule arrangement was typical of the Mesozoic peltasperms less frequently occurring in the Permian calliptrids that are presently considered as archaic peltasperms. The ovules were platyspermic, with a long reflexed micropyle tube, a free nucellus and a cutinised megaspore membrane.

The Corystospermales were widespread in the Triassic of the southern continents producing a great diversity of growth forms (Anderson & Anderson, 1983, 1985). The stem in the larger forms (*Rhexoxylon*) shows a medullosalean vascular structure (Archangelsky, 1968b). The leaves were essentially like those in peltasperms (see above), but forking in the proximal third, in the derived forms bilobed, flabellate or entire. The venation was typically open, the ultimate segments were occasionally single-veined. The stomata show an irregular or, less commonly, a radial or tetracytic arrangement of proximally papillate or ridged subsidiary cells. The pollen organs were pinnate or strobilate, with forking peltate sporangiophores bearing free sporangia in rows or radial clusters. The pollen grains were bisaccate (Taylor & al., 1984). The ovulate organs were loose gynoclades, their lateral branches forking, terminating in a pair of incurved, indistinctly lobed cupules, the latter containing a solitary ovule with a protruding bilobed micropyle (Thomas, 1955). The nucellus was thinly cutinised and lacking cutinized megaspore membrane.

The Pentoxylales are known from the Jurassic and Lower Cretaceous of India and Australia (Rao, 1974; Harris, 1983; Drinnan & Chambers, 1986). The group received its name due to the vascular system with a ring of typically five, rarely three to nine, primary bundles producing pycnoxylic secondary wood with uniseriate rays. The tracheids showed hexagonal bordered pits, as in bennettites. The shoots were dimorphic, with sparse or dense leaf scars. The leaves were simple taeniopteroid or pinnate with taeniopteroid pinnules, their midribs formed of up to nine separate bundles; the lateral veins were simple or forked and looping, terminating in the marginal hydathodes. The leaf blades were pubescent, hypostomatic. The stomata were typically tetracytic or with 6-7 radial subsidiary cells. Both pollen and ovulate organs were clustered on different short shoots, the former delicate, simple or forked, with free sporangia, the latter strobilate, with erect ovules in five oblique rows, coalescing at maturity into a fruit-like body. The pollen grains were monosulcate asaccate with a granular infrastructure and a lamellate nexine (Osborne & al., 1991). The ovules seem to have been bitegmic (?), with a massive sarcotesta and a cutinized sclerotesta, the latter forming a micropyle. The nucellus was free, the megaspore membrane was not observed.

The Nilssoniales were ubiquitous in the Mesozoic, surviving to the terminal Cretaceous and usually assigned to cycads, but they produced the branching long shoots and the deciduous spur-shoots bearing clusters of simple, lobed or pinnate, rarely bipinnate, leaves (Krassilov, 1972c; Kimura & Sekido, 1975). The leaf blade was thick, flat or involute, occasionally serrate. The venation was typically open pinnate, with simple lateral veins. Some associated foliar morphotypes had reticulate venation (*Ctenis*), or their pinnae were supplied with two basal veins giving off subparallel branches (*Heilungia*). The pollen cones were cycad-like, with scaly sporangiophores. The pollen grains were sulcate asaccate, with an alveolar infrastructure. The ovulate cones were loose, pendent, with peltate ovuliphores bearing two ovules. In distinction from cycads, even the immature cones were open and the ovules occurred abaxially between the involute margins of a distal pelta (Harris, 1964). The ovules were platyspermic, with resin bodies in the sarcotesta. The nucellus was free apically, forming a pollen chamber crowned with a short nucellar beak. There was a cutinized megaspore membrane.

The Cycadales include ten extant and a few Tertiary genera, whereas the pre-Tertiary records are problematic except, perhaps, for the Late Cretaceous *Cycas* showing cuticular structures peculiar for the genus (Krassilov, 1979). The stem is thick cylindrical or tuberous subterranean, unbranched (pachycaul), armoured with petioles of shed leaves, with an apogeotropic coralloid root mass. Occasional branching is due to wound meristems or adventitious buds. The primary body is formed by the apical meristem as well as by the separate lateral meristem and diffuse growth, the contribution of the secondary tissues being less significant. The leaf traces are girdling the stem. The leaves are produced in clusters (in *Stangeria* singly) once a year or biannually, alternating with the cones. They are 1-pinnate stipulate, conduplicate or involute in bud, in *Bowenia* 2-pinnate fern-like and circinnate in bud (Stevenson, 1981), with the pinnules linear, lanceolate, rhomboid or flabellate, pubescent. The venation is open or anastomosing, rarely with a series of marginal loops (*Stangeria*). The epidermis shows alternating zones of longitudinal and transverse-oblique cells (occurring in the Cretaceous species already), the cell walls straight or sinuous. The stomata occur in groups or files, slightly or deeply sunken, sometimes with encircling cells dome-like overhanging the guard cell - subsidiary cell complex (Pant & Nautiyal, 1963).

The reproductive organs are strobilate, diclinous, but occasionally with a few sporangia in ovulate strobili. The pollen cones are large, with numerous (up to 700) sporangia abaxial on the shortly petiolate or sessile, entire or, rarely, bilobed scales. The sporangia occur in dense masses, grouped by 2-5 in inconspicuous sori, free or proximally fused in half-synangia (Smith, 1907), with thickened cells at the apex and along the dechiscence line and with stomata, a feature shared with peltasperms (*Antevsia*) alone. The pollen grains are produced in great quantities, sulcate asaccate or with rudimentary sacci, transferred by insects (Tang, 1987). The pollen wall infrastructure is alveolate. The ovulate cones are very large, developmentally apical on the stem, seldom lateral, in *Bowenia* on short lateral branches, shifted to the leaf axils or, in *Cycas*, forming an imbricate cover over the stem apex that resumes growth after shedding the ovuliphores. The latter are scaly, with sterile apophyses peltate or flared and lobed, in *Cycas* leaf-like (Chamberlain, 1966b). The ovules are typically two per scale, inverted, pendent at maturity, in *Cycas* from two to eight in two rows, radiospermic. The integument is thick three-layered, with stomata and hairs, the latter sometimes clustered around the micropyle. The nucellus is strongly cutinized in the upper third alone, but with a delicate cuticle extending down to the base. The pollen chamber is crowned with a nucellar beak. The vascular system is double, the peripheral in the integument and the interior shared by the nucellus and integument (Pant & Nautiyal, 1963). The megaspore tetrad is linear. There is a conspicuous megaspore membrane.

The megagametophyte remains nuclear until a 100-nucleate stage, with two to four apical archegonia. The pollen tubes are haustorial. Both sperms may function in fertilization (Rao, 1974). Polyembryony is common. The proembryo remains nuclear until a 64-100-nucleate stage. The cotyledons are usually two (in *Ceratozamia* solitary), haustorial, their apices remaining in the seed.

The Ginkgoales were diverse in the Mesozoic, with the only extant - cultivated - species, a deciduous tree with dimorphic shoots. The long shoots are pycnoxylic, whereas the short shoots are manoxylic. The leaves are long petiolate flabellate, on the long shoots lobed, on the short shoots notched, with short resin ducts between the veins. The venation is flabellate dichotomous, arising from two basal veins running along the margins, mostly open, with infrequent but constantly occurring anastomoses (Arnott, 1959. The stomata are provided with radial subsidiary cells, papillate. The juvenile leaves are pubescent, shedding hairs at maturity.

The trees are dioecious. The pollen organs are strobilate, pedicellate, occurring on the short shoots in a subapical ring peripheral to the apical sterile leaves and are encircled by the glandular bracts. The sporangiophores are stalked, peltate, with an apical gland, bearing two or, rarely, 3-4 abaxial sporangia The pollen grains are sulcate asaccate or with residual sacci (Herzfeld, 1927), and with alveolar infrastructure. The gynoclades consist of a long, slightly flattened stalk, distally expanded and bearing two erect arillate ovules. They are borne in the apical zone of the short shoots intermingled with sterile leaves (and are sometimes described as axillar to the leaves but topologically equivalent to them) and surrounded by two circles of bracts, the outer sessile, the inner petiolate. The leaf petioles and the ovulate stalks are identical externally, but the latter are supplied with four bundles instead of two. Anomalous ovuliphores bear a solitary ovule or several ovules on short branches, or they are leaf-like (Favre-Duchartre, 1956) and probably chimeric, arising by fusion with sterile leaves.

The ovules are typically platyspermic bicarinate, with two integumentary bundles along the keels, less commonly radiospermic 3-4-keeled, with a corresponding number of vascular bundles. An arillate collar develops later than the integument (Pankow & Sothman, 1967). The nucellus is free in the apical portion alone, scarcely cutinized, forming a broad pollen chamber with a nucellar beak. The megaspore membrane is cutinized. The megagametophyte remains nuclear until a 256-1000-nucleate stage, with two or, rarely, four apical archegonia and with a tent-pole. Fertilization occurs before or after shedding the ovules. The proembryo remains free nuclear until a 128-256-nucleate stage, but with a partial cell wall laid down after the first division (Lee, 1955). The cotyledons are normally two, as in cycads.

The extinct Mesozoic forms were either similar to the extant species or differing in the lanceolate parallel-veined or palmately dissected leaves with or without petioles as well as in the distinctly strobilate ovuliphores bearing erect or incurved ovules (Atchangelsky, 1965), the latter with free cutinized nucelli and the early embryogenesis accomplished before shedding the seeds (Krassilov, 1969).

Coniferidae (coniferids)

Mostly arboreal with pycnoxylic wood, heteroblastic, microphyllous, with the pollen and seed cones, or the latter only, compound, doubly-strobilate, including conifers and two extinct groups, cordaites and walchians.

The Cordaitales were represented by the mid- to late Paleozoic arboreal or shrubby plants with simple parallel-veined leaves and with both pollen and seed cones doubly-strobilate. In the "hydrophilic" group, the stems were creeping, with axillar branching and adventitious roots, eustelic. The primary vascular system was sympodial, with collateral leaf traces extending through two or three internodes (Rothwell & Warner, 1984). The shoots were heteroblastic, with proximal needle leaves and distal parallel-veined leaves of the *Cordaites* type; the leaves were sclerenchymatous, hypostomatic, with tetracytic stomata. The cones were epicormic, bearing bracteate fertile shoots. The sporangiophores produced terminal clusters of 4-5 tubular sporangia axillar to bracts. The pollen grains were monosaccate, with a distal leptoma. The ovuliphores were simple or branched, with terminal ovules. The latter were erect, laterally winged, with the integument 3-4- layered, vascularized. The nucellus was free, with a vascular platform at base (Serlin, 1982a).

In the "mesophilic" group, the adventitious roots were wanting, the stems showing a sparganum-type sclerenchyma. The primary vascular system was formed of leaf traces (Trivett & Rothwell, 1985). The leaves were of the *Cordaites* type. The sporangiophores bore four sporangia in rows, producing the monosaccate prepollen that showed a distinct triradiate scar (Taylor & Daghlian, 1980). The ovules were platyspermic, with the sarcotesta forming a chalazal collar. The megaspore membrane was prominent, the early megagametophyte stages were allegedly free nuclear (Grove & Rothwell, 1980).

The Walchiales were the coniferoid mid- to late Paleozoic plants with heteromorphic, scaly falcate or needly, pubescent leaves, with the stomata in single or double rows. The pollen strobili were simple, with the sporangial clusters abaxial or whorled on the stalk. The prepollen grains were saccate. The seed cones bore the axillar shoots with up to 30 sterile scale-leaves and the terminal or subterminal abaxial ovules, the latter platyspermic, mostly solitary, sometimes two or several per shoot, originally described as typically erect, occasionally inverted as in *Walchia arnhardtii* (Florin, 1938-1945), but later proved to be typically inverted (Mapes & Rothwell, 1984; Clement-Westerhof, 1984; Meyen, 1981, 1984) with a possible exception of *Moyliostrobus*. The nucellus was free, forming a pollen chamber with a nucellar beak. The megaspore membrane was cutinized.

Although traditionally assigned to conifers, those plants are here considered as a separate order on account of the prepollen and the leafy ovulate shoots that are widely held as precursory to the coniferous seed-scales (Florin, 1938-1945), but are morphologically distinct from the latter. The southern Paleozoic coniferids *Buriadia* and related genera with furcate leaves and stalked inverted ovules that are solitary on ultimate or penultimate branches are remotely if at all related to the Walchiales (Pant, 1982). *Ferugliocladus* represents another potential Paleozoic order with simple pollen cones, monosaccate pollen and the compact seed cones bearing erect axillar ovules (Archangelsky & Cuneo, 1987).

The Coniferales constitute, since the early Mesozoic and to this day, the most diverse and highly heterogeneous group of gymnosperms with 57 extant and a comparable number of extinct genera belonging to about 15 families. The popular idea of a conifer based on the common Northern Hemisphere genera is that of a tall tree with conical crown, needle-leaves, simple pollen-cone, bisaccate pollen grains, compound seed cone bearing biovulate seed-scales axillar to bracts, and winged seeds. All these characters are variable, however. Some conifers are giant trees up to 110 m high, others are shrubs or even dwarf-shrubs. The wood is pycnoxylic, the branch systems are radial, less commonly flattened, rarely forming pinnate cladodes. The shoots are often heteromorphic, with the short shoots deciduous or sometimes persistent, shedding their leaves, commonly heteroblastic, with the basal scale-leaves and the distal foliage leaves having different epidermal structures. Phyllotaxis is helical, less often decussate, in the latter case with more or less prominently heteromorphic lateral and facial leaves. The leaf traces are single, in the decussate leaves sometimes double, derived from different axial sympodia (Nambudiri & Beck, 1968). The leaves are shortly petiolate or sessile, often decurrent, sometimes basally connate with each other and with the axis, rarely double, as in *Sciadopitys*, needly or falcate and 3-4-angular in cross-section or flat bifacial, rarely bilaterally flattened, as in *Acmopyle*, occasionally forked, single-veined or with two or many parallel veins. Ontogenetically, linear leaves usually precede scale leaves (Laubenfels, 1953). The stomata are mostly perigenous, reportedly also mesoperigenous, or the polar subsidiary cells mesogenous (Johnson & Riding, 1981).

The cones are normally diclinous, but with sporangia sometimes occurring in seed cones (Tosh & Powell, 1986). The pollen cones are compound in the Triassic Voltziaceae, that are often considered as a stem group, simple or rarely compound in the modern genera in which they are sometimes gathered in inflorescence-like aggregates. The sporangiophores are peltate hypo- or perisporangiate, with the sporangia aligned with a stalk, pendent at maturity. The sporangia are developmentally epidermal or, in the Taxodiaceae, Cupressaceae, Taxaceae and Cephalotaxaceae, hypodermal. The pollen grains are sulcate or monoporate, with a distal papilla, rarely inaperturate (*Agathis*), occasionally with a triradiate scar and prepollen-like (Cranwell, 1961). The pollen wall infrastructure is typically alveolar in the saccate forms, granular in the asaccate and some protosaccate forms (Zavada, 1985), occasionally with irregular columellae (Klaus, 1979). The seed cones are terminal or lateral on branches of different orders, occasionally proliferating, usually solitary, but sometimes gathered, like pollen cones, in inflorescence-like aggregates. Typically, the cone scale consists of a bract and an axillar seed-scale, the latter bearing inverted ovules. The relative development of the bract in respect to the seed-scale, the degree of their fusion and the number of ovules are taxonomic characters of familial rank, although their alternative states can co-occur in some genera, such as *Athrotaxis*. Forms with solitary ovuliphores are usually considered destrobilate, yet in a number of genera they can be primary (Sinnott, 1913). Erect ovules, occurring in such forms alone, can be secondary - pseudoorthotropous - changing orientation during their development, as in *Dacrydium*. In the Podocarpaceae, the ovule is enclosed in the seed-scale derived epimatium that sometimes fuses with the integument and, in *Dacrycarpus*, also with the bract. In *Dacrydium*, the epimatium is short, arillate and not unlike the arillus of the Cephalotaxaceae. In *Araucaria*, the ovule is imbedded in the cone scale, with the apex of the seed-scale forming a membranous outgrowth, or ligule. In some araucarias, the winged scale with seed in it serves as a fruit-like dispersal unit.

The bract - seed-scale complex is typically supplied with two groups of vascular bundles, with the xylem adaxial in the bract and inverted in the seed-scale, the latter bundles mostly terminating in the chalaza or, rarely, extending into the integument, as in *Amentotaxus* that has 8-14 integumental bundles (Chen & Wang, 1985). The nucellus is fused proximally to the integument, less commonly free. The megaspore tetrad (triad) is linear, seldom tetrahedral. The megaspore membrane is variable, thick in podocarps (Quinn, 1966). At anthesis, the megagametophyte is still nuclear. The tent-pole occurs in the Podocarpaceae, Cephalotaxaceae and Taxaceae. In the former family, the archegonia are large, up to 2 mm, in the latter two they are strongly reduced, with their initials sometimes directly functioning as eggs (Georgiev, 1985). In the Taxodiaceae and Cupressaceae, and occasionally in other families, the archegonia form the chalazal and/or micropylar, as well as lateral, complexes. The neck cells are reduced and, in the latter family, the ventral canal cell tends to be lost. This tendency parallels the reduction of the prothallial cells. Occasional fusions of the ventral canal nucleus with the egg, as reported in *Keteleeria* and *Pseudotsuga* (Konar & Oberoj, 1969), is of interest as precursory to double fertilization.

The pollination mode varies from the one conventional for gymnosperms to the extraovular pollen germination, as in angiosperms. The micropyle is often secretory, trapping pollen grains that are then sucked in with pollination drop secreted by the nucellus. In the Pinaceae, the micropyle or one lobe of it is often papillate, stigmoid (the Narbenmicropyle: Kügler, 1970). Extraovular pollination occurs in the Araucariaceae and, sporadically, in other families, correlating with erosive siphonogamous pollen tubes and with reduced micropyle that leaves the nucellar beak exposed. In *Araucaria*, the pollen grains settling on the cone scale concentrate in the median groove and are directed to the ligule, germinating on the membranous fringe of the latter (Fiordi, 1984). Both sperms may function in the Podocarpaceae and Cupressaceae and, less regularly, in other families, with two zygotes per archegonium reported in *Agathis* (Eames, 1913). Cleavage polyembryony is common. The proembryo cells are tiered. The cell walls are laid down at various stages, in *Sequoia* at the first division of the zygote. The dicotyledonous condition might have been derived by fusion from the 6-8-cotyledonous one that is considered primitive (Laubenfels, 1953). The cotyledons are occasionally lobed, with two veins, in *Agathis* with many veins like in the foliage leaves.

All attempts at splitting conifers into more natural groups on the basis of comparative morphology alone are deterred by the mosaic combinations of the putatively primitive and advanced characters. *Araucaria*, for instance, combines the nearly complete fusion of bracts and their axillar ovuliphores, the adaptations for extraovular pollen germination and the fruit-like dispersal units with the largest number of prothallial cells (40), the primitive archegonia, and the prephanerogamous shedding of seeds at a fairly early stage of embryogenesis.

On the basis of fossil links, a number of modern families can be traced back to the Voltziaceae. The latter appeared in the Permian and diversified in the Triassic. The Permian *Pseudovoltzia* had dimorphic, scaly and bifacial, leaves and the entire bracts that were fused basally to the lobed seed-scales, the latter bearing three inverted ovules (Schweitzer, 1963). The seed-scale was supplied with inverted bundles. It did not look like a shoot and bore no resemblance to the axillar shoots of the contemporaneous walchias. In the Triassic voltziaceans, either the seed-scale or the bract morphologically prevailed in the cone-scale complexes. The first appearing forms with needles on spur-shoots, as well as with lanceolate multinerved leaves, were parallely trending from seed dispersal to cone-scale dispersal, the latter assisted by the bract modified into a multinerve wing, as in *Cycadocarpidium*. The Late Triassic - Early Jurassic transitional forms, such as *Schizolepis, Elatides, Doliostrobus, Metaia*, etc., link them to the modern Pinaceae, Taxodiaceae and Podocarpaceae (Townrow, 1967; Harris, 1979), while the coexisting Araucariaceae and Taxaceae were as distant from the voltziacean cluster as they are today. Their independent origin from different Paleozoic coniferoids thus seems fairly probable. In the Jurassic, a number of conifer lineages underwent parallel modifications of their seed-scale complexes under the influence of semenivory and zoochory. Remarkably, flashy seeds appeared simultaneously in the conifers, ginkgoaleans and cycadophytes.

Proangiospermidae (proangiosperms)

The group is defined primarily on the basis of the ovules enclosed in the pistil-like cupulate structures that are strobilate, apically clustered on dwarf shoots or whorled, flower-like, typically subtended by perianth bracts. In distinction, the pteridosperms have non-strobilate ovuliphores, while in the cycadophytes, the ovuliphores are scaly or, if cupulate, then open, enclosing the ovules at early developmental stages alone, with fertilization never occurring inside the cupules.

The scope of the group is not yet finally determined. The bennettites and gnetaleans are included on account of their derived forms, such as *Baisia* that is distinctly cupulate (Krassilov & Bugdaeva, 1982), or *Eoantha* having flower-like gynoecial structures (Krassilov, 1986). Whether the whole orders or the derived forms alone are proangiospermous, is left an open question at the moment.

The Bennettitales were dominant in the Mesozoic, declining through the Late Cretaceous up to the eventual extinction at about the Cretaceous/Tertiary boundary. The early forms were leptocaul, with dichopodial lateral branch systems. The geologically younger forms were pachycaul as well, columnar, barrel-shaped or tuberous, sparsely branched, densely covered with multicellular ramenta (Fig. 3). The vascular system was eustelic, the primary xylem consisting of helical tracheids, the leaf traces were U-shaped, divided into collateral bundles; the secondary xylem was manoxylic but denser than in cycads, consisting of tracheids with hexagonal bordered pits. The leaves were simple taenioid, entire or lobed, or pinnate, in the pachycaul forms pinnate, petiolate, often swollen at the base shed at the abscission layer leaving an armour of persistent petioles on the stem. The pinnae were spread in one plane or imbricate and decurrent, linear, lanceolate or lingulate, rarely flabellate (Ash, 1976), sometimes with basal auricules. The venation was mostly open, rarely reticulate. The cuticle was thick, showing sinuous, rarely straight, cell walls. The stomata were paracytic, developmentally mesogenous (Krassilov, 1978a, 1984), laterocytic on the leaves, tetracytic on the floral bracts. Vessels with porous and scalariform perforations occurred in the leaf bundles (Krassilov, 1982b).

The reproductive organs were flower-like, diclinous or monoclinous, the latter perhaps protandrous, developing in the forks of the lateral branches or, in the pachycaul forms, in the leaf axils, immersed in the ramenta and supplied with vascular bundles departing from two or more leaf traces (Delevoryas, 1968), typically consisting of a fleshy receptacle bearing perianth bracts, leafy sporangiophores and a gynoecium of stalked ovules intermingled with peltate interseminal scales. The receptacle was conical, with a more or less prominent apical column ending in a "corona" of scales (Harris, 1969), sometimes expanded in a funnel-shaped structure interpreted as a vestigial second floral node (Krassilov, 1991). The perianth bracts arose in two circles and were densely pubescent, the outer thick and stomatiferous, the inner thin, lacking stomata,

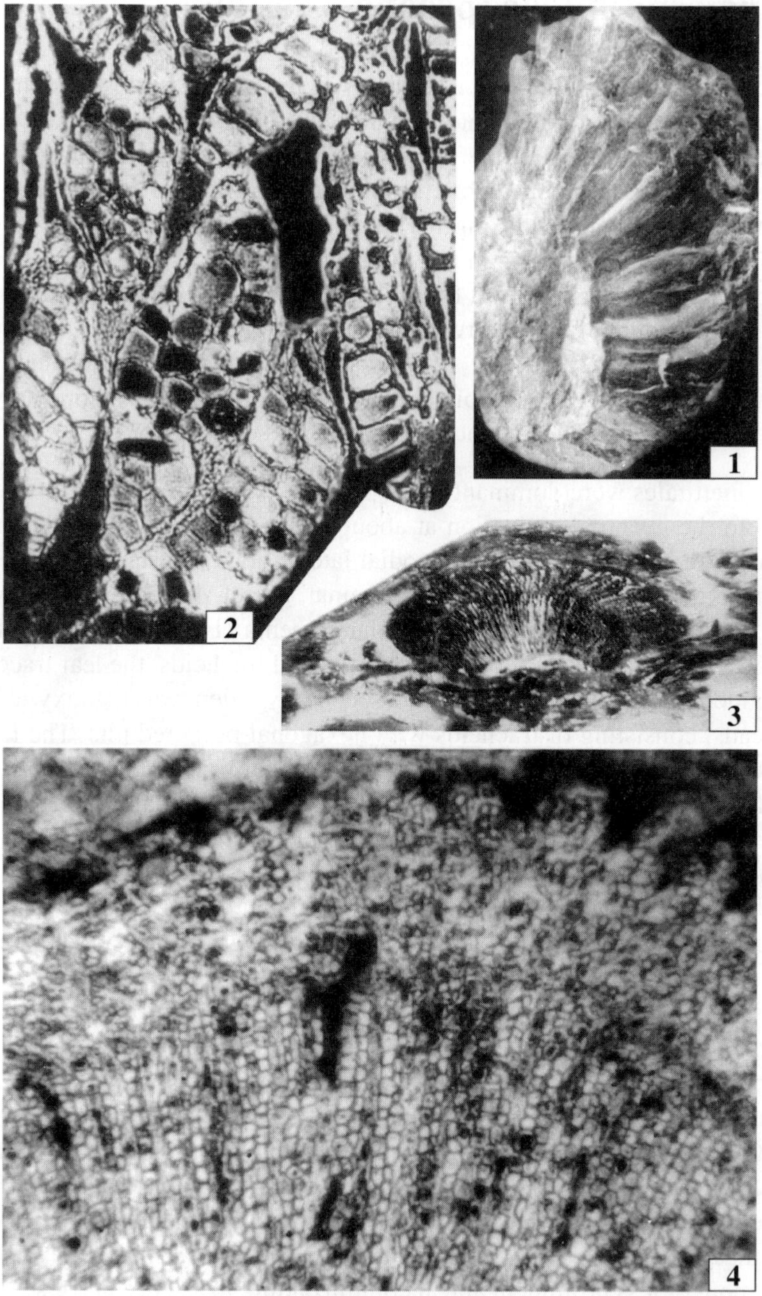

Fig. 3. *Cycadeoidea bikinensis* Krassil., a pachycaul bennettite from the mid-Cretaceous (Albian) of Sikhote-Alin Mts., Far East, Russia (Krassilov, 1990): (1) stem armoured with persistent leaf petioles, x,1/3, (2) ramenta of multicellular trichomes, x200, (3, 4) leaf trace, cross-section, x15 and 80.

perhaps attractively coloured (Harris, 1973), covering young gynoecia and sometimes even mature gynoecia, as in the cleistocarpous flowers (Gupta, 1958). Occasional bracts proliferated in miniature pinnate blades betraying their phyllodic origin (Harris, 1969; Krassilov, 1982b). The sporangiophores were basically pinnate, circinnately coiled or bent, with conduplicate pinnae, often proximally connate and then forming a cup-like structure, sometimes apically connate in a massive dome-like structure that prevented their unfolding at maturity (such structures would suggest self-pollination unless they were shed as a whole). The pinnae bore bivalved synangia or the latter developed directly on the interior wall of the cup, the proximal ones often replaced by glands, these perhaps secretory, filling the cup with nectar and suggestive of insect pollination. The pollen grains were large, sulcate, asaccate, with alveolar or granular infrastructure (Taylor, 1973).

The gynoecia were densely covered with interseminal scales, their polygonal apophyses strongly cutinized, adpressed, forming an armour with small holes through which the micropyles were protruding. The ovules were multiple or, in the aberrant forms like *Vardekloeftia*, a few within the interseminal armour, erect on a simple or occasionally forked stalk, the latter externally like the stalk of an interseminal scale but supplied with a single vascular bundle scarcely if at all divided in the chalaza, in contrast to up to six vascular bundles in the interseminal scales (Sharma, 1970), radiospermic, somewhat flattened, ribbed, sometimes arillate, with the arillus pedicellar, scarcely extending beyond the chalaza, rarely replaced by bristles (Seward, 1917). The integument was either single, poorly cutinized, or sometimes double, with the external cover apparently formed of interseminal scales and apically similar to their apophyses (Harris, 1932). The nucellus was free and thickly cutinized, less typically free in the distal third alone. The megaspore membrane was not reported. The megagametophyte early stages were free-nuclear (Crepet & Delevoryas, 1972). The embryo differentiated before shedding the seeds, with two large cotyledons. The seeds were sometimes provided with a chalazal or micropylar pappus of long bristles (Krassilov, 1973b, 1982b).

The Gnetales include three extant genera that are fairly insulated, perhaps representing different orders. Among them, *Gnetum* is most angiosperm-like in the life forms comprising tropical lianas and, in a single species, small axillary branched trees. The shoots are dimorphic heteroblastic. The lower leaves are scaly, the foliage leaves are lauroid, entire or lobed, borne in decussate pairs, but pinnately spread in one plane, basally connate and decurrent, with areolate venation as in typical angiosperms. The vascular system consists of axial and cortical bundles, with the leaf traces departing from a half of the axial bundles at each node and extending through two internodes. The foliage leaves are supplied with 3-5 traces forming 5-11 separate bundles that run parallel in the midrib sequentially diverging as lateral veins (Rodin, 1966). The tracheary elements of the primary xylem with helical, annular or reticulate thickenings, those of the secondary xylem with bordered pits. Characteristic of the vascular system are the vessel elements with oblique pergoration plates, porous or scalariform (Mouhammad & Sattler,1982), and the sieve tubes with likewise oblique sieve plates, the albuminous cells resembling the sieve element companion cells of angiosperms. The stomata are paracytic, reportedly perigenous or mesogenous (Maheshwari & Vasil, 1961).

The reproductive organs are produced in terminal, axillar or, sometimes, epicormic panicles of spicate strobili supplied with the central collateral and the peripheral concentric vascular bundles. The spikes are monoclinous, functionally diclinous, bearing decussate bracts at the nodes. The bracts, except for the basal ones, are fused in the sheathing collars. The sporangiophores are produced in the nodal zones in several whorls above the collar, are followed by a whorl of ovules, and are enveloped in the cupular perianths - occasionally two per perianth - protruding at maturity, bearing apical synangia of four or fewer sporangia. The pollen grains are distally leptomate, with a granular-columellate infrastructure (Orel & Kupriyanova, 1986). The ovulate spikes bear a single whorl of ovules in the nodal zone axillar to the collar bracts or fused with them, proximally sometimes replaced by sporangia, in *Gnetum africanum* sunken in the nodal cushion (Pearson, 1929), supplied with the vascular bundles departing form the collar bract traces and additionally from the axial bundles. The ovules are erect, radiospermic, enclosed in the cupule, or "perianth", that sometimes contains two ovules or also sporangia or even an aborted spike (Keng, 1965), bitegmic, the inner integument forming a micropyle tube with a fringe, or sometimes with two fringes, or with hairs overhanging the outer integument. The integuments develop in acropetal succession (Takaso & Bouman, 1986), the inner laid down as an annular ridge, the outer as two decussate knobs. The vascular bundles extend to the level of splitting of the nucellus from the inner integument.

The pollen grains sometimes germinate in the micropylar canal that is later sealed with papillae. The megaspore tetrad is linear. The megagametophyte is tetrasporic or monosporic, nuclear until a 1000-1500-nucleate stage. The archegonial initials seem directly functioning as egg cells that differentiate simultaneously with the entrance of up to four pollen tubes, both their sperms functioning in fertilization, producing two to six zygotes. The cell wall formation proceeds in acropetal succession and is accompanied by fusion of the nuclei. The seeds are endowed with a fleshy sarcotesta formed of the cupule (thus non-homologous to those of *Ginkgo* and cycads). The embryo produces two cotyledons and an haustorial appendage, or feeder, supplied with two series of vascular bundles.

Ephedra is an ubiquitous genus of rhizomatous shrubs or, in a single species, small tree, with green branchlets produced yearly, deciduous in dry climates. The shoots are articulate, with two, less commonly 3-5 nodal leaves, eustelic, with a double system of large and small bundles, the latter disposed more centrally, both coalescing in the nodal girdle, alternately producing leaf traces, typically two per leaf, and the axillar bud bundles above them. The leaves are sheathing, their free blades reduced, or the sheath reduced, the leaves then spreading directly from the nodes (Pearson, 1929). The tracheary elements are tracheids and vessels with bordered pits and helical to reticulate thickenings. The vessels are tracheid-like, with numerous separate or confluent porous perforations. The stomata occur in grooves on the stem, rare on leaves, sunken, with radial subsidiary cells. The strobili are spicate, functionally diclinous, axillar to leaves or terminal. The pollen spikes bear 2-3 bract whorls of which the proximal ones are sterile, the distal supporting axillar sporangiophores. The latter are enclosed in the cupules formed of two bracteoles, often containing two or more sporangiophores, their stalks often connate, sometimes forked, bearing 6-8 terminal synangia with a porous aperture. In some species, the pollen spikes produce ovules functioning as nectaries. The pollen grains are ribbed inaperturate, with the tectum thin between the ribs. The pollen wall infrastructure is granular (Zavada, 1984a).

The ovuliferous spikes bear bracts in 2-3(4)-membered whorls, the distal whorl supporting a few axillar ovules or the ovules are pseudoterminal, the proximal whorl sometimes containing pollen organs. The ovules are either solitary or, occasionally, two per cupule and accompanied by sporangia. The cupule is formed of 2-4 connate bracteoles, is supplied by a respective number of collateral bundles and is therefore bilateral and occasionally bilobed or 3-4-angular. The pseudoterminal cupule becomes bilateral through the loss or fusion of the primordial lobes. The integument (the inner seed coat) is thin, membranous, laid down as an annular primordium, proximally fused with the nucellus, forming a long micropylar tube. In seed, the cupule hardens, while a fleshy sarcotesta is formed of the bracts. The megagametophyte is monosporic, initially nuclear, as in *Gnetum*, but later forming archegonial complexes, as in conifers (Singh & Mageshwari, 1962). Archegonia produce 30-40 neck cells, as in araucarias. The egg cell and the ventral canal cell are not divided by a cell wall, or the cell wall is reduced. Sperms couple one with the egg, the other with the ventral canal cell nucleus (Herfeld, 1922; Khan, 1940; Moussel, 1978). The embryogenesis starts as free-nuclear. The cotyledons are two or three.

Welwitschia is a monotypic genus of a peculiar pachycaul life form with an obconical stem and two ribbon-shaped leaves arising from the rim of a broadly concave apex. The vascular system is segmented protostelic, with the partial stelae differentiated as leaf traces alternating with the bundles that terminate in the annular ridge of the apex. The venation is parallel "goniopteroid", with the short side branches coalescing to give rise to interstitial veins (Takeda, 1913). The stomata occur on both sides, are sunken, paracytic, developmentally mesogenous or mesoperigenous. The reproductive shoots are dichasial, borne on the annular ridge, their terminal branches are spicate, with empty connate bracts at the base. The pollen spikes bear flower-like spikelets axillar to bracts, or glumes. The spikelets are radial but slightly compressed. The perianth consists of two decussate pairs of bracteoles the inner of which are connate to half their length. They form an open cupule that contains six sporangiophores arranged radially around the apical ovule, the latter extended into a papillate corona, perhaps glandular. The sporangiophore stalks are connate basally, bent toward the central ovule, each bearing 3-locular synangia. The pollen grains are ribbed, as in *Ephedra,* but with a single or several colpous apertures between the ribs. The ovulate spikes bear large spoon-like bracts subtending the cupular ovules with rudimentary bracteoles at the base. The cupule is supplied with two vascular bundles forming a chalazal ring. The micropyle tube is long, fragile. The axillar complex develops into a samara with membranous lateral wings. The megagametophyte is monosporic or tetrasporic, nuclear until about a 1000-nucleate stage, then divided into multinuclear cells in which multiple nuclei fusions take place (Martens, 1971). The micropylar cells form the ascending tubes that meet pollen tubes. Cell walls are laid down after the first division of the zygote. The embryo produces a haustorial feeder, like in *Gnetum*, and two unequal cotyledons with the axillar buds that overtop the primary apex.

Typical gnetalean characters appear in the Triassic genera *Dinophyton* and *Sanmiguelia* that can be considered as protognetalean. They can then be traced through the Jurassic *Heerala* to the even more definitely gnetalean Cretaceous *Leongathia, Drewria* and *Eoantha* and the recently found pollen inflorescences of the Baisian Assemblage. These proangiospermous plants will be described in the next chapter. Gnetophytes are thought to be represented by the diverse polyplicate "ephedroid" pollen grains of the *Ephedripites, Equisetosporites, Singhia, Steevesipollenites, Gnetaceaepollenites* and other types (Osbon & al., 1993). However such pollen grains were produced also by the pollen organs of the typically cycadalean morphology, such as *Piroconites* (Van Konijnenburg-van Cittert, 1992). On the other hand, the typically gnetalean androclades, such as *Baisianthus*, might produce entirely different pollen grains (see below).

The Caytoniales constituted a small but widespread group of extinct Mesozoic plants reconstructed as trees or shrubs with thick heteroblastic shoots bearing scaly cataphylls and compound foliage leaves. The latter were petiolate, on prominent leaf cushions, abscissed at the base of the petiole (Harris, 1951a, 1964), with four, occasionally two, leaflets that spread palmately, but not exactly in one plane and could be separately abscissed. The two median leaflets were somewhat larger than the lateral pair, the arrangement suggesting two decussate pairs. The leaflets were lanceolate, somewhat asymmetrical, with the lamina unilateral at the base. The venation was reticulate, with the midrib consisting of separate parallel bundles that successively diverged as lateral veins. The reticulations were formed of coalescing terminal branches of the repeatedly forked lateral veins or by short transverse anastomoses, the areoles decreasing toward the margins. The stomata were fully exposed, with the guard cells not sunken below the level of anomocytic subsidiary cells. The pollen organ was a dichopodially branched androclade, with the lateral branches forked, bearing elliptical synangia terminally in sessile radial clusters or laterally on short stalks twisted to one side. The synangia were radially symmetrical, consisting of four tubular sporangia, occasionally bilateral, with the sporangia unequally developed (Krassilov, 1977b), adnate from end to end but splitting in the middle at maturity, dehiscing by an adaxial slit. The pollen grains were monoprotosaccate, with a bilobed saccus. The pollen wall infrastructure was alveolate (Zavada & Crepet, 1986). The gynoclades were pinnately branched with the lateral ovuliphores bent toward one side, consisting of a short stalk bearing a spherical cupule. The latter was fleshy, strongly cutinized, lacking stomata, vascularized, with the tracheary elements showing bordered pits (Reymanówna, 1974). A slit-like exostome occurred at the base of the cupule, was subtended by a short appendage ("lip"), and was connected to the ovules by a series of channels (Thomas, 1925). The ovules were numerous, clustered at the base of the cupule, with their micropyles facing the exostome, bitegmic, with the inner integument not reaching the micropyle (Krassilov, 1984a). The nucellus was free, strongly cutinized. The megaspore membrane was not observed. The embryogenesis was well advanced before the seeds were shed (Krassilov, 1977b).

The Czekanowskiales (Leptostrobales) were a dominant group of Mesozoic plants in northern Asia, (formerly assigned to ginkgophytes) arboreal with dimorphic shoots. The spur-shoots were tuberous, deciduous, bearing scaly cataphylls and linear foliage leaves. The latter were ribbon-shaped, parallel veined, bifacial, hypostomatic, in the *Phoenicopsis* group sometimes amphistomatic, forked, single-veined, bifacial or, in the *Czekanowskia* group, sometimes triangular in cross section, with the stomata on three facets. The stomata were tetracytic or cyclocytic, with the subsidiary cells petalocytic, unequal, strongly cutinized, often papillate. The pollen organs were catkin-like strobili, with the lateral sporangiophores arranged spirally or pseudowhorled, consisting of a slender, apically reflexed stalk bearing an adaxial synangium. The latter was 4-lobed, spreading at dehiscence, each lobe vascularized by a branch of a single pedicellar bundle. The gynoclades were terminal on short shoots, with scale leaves at base, loose, branching spirally, with the lateral ovuliphores distant or overlapping, consisting of a stalk and a terminal bivalved cupule (Harris, 1951b; Krassilov, 1969). The valves were

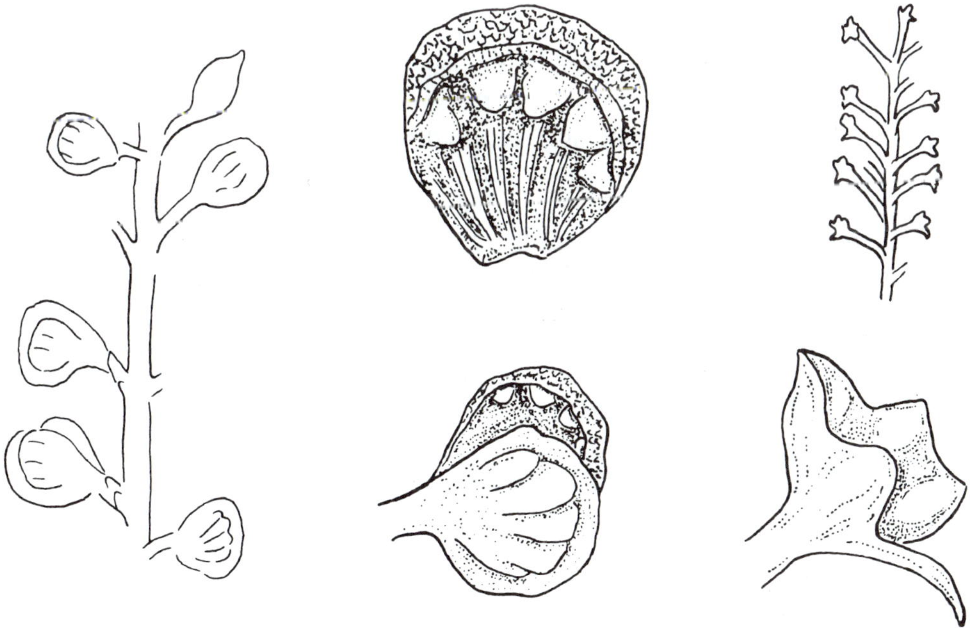

Fig. 4. Czekanowskialean reproductive structures, from left to right: *Leptostrobus* gynoclade, bivalved cupule and valve showing a stigmatic marginal flange and ovules, *Ixostrobus* androclade and synangium (after Krassilov, 1969, 1982).

gaping at maturity, externally ribbed and strongly cutinized, with stomata, internally thinly cutinized, with a broad sutural flange. In the advanced species, the latter was internally papillate, apparently stigmatic (Fig. 4). The ovules were 5-8 per valve, submarginal, inverted, shed from the ripe cupules.

Selected fossil links

Phylogenetic links are the forms either showing intermediate character states in the character polarity clines or combining the characters typical of the taxa that replace each other in a geochronological series, or chronocline. The latter forms are due to mosaic evolution, that is, the discrepant evolution rates of different functional units. A few such fossil links are selected for further discussion of the gymnosperm phylogeny and of the origin of proangiosperms.

Cycandra Krassilov & Delle in Krassilov & al., 1996; Middle Jurassic of the Caucasus (Plates 1,2).

The genus is based on the pollen cones that are about 20 cm long, compact, bearing spirally arranged, imbricate, peltate sporangiophores. The peltae of the latter are rhombic, with a sterile marginal flange that abaxially shows cyclocytic stomata and is sharply delimited from the fertile central part, the latter containing a large adaxial synangium of multiple, concentrically arranged, vertically aligned sporangia. The synangium is covered with a delicate cuticle the presence of which is indicative of a hypodermal development of the sporangia. On the surface, the synangium shows adpressed prismatic sporangial heads alone. The sporangia, exposed by maceration, are tubular, laterally adnate and proximally fused. The sporangial wall consists of two cutinized layers, the outer epidermal, forming a thickened apical head traversed by a short opening slit, the inner endothecial, with an apical beak often filled with pollen grains that are produced in great quantities, passing through the bottle-neck-like sporangial beak before dispersal. The pollen grains are anasulcate, asaccate, with alveolar infrastructure.

Although resembling the pollen cones of cycads and nilssonias (*Androstrobus*), *Cycandra* differs in the laterally adnate sporangia forming a giant synangium the concentric structure of which reminds of synangial structures occurring in the advanced medullosalean pteridosperms. The dehiscence mechanism of the sporangia, with an apical slit and an endothecial beak under it, is unique in cycadophytes supposedly mediating a gradual dispersal of pollen grains from the long-living giant pollen cones.

Semionogyna Krassilov & Bugdaeva, 1988c; Lower Cretaceous of Transbaikalia (Plate 3).

The genus is erected for the large spicate gynoclades with a stout axis bearing ovules subtended by bracts, the latter arranged spirally, spreading radially, narrow lanceolate, expanded and decurrent basally, adhering to the ovules. Several distal bracts are empty.

Fig. 5. *Semionogyna bracteata* Krassil. & Bugd., a protocycadean plant from the Early Cretaceous of Transbaikalia: ovulate spikes with juvenile and ripe ovules subtended by the bracts, with distal bracts empty, compared to extant *Cycas revoluta* (right), (after Krassilov & Bugdaeva, 1988c).

Young ovules are shorter than bracts and are protected by the bract bases, protruding at maturity, with the bracts pendent and eventually shed. The ovules are large, sessile, erect, radiospermic. The integument is thick, strongly cutinized, with stomata. The nucellus is free, cutinized. The associated pollen cones are loose, bearing spirally arranged sporangiophores that are stalked, each with a head of the proximally fused sporangia. The associated leaves are 2-pinnate, with interstitial pinnules.

In this genus, the ovules are definitely cycadeoid. In extant *Dioon*, the ovules sometimes show basal appendages (Sabato & De Luca, 1985) that may correspond to vestigial subtending bracts, while the "pinnae" of the leafy apex could be formed of empty distal bracts (Fig. 5). At the same time, the seed cones resemble *Sobernheimia*, a rare morphotype of the Permian ovuliferous organs associated with callipterids (Kerp, 1983). In addition, the associated pollen cones and leaves are similar to those of the typical callipterids.

Dirhopalostachys Prynada ex Krassilov, 1975a; Late Jurassic of the Bureya Basin, Far East, Russia (Plate 4).

The gynoclades are large, racemose, with the lateral ovuliphores spirally arranged, consisting of a stalk and a pair of terminal cupules. The latter are connate at base, distally spreading at an acute angle, follicular, laterally compressed, with a long, reflexed beak and with the abaxial suture extending along the beak. The cupule is ribbed, with the ribs diverging from the suture, the cupular wall is pubescent, containing resin ducts, lacking stomata. The beak is ribbed, abaxially pitted, with the suture marked by a median groove, gaping in the ripe cupules (Fig. 6). The ovules are solitary, filling the

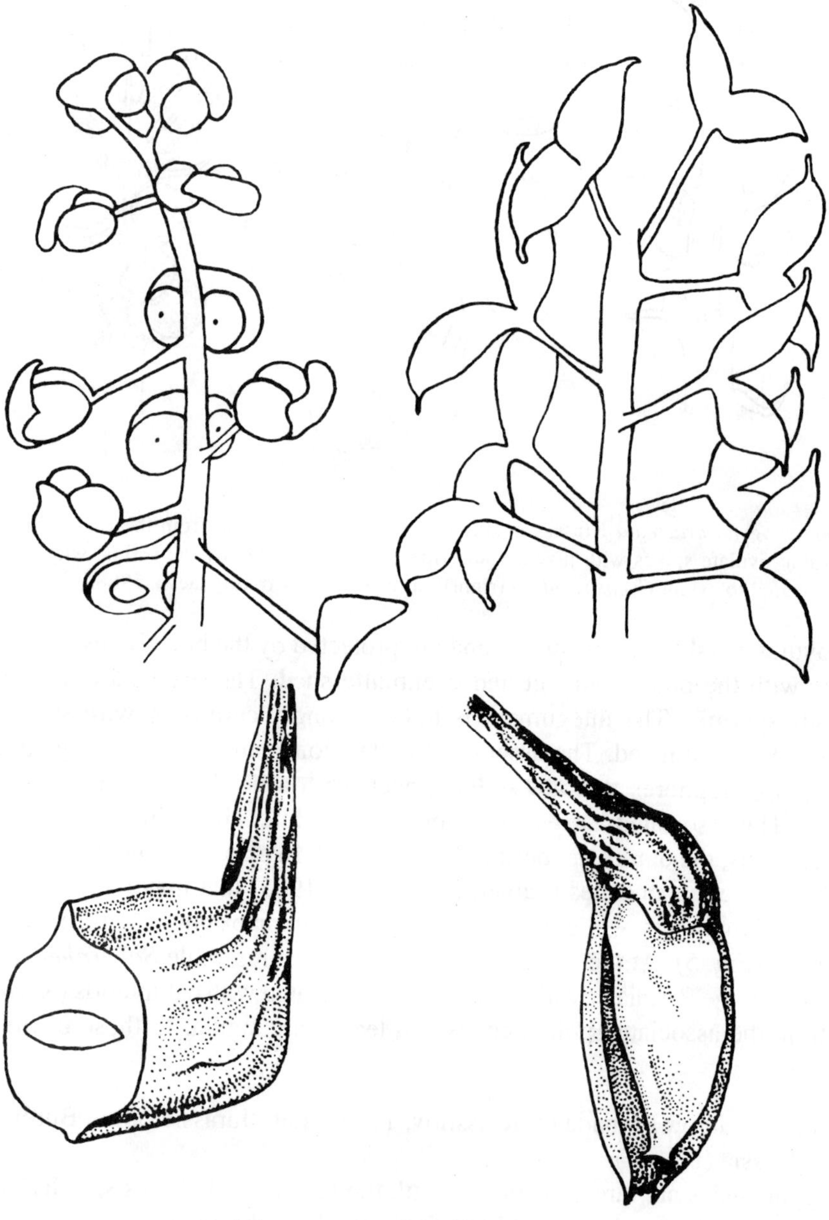

Fig. 6. *Dirhopalostachys rostrata* Krassil., a proangiosperm from the Late Jurassic of Burya Basin, Far East, Russia: gynoclade bearing paired cupules (right) and split cupules showing a soliltary anatropous ovule and a rugulate beak (after Krassilov, 1975) compared with a supposedly progenitorial *Beania* gynoclade (left, after Harris, 1964).

locule except in the basal portion, inverted, platyspermic. The micropyle is short, indistinctly lobed. The integument is single, glabrous, thinly cutinized. The nucellus is free. The megaspore membrane was not observed. The shed cupules are empty, filled with rock matrix. The pollen grains sticking to the cupule are anasulcate, asaccate, of the *Cycadopitys* type. The associated leaves are simple, lobed, of the *Nilssonia* type.

The paired cupules of *Dirhopalostachys* seem derivable from the peltate ovuliphores of the Nilssoniales (*Beania*) by involution of the peltae bearing two inverted ovules (Krassilov, 1975a). At any rate, they evidence the possibility of the involute follicular cupules arising in cycadophytic plants. Whether the pollen grains entered the ovules or germinated on the pitted beak surface, is still to be learned, but the latter possibility appears more likely because the suture of the beak was sealed in unripe cupules. Externally at least, the paired cupules resemble the gynoecia of hamamelid angiosperms and, in particular, the Cretaceous *Trochodendrocarpus* (see below).

Irania Schweitzer, 1977; Early Jurassic of Iran.

The fertile shoot bearing proximal pollen strobili and a distal gynoclade. The pollen strobili are compact, consisting of peltate sporangiophores with two or more sporangia aligned with the stalk. The ovuliphores are distant, stalked, bearing a terminal bilobed cupule, the latter dehiscing into two valves with a flat marginal flange (Fig. 7).

Irania is interesting as an early example of monoclinous structure, although the interpretation is not yet conclusive. Probably related to *Irania* are the Early Cretaceous *Ktalenia* gynoclades bearing bracteate cupules (Taylor & Archangelsky, 1985).

Fig. 7. *Irania,* a Jurassic proangiosperm: reconstruction of a fertile short shoot bearing a central gynoclade and peripheral androclades, and part of the gynoclade with bivalved cupules (modified from Schweitzer, 1977).

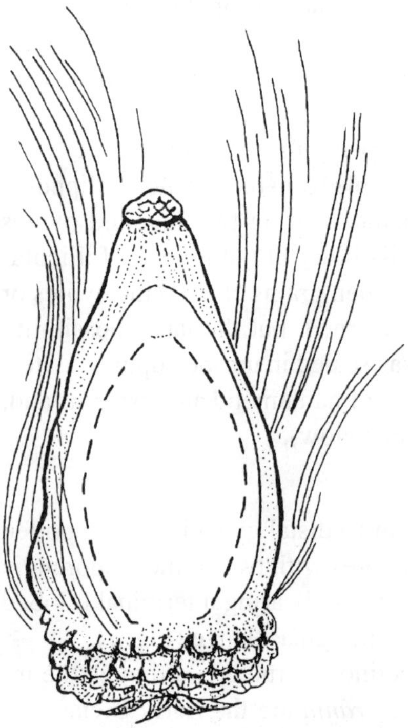

Fig. 8. *Baisia hirsuta* Krassil., a proangiospermous plant of bennettitalean affinities of the Baisian Assemblage, Early Cretaceous of Transbaikalia, reconstruction (after Krassilov & Bugdaeva, 1982).

Baisia Krassilov in Krassilov & Bugdaeva, 1982; Baisian Assemblage, Early Cretaceous of Transbaikalia (Plates 5-7).

The genus comprises dispersed cupulate ovules that are apical on a short rounded-triangular receptacle that is persistent in disseminules, and is bearing and is minute bracts and tufts of long bristles (Fig. 8). The cupule is flask-shaped, apically thickened in a hairy corona, thick, utriculate, with the exterior epidermal cells large, thick-walled, tabulate, scarcely cutinized, the stone cells large, porous, and the interior epidermal cells thinly cutinized, papillate. The ovule is solitary, erect, basal, not filling the locule. The integument is thin, cutinized, forming a short micropyle. The nucellus is free, with a low chalazal platform. The pollen chamber is scarcely differentiated. The pollen grains trapped in the nucellar beak are anasulcate, asaccate, with alveolar infrastructure.

The taphonomy of *Baisia* that came in large numbers from lacustrine deposits suggests an aquatic or water-side plant dispersing the wind and/or water-borne "achenes" while its vegetative parts scarcely reached the deposition site. The most probable candidate for the leaves is *"Dicotylophyllum" pusillum* (Vakhra-meev & Kotova, 1977), small, petiolate, with a lanceolate blade and with irregular eucampt-odromous venation. An additional specimen recently found by Dr. Eugenia Bugdaeva (Fig. 9) is more bract-like, thick with an indistinct immersed lateral veins, pubescent and comparable to bennettitalean bracts, except that the venation make it angiosperm-like.

The baisian disseminules are superficially similar to cyperaceous achenes, the analogy extending not only to the tufts of bristles but also to the large tabular surface cells. However, the pollination was obviously gymnospermous. The persistent hairy receptacles were common in bennettites in some lineages of which the number of ovules was reduced to several or even two, the ovules sometimes enveloped in a cupule formed of interseminal scales. *Baisia* might have been an extreme form with a single cupulate ovule developing from the apex of the receptacle. The vestigial corona, the structure of the ovules, and the associated pollen grains are definitely in favour of the bennettitalean affinities.

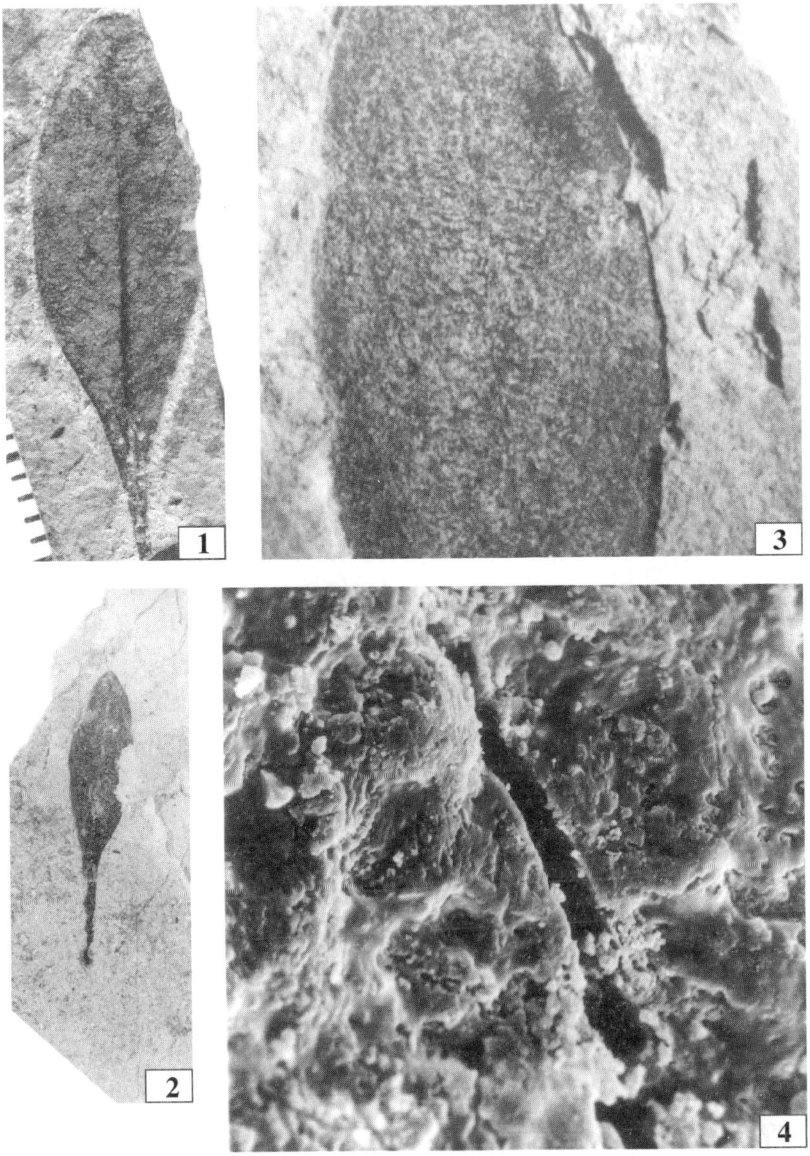

Fig. 9 Angiosperm leaves from the Baisian Assemblage, Early Cretaceous of Transbaikalia: (1) *"Dicotylophyllum" pusillum* Vachr. (Vakhrameev & Kotova, 1977), x3 (courtesy of A.B. Herman), (2-4) another specimen found by Dr. Eugenia Bugdaeva, perhaps of the same species, but more bract-like (2), x2, with indistinct oblique lateral veins (3), x4, and round structures (hair bases?) on the coating mineral film (4), SEM, x630.

Dinophyton Ash, 1970, emend. Krassilov & Ash, 1988; Late Triassic of Arizona (Plates 8, 9).

The shoots are copiously branched, heteroblastic, with scaly and needly leaves, the latter decurrent, round in cross-section, occasionally forked or even pinnate, pubescent and papillate, with two veins converging toward the apex. Occasional specimens with involute pinnae-like leaves suggesting a transitional caulome/phyllome nature of the leafy shoots (Fig. 10). The stomata occur both on the shoot axis and on leaves, are sunken, mostly tetracytic. The pollen cones are small, with four cross-like bracts at the base and with a few sporangiophores encased in bracteoles, the latter pubescent and with stomata as on leaves. Ripe sporangiophores are protruding, shortly stalked, concave, with a serrate membranous fringe one cell thick. They bear a solitary adaxial

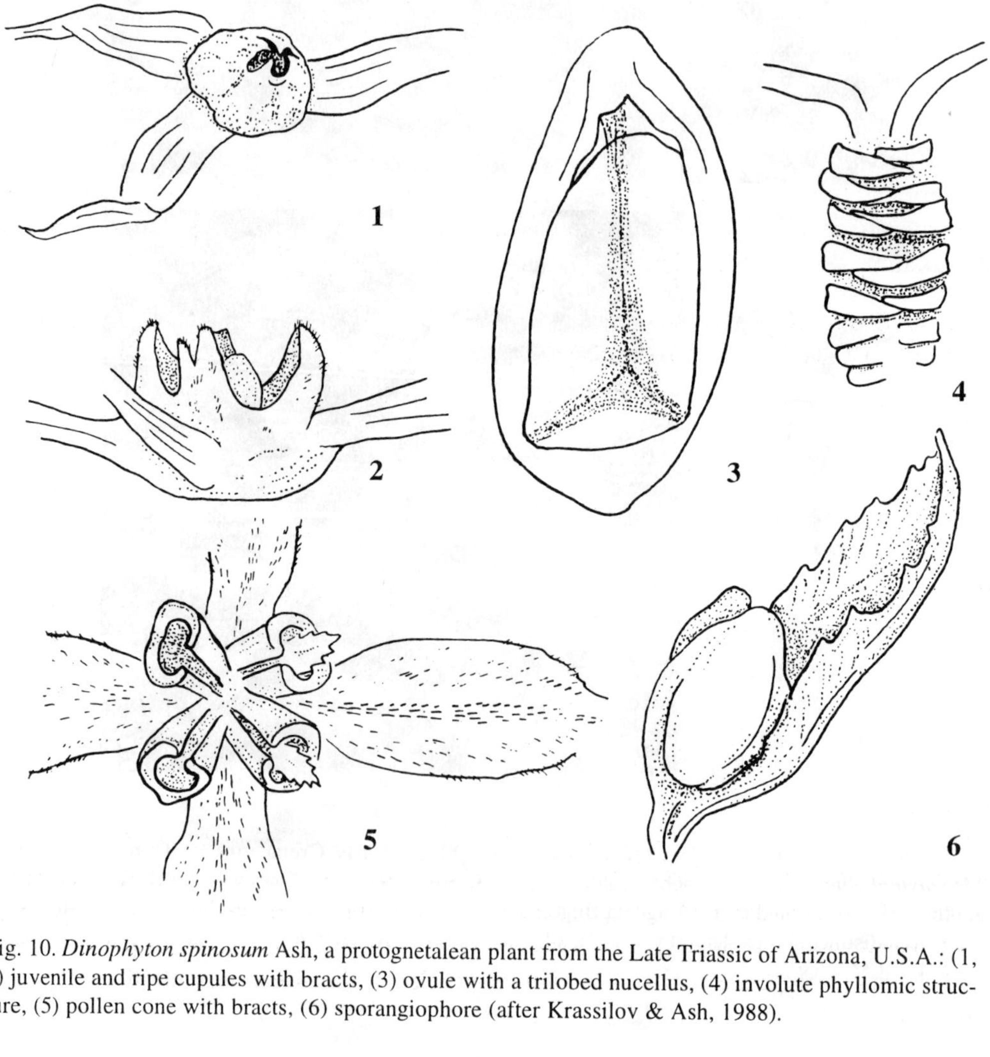

Fig. 10. *Dinophyton spinosum* Ash, a protognetalean plant from the Late Triassic of Arizona, U.S.A.: (1, 2) juvenile and ripe cupules with bracts, (3) ovule with a trilobed nucellus, (4) involute phyllomic structure, (5) pollen cone with bracts, (6) sporangiophore (after Krassilov & Ash, 1988).

sporangium or a group of three apparently fused sporangia that are basifixed, dehiscing in two valves. The pollen grains are bisaccate. The ovuliphores are 4-lobed pedicellate cupules borne on shoots of different orders, with a 4-lobed cross-like perianth formed of two decussate pairs of bracts and fused to the base of the cupule, persistent as a propeller in the fruit-like diaspora. The cupule lobes are imbricate over a slit-like exostome, spreading at maturity, pubescent, with the hairs longer and tufted on tips, cutinized, showing irregular-polygonal cells, lacking stomata, each receiving a single vascular bundle. The propeller bracts are flat, broadly attached or slightly constricted at the base, tapering toward the apex, less strongly pubescent and with the cuticle thicker than on the cupule, showing rectanguloid cells arranged in longitudinal rows. The interior cuticle of the cupule is thin with distinctly marked, large, polygonal cells. The ovule is basal, erect, 3-keeled, not filling the locule. The integument is scarcely cutinized, the nucellus is free at the micropylar end alone, forming a broad pollen chamber. The megaspore membrane is thick, showing a tetrad mark.

A phylogenetic significance of *Dinophyton* lies in the combination of the gnetalean and pteridospermous characters. The former include the decussate perianth bracts both in the pollen cones and seed organs and the 4-lobed bracteolate cupule containing a solitary erect ovule, whereas the structure of the latter is like in the hydrospermalean *Lyrosperma*, and the leafy shoots show transitional caulome-phyllome characters, as in the calamopytiacean foliar branch systems.

Sanmiguelia R.W. Brown, 1956, emend. Cornet, 1986; Late Triassic of Arizona.

Rhizomatous shrubs with slender aerial stems branching by axillary buds, eustelic, producing a relatively thick secondary wood. The tracheary elements with helical or reticulate thickenings. The shoots are heteroblastic, with spirally arranged cataphylls and foliage leaves, the latter up to 30 cm long, sheathing, plicate, with numerous, subparallel, thick and thin veins converging toward the apex, anastomosing, looping or the thin veins coalescing to restore a thick vein, with resin ducts in the sheaths. The leaf cuticle is thin, showing anomocytic or cyclocytic stomata. The pollen cones of spirally arranged sporangiophores with bisporangiate synangia dehiscing by a longitudinal slit. The pollen grains are anasulcate, with granular infrastructure. The seed organ is paniculate, bearing terminal clusters of bracteate cupules. The bracts are decussate, adnate to the cupules, heteromorphic, the lower pair dissected, the upper pair entire, pubescent. The cupule is bilobed apically and with two lateral pocket-like appendages, biovulate. The ovules are basal anatropous, some containing embryo with two heteromorphous cotyledons.

Sanmiguelia is remarkable for a monocot aspect of its plicate leaves and paniculate inflorescences. The "goniopteroid" venation pattern, as in *Welwitschia,* the decussate pericupular bracts, the pocket-like bracteoles (?), and the cupule morphology suggest a gnetalean affinity which is, however, inconsistent with the allegedly anatropous ovules. The alternation of large and small vascular bundles in the apparently phyllodic leaves finds its analogy in the axial vascular system of *Ephedra*.

Hirmerella Hoerchammer, 1933, emend. Jung, 1967; Jurassic of Central Europe (Plate 10).

The generic name was corrected by Jung (1967) from a grammatically incorrect *Hirmeriella*; the preoccupied *Cheirolepis* Schimper and the substituting *Cheirolepidium* Takhtajan are synonymous to *Hirmerella*.

Trees or shrubs, sometimes preserved in situ in littoral deposits (Oldham, 1976; Francis, 1983). The wood is coniferoid, with pinoid or cupressoid pitting of tracheary

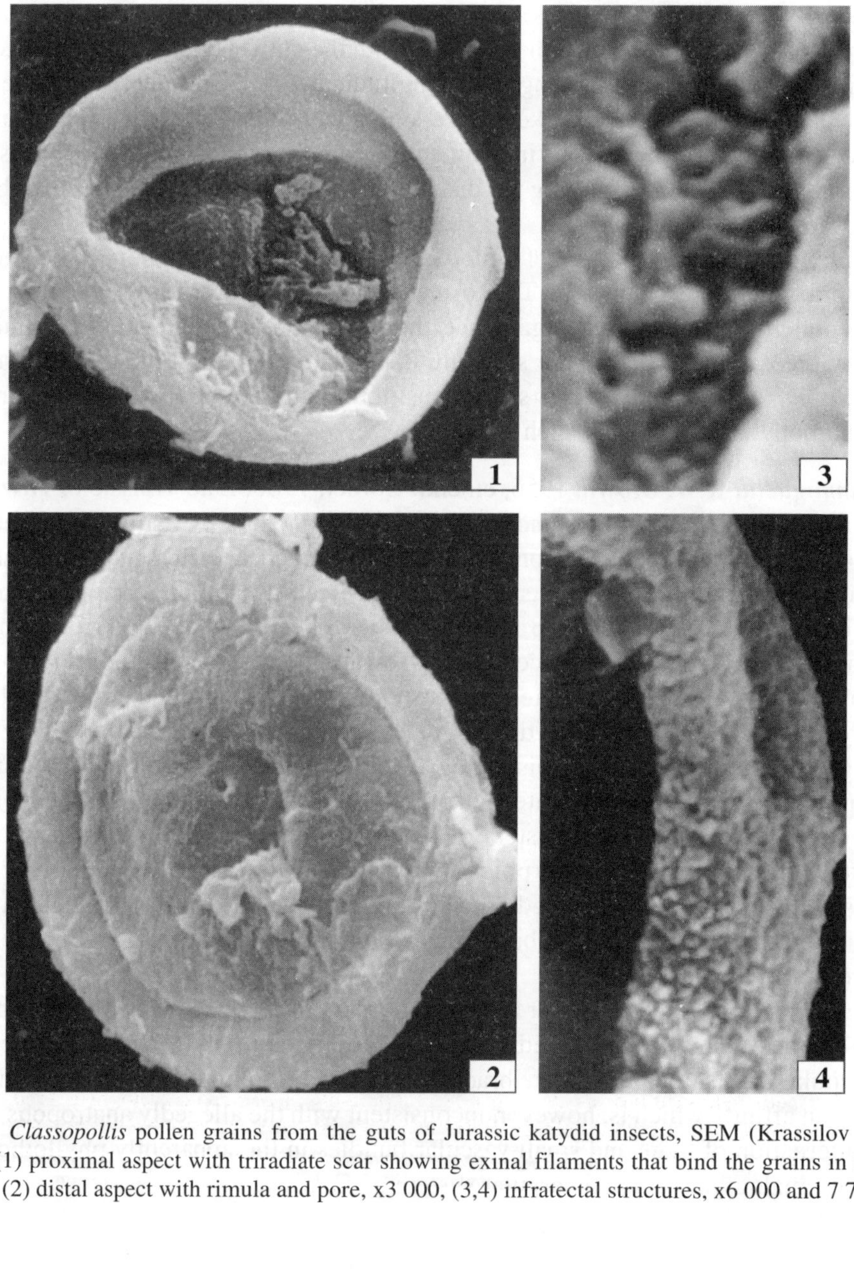

Fig. 11. *Classopollis* pollen grains from the guts of Jurassic katydid insects, SEM (Krassilov & al., 1997): (1) proximal aspect with triradiate scar showing exinal filaments that bind the grains in tetrad, x3 000, (2) distal aspect with rimula and pore, x3 000, (3,4) infratectal structures, x6 000 and 7 700.

elements that are sometimes "septate" (Alvin & al., 1978). Branching is axillar. The leaves are scaly, sometimes dimorphic, spiral or decussate or verticillate, decurrent or sheathing, sometimes deciduous or the short shoots (branchlets) abscissed, xeromorphic with a thick cuticle, often pubescent and papillate. The stomata are sunken, cyclocytic. The pollen cones are bearing peltate sporangiophores, with sporangia 6-8 in pendent clusters, sometimes synangiate (Archngelsky, 1986a). The pollen grains (*Classopollis*) sometimes in permanent tetrads, interconnected by sexinal filaments, with an equatorial sac-like girdle, a distinct triradiate mark, an incipient subequatorial zonosulcus (rimula), and a distal porous leptoma (Taylor & Alvin, 1984), the latter showing a microporous membrane and perhaps aperturate (Krassilov & al., 1997a). The pollen wall infrastructure is columellate-rugulate, sometimes granulate (Fig. 11). The seed cones with bracts subtending axillar ovuliphores that were usually considered as seed-scales but were reinterpreted by Krassilov (1982a) as lobed cupules with adnate lateral bracteoles. They are basally attached to the bracts, but are shed separately as samaras, with the bracts remaining on the axis (Fig. 12). The cupules are thickly cutinized, with stomata as on leaves. The locule is lined with a thin cuticle showing sinuous cell walls. The ovule is solitary, erect, adpressed to the locule. The integument is thinly cutinized. The nucellus is free, cutinized, containing a thick megaspore membrane or, occasionally, two megaspores. Probably entomophilous, with the pollen germinating on the cupule.

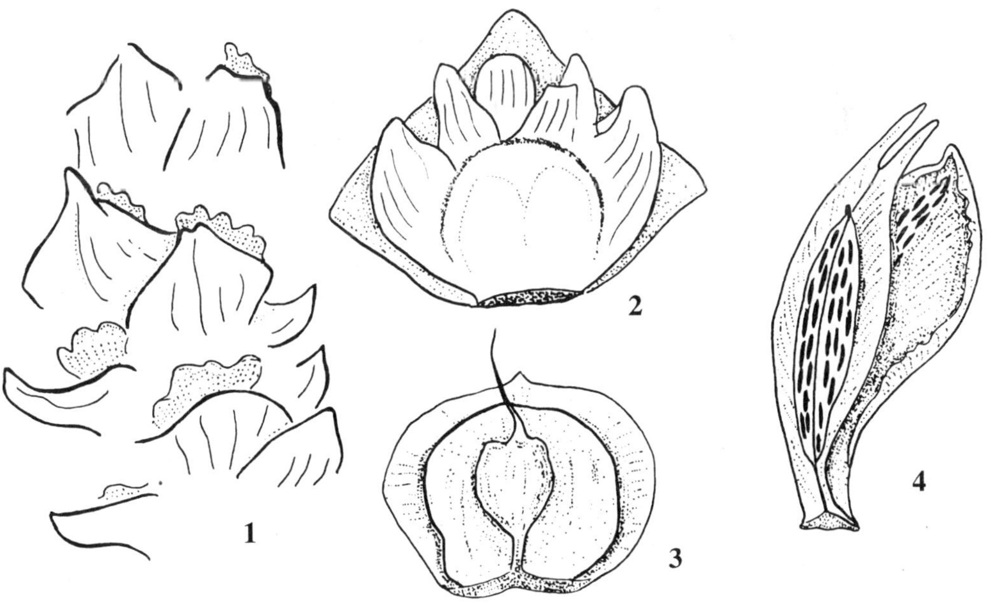

Fig. 12. Gnetophyte seed-bearing structures: (1,2) *Hirmerella*, a Mesozoic gnetophyte, seed cone and a winged samaroid cupule axillar to bract, (3,4) samaroid bracteate cupules of extant *Welwitschia* and the Jurassic *Heerala -Angarolepis* plant (reconstruction based on dispersed samaras and bracts: Krassilov & Bugdaeva, 1988a).

Hirmerella is usually assigned to conifers and is actually similar to the Permian coniferoid genus *Majonica* (Clement-Westerhof, 1987), but the cupules show an interior cuticle, thus differing from epimatia of conifers with embedded ovules. On the other hand, the bracteate cupules with adnate lateral bracteole-like appendages that are transformed into wings in the samaras resemble those of *Welwitschia* (sometimes also having two functional megaspores per nucellus). The decussate sheathing leaves and the "septate" tracheary elements are additional gnetalean characters. The pollen grains are remarkable in showing incipient angiospermoid structures.

Heerala Krassilov & Bugdaeva, 1988a; Middle Jurassic of Siberia (Plates 11,12).

The genus is based on *Ephedrites antiquus* Heer, 1978, the large samaras with a bilobed membranous wing, a spindle-shaped central body with conspicuous resin ducts, a median suture line and a callous scar at the base. The associating spoon-like bracts with identical resin ducts might have been attached at the basal scar. They are petiolate or sessile, bluntly pointed or acuminate, showing cyclocytic stomata, sometimes bearing a triangular appendage, probably the attached prophyll of an axillary ovuliferous shoot. The bracts and samaras are like those of *Welwitschia*, also having conspicuous callous attachment scars at the base. Even the blunt and acuminate bract variants have their analogues in the respectively distal and proximal bracts of *Welwitschia*. The associated pollen cones are bearing spirally arranged peltate sporangiophores with the radially arranged sporangia adpressed to the peltae, dehiscing by an apical pore or a short slit. The pollen grains are anasulcate asaccate. All that is presently known of the vegetative parts are the detached dagger-shaped leaves that are basally expanded, apparently sheathing. The leaf blade is membranous, with a thickened median zone, variably supplied by 1-3 veins as in extant *Ephedra chinensis* (Foster & Gilford, 1974).

Drewria Crane & Upchurch, 1987; Early Cretaceous of eastern North America.

Branching shoots with decussate sheathing leaves. The venation is "goniopteroid". The seed cones are borne in terminal or lateral dichasia, spicate. The ovules are subtended by a pair of bracteoles. The associated dispersed pollen grains are ribbed, ephedroid.

Drewria shows a remarkable combinatin of a herbaceous life-form with the *Welwitschia*-like leaf venation and gynoclade morphology.

Leongathia Krassilov, Dilcher & Douglas in Krassilov & al., 1997b; Koonwarra Fossil Bed Assemblage, Early Cretaceous of Australia (Plates 13-15).

Slender articulate shoots with the distal internodes much shorter than the proximal and the distal leaf whorls crowded or even conjugate. The shoot axis is longitudinally ribbed, with a transverse rugulate structure reflecting the sclerenchymous fibres. The nodes are typically 4-leaved, occasionally 2-3- leaved, with up to 8 leaves in the terminal whorls. The leaves are decussate, spreading at acute angles or almost horizontally or pendent, lanceolate, acute, about 2 mm long, connate at the base, sheathing about 1/3 to

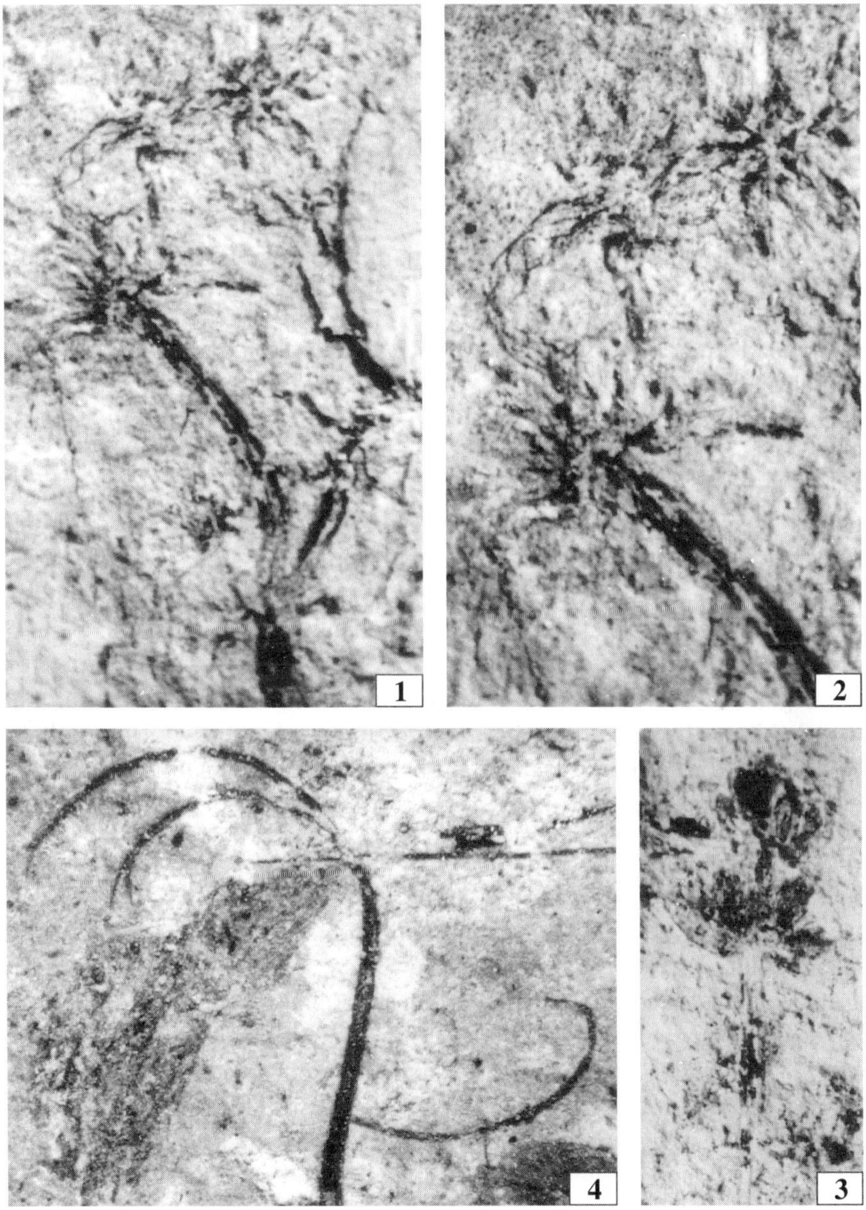

Fig. 13. *Baisianthus*, a gnetalean androclade of the Baisian Assemblage, Early Cretaceous of Transbaikalia (Krassilov & Bugdaeva, 1997): (1, 2) branching shoot bearing successive whorls of the distally crowded cupulate sporangiophores on the central axis and a terminal whorl on a side branch, with linear bracts at the nodes, x8 and 10, (3) shoot fragment with two sporangiophore whorls, x7, (4) axis with tendril-like appendages found with androclades, possibly of the same plant, x20.

1/2 of internode or with the sheaths reduced and then directly inserted upon the node. The stomata are densely crowded in the grooves between the ribs, sunken, encircled by a thick annular ridge. The cortex traversed with a sclerenchymous network. The vascular system consist of alternating large and small bundles. The leaf traces are paired, departing from the nodal girdles. The tracheary elements are pitted, with reticulate and helical wall thickenings. The pits occur in two rows or occasionally crowded and some confluent forming porous perforations at the poles.

This fossil is the first evidens of the ephedroid gnetaleans existing in the Early Cretaceous. It deffers from extant *Ephedra* in more numerous leaves per node and in the leaf to vascular bundle ratios.

Baisianthus Krassilov & Bugdaeva, 1897; Baisian Assemblage, Early Cretaceous of Transbaikalia (Plates 16-19).

Branched androclades bearing flower-like whorls of bracteate cupules that are subtended by bracts. The cupules enclose sporangiophores that are bearing synangia on the short lateral branches and terminally. The synangia are 2-4-sporangiate, showing various degrees of sporangial fusion, the terminal sporangia sometimes free, producing monosulcate pollen grains in tetrahedral tetrads. The pollen grains are fusiform, cavate, with sexine loosely attached to the nexinal body and easily separable. The tracheary elements of the axis are septate, with pits and helical thickenings that are occasionally transformed into oblique perforation plates.

These organs are basically similar to the androclades of *Gnetum,* but of a peculiar branching habit, with the fertile whorls more flower-like and with the pollen grains perhaps derived from a monosaccate morphotype. The associating vegetative organs with tendril-like appendages (Fig. 13) might belong to the same plant.

Eoantha Krassilov, 1986; Baisian Assemblage, Early Cretaceous of Transbaikalia (Plates 20-22).

Flower-like gynoecial structure of four ovuliphores surrounded by a perianth of linear bracts, radially spreading in the open flowers, with the floral axis protruding and bearing apical bracts that supposedly representi a vestigial second floral node. The ovules are single per the gynoecial lobes, orthotropous, with the nucellus trapping ribbed *Ephedripites*-type pollen grains in the beak and with a conspicuous granular megaspore membrane. A recently found specimen (Fig. 14) is attached by a slender stalk to the axis with the vascular bundles consisting of pitted fibres and tracheary elements, the latter showing distant pits and the helical to reticulate thickenings.

Similar vascular elements have been observed in the vein bundles of the associating parallelinerve graminoid leaves that might belong to the same plant. The leaves (Plates 23, 24) show short cross-veins and longitudinal rows of intercostal stomata that are stretched as if the blade growth continued after the stomatal pattern was laid down.

The gnetalean affinities of *Eoantha* are evidenced by the basically decussate, although spreading in one plane, gynoecial lobes, orthotropous ovules and characteristic

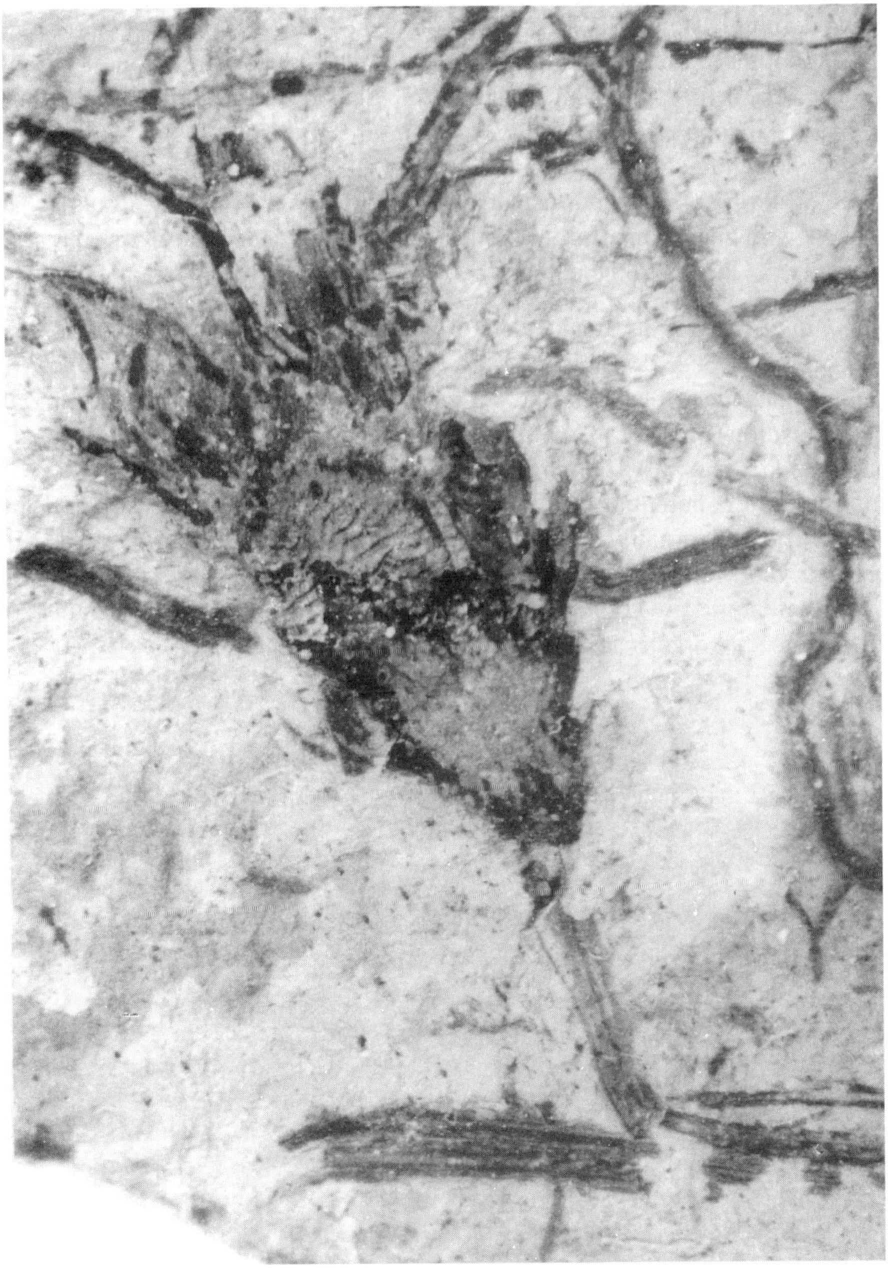

Fig. 14. *Eoantha zherikhinii* Krassil., a proangiospermous gnetalean plant of the Baisian Assemblage, Early Cretaceous of Transbaikalia (Krassilov, 1986): recently found attached flower (Krassilov & Bugdaeva, 1997), x15.

pollen grains. At the same time the gynoclades are reduced to a single fertile node, thus becoming definitely flower-like. Leaves attributed to the *Eoantha* plant on the evidence of association and vascular similarities suggest progenitorial relations to graminoid monocots. Some incompletely known Mesozoic plants, perhaps related to *Eoantha*, are represented by dispersed ovules with *Eucommiidites* pollen grains in the pollen chamber. The pollen grains are characteristic in having exinal slits parallel to the sulcus (Pedersen & al., 1988). The presence of a megaspore membrane was seen as a character inconsistent with gnetalean affinities of these seeds (Reymanowna, 1968), but presently *Eoantha* gives evidence to the contrary.

Vitimantha Krassilov & Bugdaeva, 1997; Baisian Assemblage, Early Cretaceous of Transbaikalia (Plates 25-27).

Longly pedicellate flower-like structure with a perianth of four leafy bracts showing reticulate venation (Fig. 15) and with four elongate follicles or cupules that are marginally papillate, occasionally trapping ephedroid pollen grains.

This structure is potentially of a considerable phylogenetic interest resembling an angiosperm flower (notably of *Kingdonia*) with leafy reticulinerve tepals and the apocarpous gynoecium of ascidiform follicles. The latter might have stigmatic marginal crests. These features are combined with the typically gnetalean pollen grains. Yet the supposed gynoecial organs are poorly known.

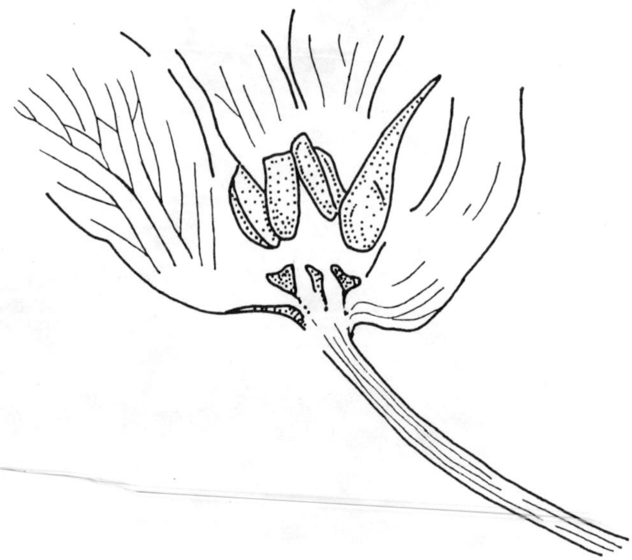

Fig. 15. *Vitimantha*, a preflower of the Baisian Assemblage, Early Cretaceous of Transbaikalia (Krassilov & Bugdaeva, 1997), showing four bracteate tepals with anastomosing veins and follicular structures (polyplicate pollen grains were found stuck to the margin of the right follicle).

Caytonia sewardii Thomas, 1925; Middle Jurassic of Yorkshire, England (Plates 28-30).

This species of caytonialean ovuliferous structures was described as foliar, with the cupules formed of involute lateral pinnae, their tips still conspicuous as short "lips" over the exostome, or "mouth" - a slit between the lip and the stalk (Harris, 1951a). I have been able to study a few specimens of this species (courtesy of Professor T.M. Harris), suggesting a somewhat different interpretation. With SEM, I have observed the mouth situated on the abaxial side of the lip turned away from the stalk rather than at the stalk to lip junction. This makes the involute pinnae model improbable. The lip appears decurrent on the pedicel and adnate to the base of the cupule in the manner of a reduced subtending bract (Fig. 16). Then the exostome could have been formed along the cupule to subtending bract junction at the site of their fusion meristem. The ovules have been found in a tight cluster rather than scattered parietal as in the classical reconstruction, their micropyles of bulging cells adpressed to each other and nearly at the same level, suggesting an original clustering rather than a depositional compression. The micropyles are facing the exostome and are connected to it by a series of the so-called channels that sometimes look like funnicles attached near the micropyle, the ovules thus being anatropous, their funnicles modified as hollow tubes acquiring a function of conducting channels for pollen grains entering the exostome. A few grains have actually been found stuck midway in the channels. The ovules are flattened, with a raphe as a median ridge

Fig. 16. *Caytonia*, gynoclade of stalked cupules and a suggested prototype of the cupule (center) with a subtending bract transformed into "lip" of the mouth (right) (after Harris, 1951a; Krassilov, 1977, 1984).

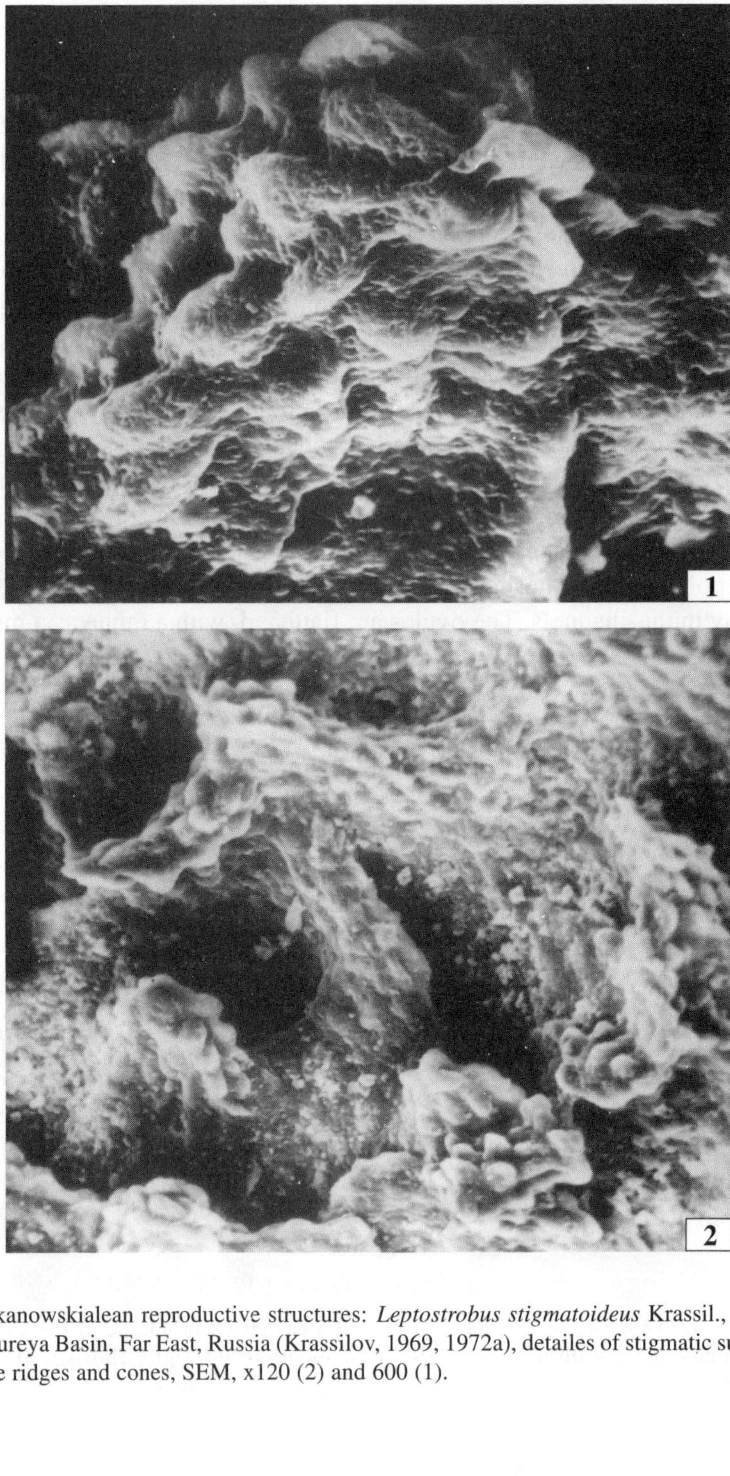

Fig. 17. Czekanowskialean reproductive structures: *Leptostrobus stigmatoideus* Krassil., from the Late Jurassic of Bureya Basin, Far East, Russia (Krassilov, 1969, 1972a), detailes of stigmatic surfaces, pitted, with papillate ridges and cones, SEM, x120 (2) and 600 (1).

conspicuous also on the nucellus that is free and cutinized, showing a basal platform, apically forming a poorly differentiated pollen chamber.

An enigmatic feature of the ovules is the so-called spotted layer previously interpreted as consisting of eleuron cells (Harris, 1951a). In cleared ovules, it can be seen as a well-differentiated coat two cells thick, thinly cutinized, enveloping the nucellus and wedging out at the level of the pollen chamber. Both topologically and morphologically, it corresponds to the inner integuments of bitegmic angiosperm ovules. Its spotted appearance can be due to tannin accumulations that commonly occur in the inner integuments of angiosperms, rather than to eleuron grains.

Leptostrobus stigmatoideus Krassilov, 1972b; Late Jurassic of the Russian Far East (Plates 31-32).

This species of czekanowskialeans deserves special mention on account of its stigmatic cupules. It is based on the gynoclades consisting of a relatively thin axis bearing loosely spirally arranged bivalved cupules that appear oriented vertically, with their median plane along the axis. The cupules are shortly stalked, occasionally with a stout cylindrical stalk bearing rudimentary scales, as in the leafy short shoots of the same plant. This is taken as evidence of their short shoot derivation, the cupular valves then corresponding to a pair of ovuliferous scales that are marginally fused, as in paracarpous ovaries. In this interpretation, the marginal flange might have evolved as a product of the fusion meristem activity. The valves remain adnate to 1/3 their length, distally gaping at maturity. Their outer wall is plicate with strong flabellate ribs, strongly cutinized, with stomata, whereas the interior cuticle is delicate but fairly resistant to maceration, lacking stomata. The ovules, four per valve, are submarginal on interior ribs. The sutural crests are smooth externally, pitted internally, microscopically with a network of steep ridges bordering rounded depressions and covered with simple or bifid papillae (Fig. 17), the latter analogous to the typical stigmatic papillae of angiosperms. The crests intrude the locule as a thin ledges 1.5 mm wide overhanging the ovules. This interior ledges, also having its analogues in angiosperms, is densely papillate, but lacks a reticulate ridge structure of the outer flange. The spherical bodies sticking to the papillae could be deformed pollen grains, but none of them is sufficiently well-preserved. In distinction from the mid-Jurassic species lacking stigmatic flanges (Harris, 1951b), no pollen grains have ever been observed inside the cupules of *Leptostrobus stigmatoideus*.

Gymnosperm phylogeny

While the above gymnosperm and proangiosperm divisions represent the distinctive adaptive types, their phylogenetic relations are to be inferred on the basis of homology of their respective morphological structures. Discrimination of homology and homoplasy is a much debated problem of apparently no once and for all solution. The cladistic techniques provide a cumulative assessment of homoplasy at most (Farris, 1989). Homology is more often inferred from phylogeny than vice versa, while phylogeny is based on oblique reasoning that is rooted in the contemporary evolutionary philosophy.

Insofar as phylogeny primarily conveys our understanding of homology, it cannot be more objective than the latter. The objective elements in it are chronological relationships as well as morphological links, that is, the intermediate forms that show character states midway in the polarity gradient or the mosaic forms that combine the typical characters of different taxonomic units (see above). A phylogenetic reconstruction would then start with assigning the operational units - gymnosperm orders in our case - to successive geological age groups reflecting their first appearances in the fossil record (not reflecting the overlaps of their stratigraphic ranges). The relations within and between the age groups would then be inferred on the basis of intermediate and/or mosaic forms.

Members of a single age group are likely to have independent origins or, if connected by intermediate forms, a common origin, that is, their relations are fraternal rather than progenitorial. On the other hand, the members of successive age groups, if connected by intermediate or mosaic forms, are likely to have progenitorial relations. Presently, the chronological relations between the major gymnosperm units are mostly based on a sufficiently high stratigraphic resolution and are fairly reliable. Discovery of linking forms may seem accidental but, since there were intermediate habitats and discrepant evolution rates for different functional units, the intermediate and mosaic forms ought to be found.

According to their first appearances in the fossil record, the gymnosperm orders can be assigned to the following age groups:

(1) Late Devonian - Early Carboniferous: Hydrospermales, Lagenostomales, Trigonocarpales (Lyginopteridales, Medullosales of the axial anatomy classification);

(2) Mid-Carboniferous: Callistophytales, Cordaitales, Walchiales;

(3) Late Carboniferous - Permian: Glossopteridales, Vojnovskyales, Gigantopteridales, archaic Peltaspermales (callipterids), archaic Coniferales (voltziaceans);

(4) Early Mesozoic: advanced Peltaspermales, Corystospermales, Nilssoniales, Ginkgoales, Czekanowskiales, Caytoniales, Bennettitales, archaic Gnetales (protognetaleans, such as *Dinophyton*);

(5) Late Mesozoic: Pentoxylales, Cycadales, modern Coniferales, modern Gnetales, their derived proangiospermous forms, such as *Eoantha*, etc.

The gymnosperms are thought to be rooted in progymnosperms of which both the heterosporous archaeopterids and the homosporous aneurophytes are considered as potential ancestors (Rothwell & Erwin, 1987). Moreover, there were herbaceous plants of a simpler axial anatomy but having elaborate cupule-like sporangial clusters, as in *Lenlogia* (Krassilov & Zakharova, 1995). Early seed plant diversity is insufficiently known, however, for establishing the progenitorial relations otherwise but in a very general form. The possibility of the gymnosperm anatomy and seeds originating in different lineages then combined by horizontal gene transfers in the Devonian plant communities cannot be excluded.

In the age group (1), the Hydrospermales with 2-many-ovulate cupules, sometimes containing sporangia (Long, 1977a), appeared somewhat earlier than the other two orders. That this type of cupules might have been primary is scarcely deducible from their morphology: it is just a geochronological fact.

The Hydrospermales merge with the Lagenostomales - their respective axial organs are assigned to a single order, Lyginopteridales, while the ovules are of the same basic type and are linked to the Trigonocarpales (Medullosales) by the forms with an intermediate callamopytialean-medullosalean anatomy, such as the Late Mississippian *Questora* (Mapes & Rothwell, 1980).

Both the Lagenostomales and Trigonocarpales show a tendency of a not quite synchronous transition of the pollen receptive function from the nucellar apex with its excessively developed pollen trapping and secretory structures, lagenostome and salpynx, to the integumental micropyles, with the lobed micropyles as a transitional stage. They differ, however, in the relative development of the inner, integumental, and the outer, cupular, coats of their ovules, a variation that can be traced up to the present-day angiosperms. While in the Lagenostomales the micropyle was primarily integumental, in the Trigonocarpales it might have been of a cupular origin, with the primary integument reduced and adnate to the nucellus, the latter thus transformed into a chimeric structure assuming the vascular system of integumental origin, hence non-homologous to the nucelli of the lagenostomalean line.

The fraternal phylogenetic relations of Lagenostomales and Trigonocarpales imply that the ovules borne on fern-like leaves, a critical feature of pteridosperms, appeared independently in both groups in the course of their parallel evolution. Such parallel developments of the fern-like foliage on the basis of syntelomic branch systems in spore plants and early gymnosperms might have been urged by a demand of larger photosynthetic surfaces in response to either a decreased ultraviolet radiation rate (due to oxidation of atmosphere, with the concomitantly increasing water vapour content and stratospheric ozone) or the build-up of the structural complexity, canopy formation and the ensuing competition for light in the early plant communities, or both. In distinction from the exposed hydrospermalean cupules, the leaf-borne cupules were less effective in wind pollination. The concomitant development of the large synangiate polliniferous structures and the secretory apical structures of the early ovules suggest pollination by arthropods finding shelter in the foliage and not yet engaged in the leaf-cutting activities (that appeared much later).

In the age group (2), the Callistophytales embodied a mosaic of forms combining an essentially lyginopteroid morphology of their vegetative parts and synangia with the platyspermic cardiocarpoid ovules (incipient platyspermy appeared already in the lagenostomalean ovules, see Stubblefield & al., 1984) and true pollen (*Vesicaspora*) also resembling some cordaitalean morphotypes (*Florinites*). Among cordaites, *Mesoxylon* is likewise a linking form with the eustele formed of leaf traces, as in megaphyllous plants, and with compound pollen cones producing prepollen. The seeds were, however, cardiocarpoid (Grove & Rothwell, 1980), like in the advanced forms that acquired a sympodial eustele and true pollen grains (Sterlin, 1982a). The Walchiales are obviously related to cordaites, their coniferoid foliage corresponding to the lower leaves of the latter (the cordaitalean ribbon-shaped parallelinerve foliage leaves might have been phyllodic). They still produced prepollen, as in the primitive cordaites, and their ovules, though borne in more or less strongly flattened axillar shoots and sometimes inverted, were of the same cardiocarpoid type.

Neither morphological nor chronological evidence suggests derivation of the age group (2) gymnosperm orders from each other. The evidence is rather in favour of their common lagenostomalean ancestry. Remarkably, these three orders have in parallel developed the anasulcate pollen on the basis of spore-like prepollen, as well as the platyspermic seeds derived from radiospermic precursors, while the compound strobili have appeared in two of them. True pollen have independantly arised also in the Trigonocarpales, a geologically older age group (1) order that underwent a considerable modernization in the late Paleozoic. A synapomorphic group based on this obviously derived character would include the advanced members both of trigonocarps and cordaites, as well as callistophytes, otherwise the most plesiomorphic order of the age group (2).

Another common tendency in the age group (2) was the loss of cupules either by reduction or by fusion with the integument, or the cupules might have lost their identity in replacing the primary integuments that became incorporated in the nucelli, thus giving rise to crassinucellate ovules. In effect, the age group (2) seed plants were rendered true gymnosperms replacing the prevailingly chlamydospermous age group (1) pteridosperms.

In the age group (3), the Permian callipterids can be considered as transitional between the callistophytes of the age group (2) and the typical Triassic peltasperms. Their leaves and pollen organs are quite similar to those of callistophytes, the pollen grains are of the same morphotype (*Vesicaspora*), while the seed organs of the *Autunia* type, with their scaly ovuliphores bearing two sessile, platyspermic, inverted ovules, are obviously a variation of the *Callospermarion* theme (Rothwell, 1980).

The same can be said of a typical coniferous seed-scale. In other characters, conifers are similar to both callistophytes (Rothwell, 1981) and archaic peltasperms (Meyen, 1984), clustering with the latter as their sister-group. The in the archaic forms compound pollen cones (Grauvogel-Stamm, 1978), true pollen, and laminar seed-scales are against their close affinities with walchians, while the compound strobilation of ovuliphores, shared with the latter group, as well as with cordaites, appears a parallel development.

At the opposite morphological pole, the peltasperms are linked to the gigantopterids through such intermediate forms as *Tinsleya, Russelites* (Mamay, 1968, Mamay & Watt, 1971), *Protoblechnum*, etc. Both peltasperms and gigantopterids show a tendency to a consecutive fusion of the pinnules and pinnae resulting in a simple angiosperm-like leaf with a hierarchical areolate venation (Krassilov, 1995).

The glossopterid stem anatomy and ovulate structures are not yet unambiguously interpreted rendering their phylogenetic position uncertain. Or the ambiguity might be caused by polyphyletic origins. The protostelic stems (*Vertebraria*) and the dichotomously branched pollen organs with terminal sporangial clusters are archaic features found in the most primitive lyginopterids and suggestive of a fraternal relation to callistophytes. The derived characters, notably the saccate pollen grains and platyspermic ovules, might have evolved in parallel with the callistophytes, cordaites and peltasperms. The leaves appear phyllodic, as in cordaites, and some leaf forms (*Euryphyllum, Noeggerathiopsis*) are actually quite similar to *Cordaites*, the similarity extending also to vojnovskyaleans, the Angaridan counterpart of glossopterids.

Arberia, the most primitive ovuliferous organ of glossopterids, is here interpreted as a somewhat flattened, profusely branched shoot. Some of the derived ovuliferous structures might have arisen as modifications of *Arberia* (Rigby, 1978). Their fusion to subtending bracts finds its parallel in conifers, the persistent bracts apparently assisting in dispersal in both cases. However, in the advanced glossopterids, a prevailing tendency seems to have been a shortening of the ovuliferous shoot resulting in a radial arrangement of the ovules in the capitulate or campanulate, sometimes discoid, clusters. This tendency was paralleled by the vojnovskyaleans. Remarkably, both groups had protosaccate taeniate pollen grains, shared also with peltasperms and some early conifers. A variety of Permian insects fed on such pollen (as is documented directly by the pollen preserved in their gut contents, see Krassilov & Rasnitsyn, 1996; Rasnitsyn & Krassilov, 1996). Pollinivorous insects might have both promoted the parallelism by mediating horizontal gene transfers (see more on it in the next chapters) and initiated entomophily, the latter explaining also the capitulate ovuliferous structures of amassed small ovules confronting a pollination vector with a brush of adpressed micropyles.

Characteristic of age group (3) are specialized laminar ovuliphores - seed-scales - developed in parallel in most of its members, although not necessarily homologous. In peltasperms and perhaps also in conifers, they are modified foliar ovuliphores of the age group (1) pteridosperms, while in gigantopterids they might have been secondarily flattened fertile shoots, partly lacking lamina in aberrant forms, such as *Physmatocycas* (Gillespie & Pfeffercorn, 1986).

In the age group (4), the nilssonialeans are linked to the peltasperms through their ovuliferous organs (*Beania*) that are basically identical to *Autunia*. Previously assigned to cycads, they differ in the leptocaul life-form with dimorphic shoots, simple leaves and the ovules abaxial on the heads of the longly stalked ovuliphores. The pollen cones are similar to those of the extant cycads, but differ from those of Mesozoic cycads, such as *Cycandra* that produced the much more elaborate multisporangiate synangia. The simpler bisporangiate half-synangia are thus derived in the modern cycads. The czekanowskialeans are similar to nilssonias in the life-form characters including the deciduous spur-shoots, while their bivalved cupules seem derivable from a pair of marginally fused peltasperm-type peltae.

The corystosperms are linked to pentoxyleans and cycads on the basis of their essentially medullosalean stem anatomy (*Rhexoxylon*), and to ginkgoaleans on the basis of their similarly paired ovules embedded in the short collar-like cupules conceivably reducible to the "collars" of the extant *Ginkgo*. In addition, the flabellate dissected leaf morphotypes typical of the Mesozoic ginkgoleans occur in *Dicroidium*, the corystosperm foliage (Anderson & Anderson, 1983. 1985). There is also a marked stomatographic similarity between corystosperms and ginkgoaleans extending to pentoxyleans.

The cycadalean stem anatomy with characteristic girdling leaf traces first appeared in *Antarcticycas* (Smoot & al., 1985) of a corystosperm-dominated Triassic flora. The Jurassic *Cycandra*, a protocycad pollen cone, resembles the medullosalean pteridosperms in having large synangia of concentrically arranged tubular sporangia (Krassilov & al., 1996). The double vascular system in the ovules of extant cycads is still another trigonocarpalean feature. At the same time, the Early Cretaceous *Semionogyna*, a protocycadean ovuliferous shoot with sessile bracteate ovules (Krassilov & Bugdaeva, 1988c), is comparable to the Permian *Sobernheimia* of the callipterid circle (Kerp, 1983), both associating with peltasperm-type leaves and pollen organs. *Semionogyna* also shows a tendency to sterilization and planation of the distal bracts, potentially leading to a *Cycas*-like leafy ovuliphore (Krassilov & Bugdaeva, 1988c). Thus the origin of the modern cycads (that appeared not earlier than the mid-Cretaceous) remains controversial, with a possibility of genetic contributions from different phylogenetic clusters.

The extant *Gingko* is markedly different from the extant cycads in the life-form and is usually allied with coniferoid gymnosperms rather than with cycadophytes. However, ginkgoaleans share with the latter the general structure of both pollen and seed strobili (with parallel destrobilation tendencies manifested in *Ginkgo* and *Cycas)*, the pollen (Shashi & Ueno, 1986) and ovule morphologies, as well as many features of reproductive biology, including the pollen tube, sperm, archegonial and embryological similarities. In the Mesozoic ginkgoalean ovules (*Karkenia*), the nucelli were free, beaked and strongly cutinized, thus more cycad-like than in the extant species. Interpretation of the characters shared by cycads and gingko as symplesiomorphies attesting to their primitiveness is hardly warranted, for both the modern ginkgoaleans and the modern cycads appeared rather late in the geological history of gymnosperms. In them, the large ovules and their correlated reproductive features are recent developments lacking in their early Mesozoic precursors. Contrary to a widely held opinion, derived states in plant morphology are not necessarily those due to reduction. Sometimes they are those due to magnification.

In the rest of age group (4), the relatedness of bennettites and gnetaleans is attested by their basically similar ovuliferous structures with orthotropous ovules enclosed in the cupules formed of bracteoles or their homologous interseminal scales. The bennettitalean receptacles with an apical corona of sterile scales correspond to the shortened binodal strobili of extant *Gnetum scandens* (Foster & Gifford, 1974), with the distal node reduced. The bennettitalean ovules, radiospermic with residual nucellar vascular bundles, are essentially trigonocarpalean, at the same time resembling *Gnetum* in the peculiar cutinized flange at the base of the micropyle tube (e.g., in *Bennetticarpus crossospermum*: Harris, 1932). The Triassic *Dinophyton*, a peculiar protognetalean form, showed a 3-ribbed nucellus as in *Pachytesta*, a trigonocarpalean ovule. Notably, some medullosaleans had a decussate leaf arrangement (Mapes & Rothwell, 1980) characteristic of gnetaleans. At the same time, the Hirmerellaceae, another protognetalean Mesozoic group, seems to have been be related to *Majonica*, a Permian coniferoid (Clement-Westerhof, 1987).

The Caytoniales produced compound leaves with *Glossopteris*-like leaflets occasionally showing a decussate arrangement betraying their caulomic origin. Their pollen organs are essentially glossopteroid while the cupules are fused to their residual subtending bracts (lips) and seem derivable from the bracteate glossopteridalean cupules, perhaps by arrested development at a juvenile stage when some of the cupules were closed, as those described by Gould & Delevoryas (1977).

Characteristic of the age group (4) gymnosperms, irrespective of their phylogenetic position, are the deciduous short shoots, the prevailingly simple, entire or pinnatifid, often phyllodic leaves, the asaccate pollen grains, sometimes with residual sacci or folds betraying their origin from saccate forms, and the strobilate ovuliferous organs, while the leafy ovuliphores, so widespread in the Paleozoic, are totally lacking. The latter change might have been due to the appearance of leaf-cutting insects or foliophagous tetrapods, or both.

In the mid-Cretaceous, the age group (4) gymnosperms either vanished or underwent a radical modernization. The nearly synchronous appearance of fleshy seeds in cycads, ginkgo and some conifers, of fleshy cupules in caytonias (as distinct from dry cupules of the Late Triassic forms), of fleshy receptacles in bennettites and of fruit-like cones in pentoxyleans might have been related to zoochory that was then fostered by a rapid spread of birds and multituberculates. This factor, perhaps in conjunction with the spread of ovule-sucking insects, might have been impelled a better protection of ovules either by the armour of closed peltate cone scales, as in cycads and the concomitantly appearing modern pinaceous, taxodiaceous and cupressaceous conifers, or by the newly formed cupules that reappeared in several phylogenetic lines leading to the proangiospermous *Leptostrobus, Caytonia, Dirhopalostachys, Baisia, Eoantha*, etc. At this stage, angiosperms have been added to the seed plant diversity.

General tendencies and patterns

A few evolutionary generalizations are deducible from the above analysis (Fig. 18):

(1) Increase in the higher-rank diversity. About twenty orders (with two or three controversial) have been added to the initial seed plant diversity in the course of gymnosperm evolution. The numbers of orders appearing and coexisting in the successive age groups are as follows:

	Appeared	Coexisted
Late Devonian - Early Carboniferous		3
Mid-Carboniferous	3	6
Late Carboniferous - Permian	5	7
Early Mesozoic	6 (9)	8 (11)
Late Mesozoic (at the time of angiosperm appearance)	0 (3)	8
Present day	0	4

These numbers seem to indicate that the higher rank non-angiosperm seed plant diversity peaked in the early Mesozoic, at the time of the proangiosperm appearances. The subsequent decrease at the expense of proangiospermous orders might reflect evolution of the latter into conventional angiosperms.

(2) Genetic continuity. A remarkable feature of seed plant phylogeny is the persistence of genetic lines. Although some orders ceased to exist nominally, none of them disappeared without leaving descendants that in turn contributed genetic material to the next evolutionary stage up to the proangiosperm level. This was accompanied by the parallel developments, reappearances of the formerly extinguished characters, and reticulations between the phylogenetic clusters. Whether the genetic continuity was truncated by the appearance of angiosperms or the latter were inheritors of all or most of the disappearing non-angiospermous lineages, is yet to be learned.

(3) Nodal clusters. The successive radiations of gymnosperms and proangiosperms appear to have occurred by their lineages deriving in clusters from their stem groups rather than by a conventional succession of dichotomies.

(4) Parallelisms. A notable feature of angiosperm prehistory is a widespread homoplasy of derived characters simultaneously appearing in the fraternal lineages, such as the leafy ovuliphores in lagenostomaleans and trigonocarps; the anasulcate pollen and platyspermic ovules (replacing prepollen and radiospermic ovules in trigonocarps, callistophytes and cordaites); the compound strobili in cordaites, walchians, conifers and other Paleozoic coniferoids; the radial arrangement of gynoecial structures in glossopterids, vojnovskyaleans and bennettites; the taeniate pollen grains in glossopterids, peltasperms and conifers; the specialized scaly ovuliphores -

seed-scales - in peltasperms, nilssonialeans, conifers, etc.; their fusion to subtending bracts assisting in seed dispersal in glossopterids, conifers and gnetaleans; the closed cones of peltate cone scales in peltasperms, cycads and several groups of modern conifers; the destrobilation in *Ginkgo, Cycas*, *Taxus*, podocarps and other conifers as well as in the Mesozoic gnetophytes; the secondary cupules in caytonialeans, czekanowskialeans and other proangiosperms; the fleshy seeds or cupules in the modern

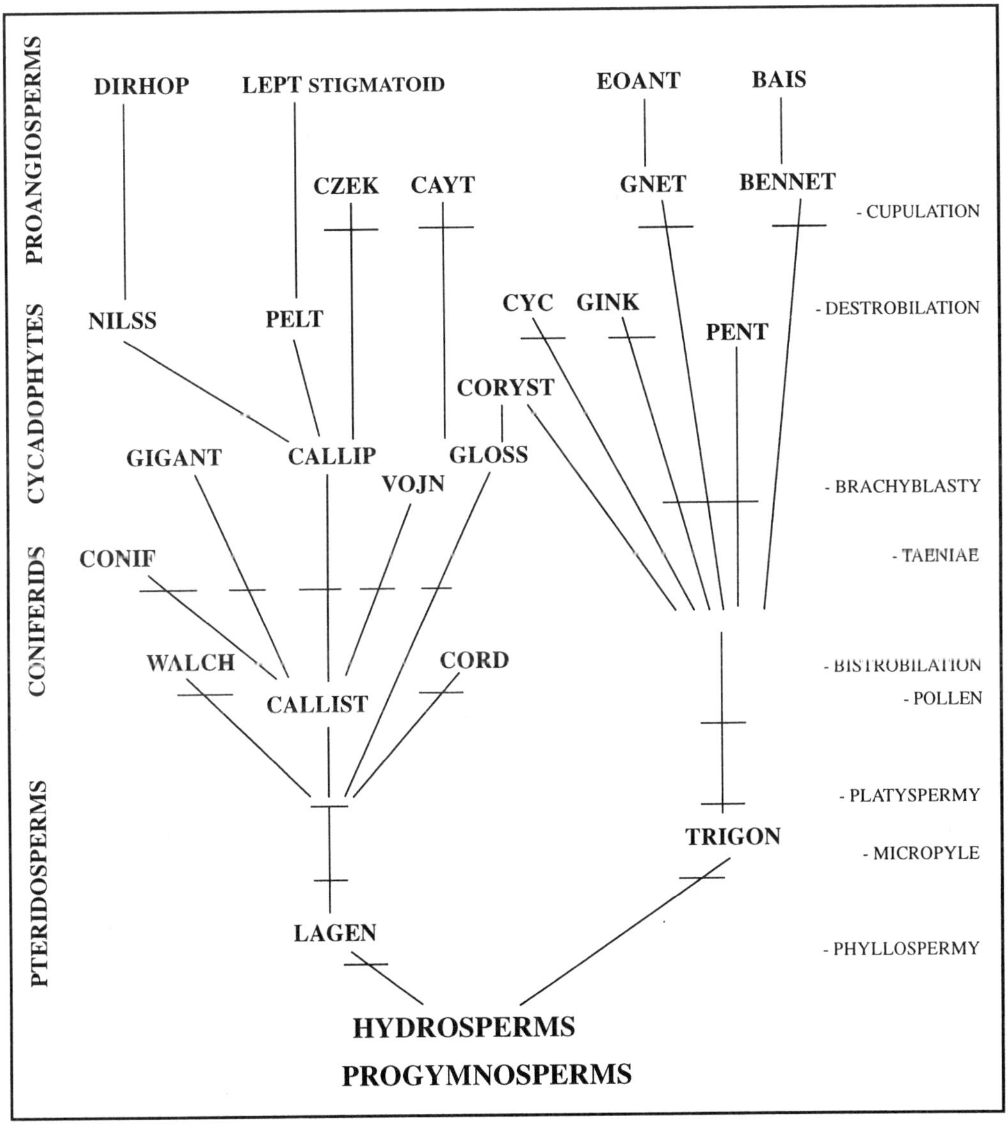

Fig. 18. Suggested phylogenetic relations in and between the gymnosperm and angiosperm grades (left), with horizontal dashes marking the characters (right) appearing in more than one clade.

cycads, ginkgo, *Caytonia*, etc.; the spread of asaccate pollen morphologies in the Mesozoic, involving nilssonialeans, cycads, ginkgoaleans, etc.

(5) Time-fashions. Of special interest among the parallelisms are those confined to particular time planes. These are the time-specific characters, or morphological "fashions", spreading among the coexisting groups of seed plants across the genetic boundaries. Among the familiar examples are the leafy ovuliphores simultaneously appearing in lagenostomalean and trigonocarpalean pteridosperms, the platyspermic ovules in most of the dominant late Paleozoic groups, and the taeniate pollen in the Permian. In distinction from diachronous parallelisms that might reflect some general evolutionary tendencies, the time-specific fashions are induced by environmental factors simultaneously affecting different seed plant lineages. Whereas some time-fashions, e.g. the fleshy disseminules in the late Mesozoic cycads, ginkgos, caytonias, podocarps, etc., are obviously adaptive, the others, such as the taeniate exine, seem far less so and are suggestive of a horizontal spread of the respective mutation (see more on this in the next chapters).

(6) Reversals. These may have been stipulated by the genetic continuity manifesting itself in the reappearances of temporarily extinguished characters while the respective genetic structures have latently persisted in the lineage. Thus, bitegmic ovules appearing already in the early seed plants then faded out of the major Paleozoic gymnosperm lineages to reappear in their late Mesozoic descendants.

(7) Cyclicity. Seed plant evolution has started with the appearance of cupulate ovules arising from sporangial clusters by involving a set, or a double set, of surrounding sterile appendages that acquired the protective function of a seed coat (or a double coat) stabilizing the internal environment in which an in situ development of the megagametophyte would take place. These compound structures functioned as dispersal units. In the early gymnosperms, the coats were still open, with prepollen settling on the sporangial apex that developed special receptive structures. Later, the receptive function was also transferred to the seed coat or coats that formed an integumental micropyle or a cupular exostome, or both. Consequently, the receptive structures of the nucellar apex lost their function and were reduced. This happened in parallel in the gymnosperm lineages of the second evolutionary stage and, likewise in parallel, a tendency to the nucellus fusion with the integument was set in. A complete fusion resulted in a thickened and vascularized secondary sporangial wall of chimeric nature, as in trigonocarps, protected by a thinner integument of cupular origin. The nucellus then proceeded upon the reduction track. Much later, the inner integument might have been restored for a special pollen-tube conducting function, as in angiosperms. Meanwhile, the reduction of the intracupular structures was compensated by the addition of a new set of peripheral structures (modified ovuliphores, subtending bracts, persistent receptacles, closed strobili) and their involvement in protection, pollination and dispersal as seed-scale complexes, bract - seed-scale complexes, secondary cupules, etc. Such new compound structures have been formed only to be engaged in the next round of integration and reduction rendering them similar to their constituent structures, as, for instance, the

compound strobili are homomorphous to their precursory simple strobili, the cupules to the integuments, and the inner integument - nucellus chimerae to simple nucelli, a phenomenon that I have called retroconvergence (Krassilov, 1995, further illustrated in the next chapters). The proangiosperms and then angiosperms appeared as participants of this incessant morphological play at a certain act of its development.

(8) Geological-scale periodicities. Remarkably, the cyclicity of the complication-reduction processes crudely corresponds to the periodicity of the geochronological scale. Insofar as most early seed plants had cupulate ovules, we can speak of the first, mid-Paleozoic, round of angiospermization followed by gymnospermization in the late Paleozoic, in turn followed by the second round of angiospermization in the late Mesozoic. A morphological distance between the cupulate structures of extinct seed plants and angiosperm gynoecia appears much shorter than between the latter and the scaly ovuliphores of extant gymnosperms. Such a zigzag course of seed plant evolution is paralleled by terrestrial vertebrates among which the mammal-like therapsids were replaced by the bird-like dinosaurs, in turn replaced by mammals.

The leafy ovuliphores simultaneously appearing in different pteridosperm lineages might have attracted pollinators sheltering in the foliage. They might also have assisted in seed dispersal. At about the Permian/Triassic boundary, they were lost in strobilation, the then prevailing tendency of ovuliphore aggregation associated with seed-scales, the then prevalent type of ovuliphores. These changes signified new pollination and dispersal environments. In the late Mesozoic time, the opposite tendency of destrobilation was accompanied by the spread of fleshy seeds or cupules, suggestive of zoochorous adaptations.

The major events of morphological evolution coincided with the most prominent global climatic changes, such as the onset of the Late Paleozoic glaciation, the ultimate Permian deglaciation and aridity, and the mid-Jurassic cooling (Krassilov, 1994). Their effect on the terrestrial ecosystems manifested itself in mass extinctions of dominant plants and animals and in the prevalence of pioneer strategists in the surviving groups. They also triggered off the widespread forestation/deforestation trends that affected the evolution both of arboreal and herbaceous life-forms as well as the pollination and dispersal environments inducing the respective changes of reproductive strategies and the corresponding morphological restructuring of reproductive sphere. Morphological evolution was thus incorporated in the biospheric processes.

Early history

The history of angiosperms might have begun well before the Cretaceous, but its major events did take place during that period. A brief overview that follows is based on my recent review (Krassilov, 1985) and the literature cited therein.

The Cretaceous

Starting as occasional newcomers into the gymnosperm-dominated Early Cretaceous floras, angiosperms became, on the global scale, a subdominant or, locally, even a dominant group of terrestrial plants to the end of the period. Their rise was paralleled by mammals that came to dominance among the smaller vertebrates (Lillegraven & al., 1979) and, in the terminal Cretaceous, were much more diverse than the then declining dinosaurs. Such replacements of the leading plant and animal groups, amply documented by the fossil record, signified a radical change in the terrestrial ecosystems. Simultaneously, the land masses stepwise acquired the familiar outlines of the present-day continents, and the earth entered the Cenozoic era. The advent of angiosperms was an integral part of this process.

The Cretaceous lasted about 70 million years, from about 130 Ma to about 65 Ma. It is divided into two nearly equal epoches about 35 million years each, corresponding to the major periodicity of the tectonic and biotic evolution, and in turn divisible into 12 geochronological ages (stages of the rock classification), each about 5-6 million years long which is an average age duration for the entire Phanerozoic (historically, the stages were based on local stratigraphic divisions that, to a certain extent, but somewhat inconsistently, reflected the global periodicities; a few ages, such as the Coniacian, are at least three times shorter than the average, actually corresponding to the average subages). The Aptian to Turonian interval, about 115-88 Ma, is informally designated as the mid-Cretaceous.

Period	**Epoch**	**Age group**	**Age**	**Ma (beginning)**
Paleogene	Paleocene		Danian	65
Cretaceous	Late	Senonian	Maastrichtian	77
			Campanian	83
			Santonian	88
			Coniacian	90
		mid-Cretaceous	Turonian	95
			Cenomanian	100
	Early		Albian	110
			Aptian	115

		Neocomian	Barremian	120
			Hauterivian	125
			Valanginian	129
			Berriasian	134
Jurassic	Late	Malm	Tithonian	141

Some stratigraphers consider periods and epoches as entirely conventional divisions of the geochronological scale. Most Phanerozoic periods, however, are of comparable durations and are marked by the boundary events that are more significant in their ranges and consequences than any intraperiod events. The Cretaceous differs from other periods, except perhaps the Permian, in having the most important events in the middle. In effect, the Late Cretaceous is very different from the Early Cretaceous both in the palaeogeography and biota. The Early Cretaceous started with the global scale rifting and faulting of continental margins that defined a tectonic style of the epoch. Thick terrestrial and paralic deposition occurred primarily in the marginal downfaulted basins, as well as in the intracratonic rift zones. The predominantly clastic marginal deposits, first studied in western Europe, became known as the "Wealden facies". They were found also in the marginal basins of eastern North America, on the Arctic coasts from Spitsbergen to northern Siberia and Alaska, in the Asiatic Far East, southern Australia, Antarctic Peninsula, etc. These deposits contain the major Early Cretaceous fossil floras described in a series of palaeobotanical monographs (Heer, 1880; Seward, 1894-1895; Fontaine, 1899; Berry, 1911; Walkom, 1919; Oishi, 1940; Krassilov, 1967; Douglas, 1969; Watson, 1969, Watson & Sincock, 1992, etc.). Rich fossil plant localities occur also in the lacustrine deposits of intracontinental rifts (e.g. Krassilov, 1982b) that traversed the Asiatic land mass from northern China across Mongolia, Transbaikalia and Amurland to the Sea of Okhotsk where they joined the West Pacific marginal fault zone.

The tectonic style changed after the mid-Cretaceous Austroalpine orogeny that affected eastern part of the Tethys belt, as well as the Pacific margins where the terrestrial volcanic ridge - back arc trough - island arc - trench systems have been laid down. These events in the fold belts were accompanied by submergence of cratonic areas worldwide.

In the Late Cretaceous, deposition continued, though on a reduced scale, in most of the earlier appearing marginal basins. However, the major depositional foci shifted to the sagging continental interiors flooded by shallow seas. The richest fossil floras recording the rise of angiosperms came from the marginal facies of intracontinental basins extending over central and northern Europe, North America, Kazakhstan, etc. (Velenovsky, 1889; Knowlton, 1930; Dorf, 1942; Vakhrameev, 1952; Knobloch & Mai, 1986, etc.). Epicontinental seas covered most of Europe, with isolated land masses, such as the Baltic and Bohemian massives forming large islands. Seaways stretched across North America from the Beaufort Sea to the Gulf of Mexico, across Asia from the Caspian Sea to the Kara Sea, and across Africa from Sirt to the Gulf of Guinea

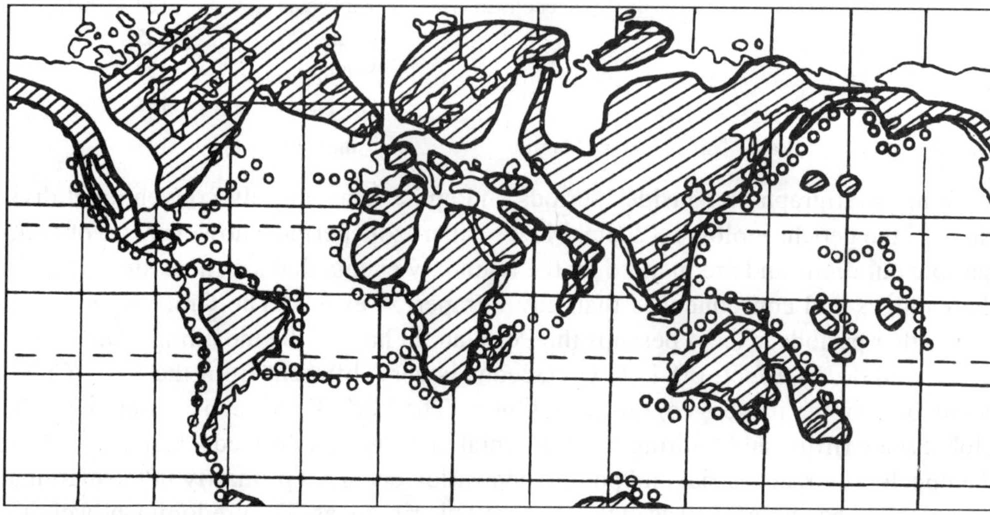

Fig. 19. Cretaceous land (stippled) and island arcs (circles) (after Krassilov, 1985).

(Fig. 19). Geographically, there were twice as many continents as at present, but the epicontinental marine passages were inconstant. In addition, large land masses emerged in the northern North Atlantic (Tule Land), the Beringian Shelf (Beringia) and the Tasman Sea (Tasmantis). They formed broad, though probably intermittent, land bridges for intercontinental terrestrial migrations.

The next global tectonic event occurred in two major stages in the western Tethys belt, notably in the Zagros and Indus-Tsangpo ophiolite zones, causing collision of the Arabian and Indian terrains to the mainland Asia at about the Campanian/Maastrichtian boundary, as well as in the circum-Pacific belt causing the Laramid orogeny and the emplacements of giant granitic bodies at about the Cretaceous/Tertiary boundary. The latter event was accompanied by emergence of cratonic areas and rifting, with abundant terrestrial basaltic accumulations, such as the Deccan traps.

Cretaceous angiosperms

Presently it seems futile to even crudely define the number of Cretaceous angiosperm species, for the major fossil floras of that age have not been revised since the initial palaeofloristic studies of the 1870's to the 1920's. Modern studies, with a few exceptions, focus mainly on selected fossils rather than on fossil floras.

The classical monographers tended to identify all fossil plant remains with the present-day plants. They widely applied generic names of extant genera, sometimes modified by adding "ites", "opsis", "phyllum", "xylon" or "carpon", just to indicate fossil material or the kind of organs. The mystery of angiosperm origin was partly due to this classification practice that produced the impression of a fairly modern aspect of Cretaceous angiosperms. The more we know of the Cretaceous angiosperms the less convincing seem their assignments to extant genera and even families. The "opsis" - "phyllum" names, although valid taxonomically, scarcely indicate taxonomic affinities implied by their etymology. Crude phylogenetic and palaeoecological errors resulted, and sometimes still elapse, from uncritical reading, e.g. of *Sapindopsis* as a representative of the Sapindales.

Of all vegetative organs, angiosperm leaves are most distinctive and, in fact, the majority of angiosperm records are detached leaves. It was argued by Berry (1911), however, that angiosperm-like leaf remains from the Lower Cretaceous, in particular, those from the famous Potomac flora, eastern North America, might actually belong to gnetophytes. This suggestion may still hold for some, but certainly not all, of the presently much better studied Potomac angiosperms (Hickey & Doyle, 1977; Upchurch, 1984). Yet only typical dicotylid leaves are distinctive, whereas the scaly, needle-like, fern-like morphotypes with open venation are shared with other plant groups and can easily be misinterpreted. For instance, the so-called *Dammarophyllum*, the linear parallelinerve leaves from the Late Cretaceous of Kazakhstan (Shilin, 1986), might belong to the same plant as the *Taldysaja* monocotylid flowers (Krassilov & al., 1983). They are similar to the *"Podozamites"* and *Krannera* morphotypes from the Cenomanian of Bohemia (Velenovsky, 1889) that are also under suspicion.

Palaeobotanists are more cautious now in comparing Cretaceous fossils, especially leaves, with extant genera. Yet recent studies of fossil reproductive structures seem to fall into the same trap as the classical "leaf palaeobotany" in assigning their material if not to extant genera, then to extant families, such as the Chloranthaceae, Platanaceae, Magnoliaceae, etc. The following review will show that in most such familial assignments the inconsistent or ambiguous features are underrated, e.g. in allocating *Nordenskioldia* in the Trochodendraceae (Crane & al., 1991; Manchester & al., 1991), a presently monotypic family bypassing the differences in the inflorescence morphology, fruit type, etc. (see below). It would be scarcely appropriate to extend familial diagnoses by introducing features based on fossil material, but lacking in the extant representatives and with no intermediate forms in sight. Those palaeobotanists who tend to neglect distinctions between the Cretaceous fossil and their superficially similar modern forms are perpetuating the "abominable mystery" of angiosperm origins by implying that a number of the present-day families are recognizable already in the earliest angiosperm records. Actually, fossil evidence seems to suggest that families of the extant angiosperm classification are scarcely applicable to Cretaceous angiosperms, and even the ordinal assignments are ambiguous. With this in mind, the following table is to be considered as indicating tentative affinities at most. The terminal Cretaceous (late Maastrichtian) records of more obviously modern alliances are not included.

Order	Plant organ	Appeared
Alismatales	?*Alismatophyllum* leaves (Berry, 1911)	Albian
	?*Liliacidites* pollen	Aptian
Arales	Fruits (Krassilov & Makulbekov, 1995)	Senonian
Arecales	Leaves, pollen (Bell, 1957; Muller, 1984)	Senonian
	Palmoxylon wood (Berry, 1916)	Senonian
Annonales	?*Triplicarpus* fruits (Velenovsky & Viniklar, 1926)	Cenomanian
Ceratophyllales	Fruits (Dilcher & al., 1996)	Aptian
Chloranthales	Androecia (Friis & al., 1986; Crepet & Nixon, 1994)	Turonian
	?*Clavatipollenites* pollen (Walker & Walker, 1984)	Neocomian
Cornales	Fruits (Krassilov & Martinson, 1982)	Turonian
Ericales	Fruits (Knobloch & Mai, 1986)	Turonian
	Actinocalyx flower (Friis, 1983)	Senonian
Fagales	*Nothofagidites* pollen	Senonian
Graminoids	*Taldysaja*, inflorescence (Krassilov & al., 1983)	Senonian
	Gerofitia inflorescence (Krassilov & Dobruskina, 1997)	Turonian
	Restionaceous pollen (Hochuli, 1979)	Senonian
Geraniales	*Sarysua* inflorescence (Krassilov & al., 1983)	Senonian
Hamamelidales	?*Lesqueria* inflorescence (Crane & Dilcher, 1984)	Cenomanian
	Klikovispermum seed (Knobloch & Mai, 1986)	Cenomanian
	Trochodendrocarpus fruits (Krassilov, 1979)	Senonian
Juglandales (Myricales)	*Dryophyllum* leaves (Jones & Dilcher, 1990)	Cenomanian
	?*Ramonicarya* fruit (Krassilov & Dobruskina, 1995)	Aptian
	Caryanthus fruit (Knobloch & Mai, 1986)	Cenomanian
	Flowers, *Normapolles* (Friis, 1983, 1985)	Cenomanian
Laurales	*Araliaephyllum, Magnoliaephyllum*	
	Proteoides, etc., leaves	Albian
	Prisca inflorescence (Retallack & Dilcher, 1981c)	Cenomanian
	Mauldinia inflorescence(Drinnan & al., 1990)	Cenomanian
	Virginianthus (Friis & al.,1994)	Albian
	Flowers (Crepet & Nixon, 1994)	Turonian
Magnoliales	Seeds (Knobloch & Mai, 1986)	Turonian
	Pollen (Muller, 19840)	Senonian
Nymphaeales	? *Afropollis* ?*Asteropollis* pollen	Neocomian
	Modern type pollen (Muller, 1984),	
	seeds (Knobloch & Mai, 1986)	Senonian
Paeoniales	*Hyrcantha* inflorescence, leaves (Krassilov & al., 1983)	Albian
Platanoids	?*Paraphyllantoxylon, Icacynoxylon* wood	
	(Serlin, 1982b; Thyan & al., 1985)	Albian
	Leaves, heads (Crane & al., 1986)	Albian
	Sarbaya head (Krassilov & Shilin, 1995), pollen	Cenomanian

Ranunculids	*Cissites, Vitiphyllum, Menispermites* leaves	Albian
	Caspiocarpus inflorescence,	
	leaves (Vakhrameev & Krassilov, 1979)	Albian
	Appomattoxia fruit (Friis & al., 1995)	Albian
	Prototinomiscium seed (Knobloch & Mai, 1986)	Turonian
Rosales	*Asterocelastrus* fruit (Krassilov & Pacltová, 1989)	Cenomanian
Sapindales	*Insiticarpus* fruit (Knobloch & Mai, 1986)	Cenomanian
	Pollen (Muller, 1984)	Senonian
Saxifragales	*Scandianthus* flower (Friis & Scarby, 1982	Senonian
	Laramisemen seed (Knobloch & Mai, 1986)	Cenomanian
Trochodendrales	?*Caloda* inflorescence (Dilcher & Kovach, 1986)	Cenomanian
	?*Protomonimia* flower (Nishida & Nishida, 1988)	Turonian
	?*Elsemaria* flower (Nishida, 1994)	Senonian
Urticales	Seeds (Knobloch & Mai, 1986)	Cenomanian
	Pollen (Muller, 1984)	Turonian

The following supraordinal grouping of some phylogenetically significant records is occasionally different from their original assignments, reflecting the present author's opinion of their systematic affinities.

Chloranthoids (piperoids)

Chloranthoid affinities have been suggested for the earliest dispersed angiosperm pollen grains *Clavatipollenites hughesii* (Couper, 1958; Walker & Walker, 1984) resembling those of the extant *Ascarina* in both the external morphology and exinal structures. Other widespread Early Cretaceous pollen morphotypes, *Asteropollis* and *Stephanocolpites*, have been in their turn compared with the extant chloranthacean genera *Chloranthus* and *Hediosmum* (Walker & Walker, 1984). *Clavatipollenites* grains have also been found stuck to small flattened unilocular one-seeded *Couperites*-type fruits from the Lower Cretaceous of North America and Portugal (Pedersen & al., 1991; Friis & al., 1994). The fruits have been compared to those of the Chloranthaceae, although the fairly similar fruit morphologies occur in a number of monocot families, such as the Juncaceae or Cyperaceae. Pollen morphologies analogous to *Clavatipollenites* and *Asteropollis* can be found also in the Cabombaceae (Nymphaeales). At the same time, pollen grains comparable with *Clavatipollenites minutus* have been obtained from the Albian flowers related to the Calycanthaceae (Friis & al., 1994).

The reportedly chloranthoid androecial remains from the upper Albian of Maryland, eastern North America, have been described as consisting of three stamens fused at the base and each bearing two opposite pairs of pollen sacs (Crane & al., 1986, 1989). However, the illustrations could also be interpreted as showing three distinct, though adpressed, stamens of an incomplete staminate flower, whereas the pollen is apparently tricolpate and not unlike that of the Cretaceous platanoids (see below) having stamens with the likewise produced connective apophyses. In this particular fossil, the platanoid affinities seem more probable than the chloranthoid, though these two groups might have been closely related (see below).

Chloranthistemon endressii from the Senonian of Sweden (Crane & al., 1989) is an androecial remain of three stamens fused by their short filaments. Their massive anthers consist of two bisporangiate thecae in the median stamen and a single bisporangiate theca in each of the lateral stamens. The connective apophyses are glandular. The dehiscent sporangia are valvate. The pollen grains are spiraperturate, a morphology occurring in no extant Chloranthaceae. In *Chloranthistemon crossmanensis,* a similarly flattened tripartite androecial structure from the Turonian of North America, the pollen aperture configuration is variable even within a pollen sac (Herendeen & al., 1993; Crepet & Nixon, 1994).

Lauroids

Lauroid leaves appear among the earliest angiosperm records and are represented by *Laurophyllum*, the entire leaf blades with eucamptodromous venation. This leaf genus overlaps with *Magnoliaephyllum*, the latter comprising larger leaves with the blade ending in a drip-tip and decurrent on the petiole, with several strands discernible in the midrib and with interstitial veins between the camptodromous lateral veins. Common since the mid-Cretaceous are the entire to apically dentate *Daphnogene*-type cinnamomoid leaves (Kvaček & Knobloch, 1967) as well as the entire to trilobed *Lindera*-like leaves described under different names, of which the etymologically unfortunate *Araliaephyllum* is validated by the principle of priority (Krassilov, 1973c). *Eucalyptophyllum* may prove to have been an early lauroid form with a marginal vein and polymorphic (paracytic, hemiparacytic or, occasionally, anomocytic to cyclocytic) stomata (Upchurch, 1984). In the geologically younger forms, including the narrow-leaved *"Proteoides"* morphotype with inconspicuous secondary veins (Krassilov, 1973c), the stomata are typically laterocytic.

Prisca Retallack & Dilcher (1981c) from the Late Cretaceous (Cenomanian) of North America is a loose panicle, the terminal branches of which are spikes of spiral follicles that develop inside calyptras. These floral structures are associated with lauroid leaves of the *Magnoliaephyllum* type and share with them the laterocytic stomatal structure. Similar calyptrate structures are described in the Cenomanian inflorescence *Mauldinia* (Drinnan & al., 1990).

The well-preserved calyptroid flowers from the Turonian deposits of the Atlantic Coastal Plain (Crepet & Nixon, 1994) are trimerous, with laminar stamens and whorled to spirally disposed follicular carpels sunken in a receptacular hypanthium. The adaxial pollen sacs contain monosulcate pollen grains. The carpels are crowned with the bulging peltate or bilobed stigmas and contain winged ovules on the marginal placentae. Between the carpels and stamens there are sterile structures described as pistillodes. These flowers are said to share eupomatoid and calycanthoid features.

Platanoids

Angiosperm wood remains first appearing in the Albian are assigned to the wood-genera *Icacynoxylon* and *Paraphyllantoxylon*, the latter widespread in the Late Cretaceous (Mädel, 1962; Serlin, 1982b; Thayn & al., 1983, 1985; Thayn & Tidwell, 1984), reaching up to 1 m in diameter in the Turonian (Cahoon, 1972). In both wood types, the names imply wrong affinities, for they show characteristic features of the wood of extant planes as well as of the Tertiary *Plataninium* that differs from *Icacynoxylon pittiense* in the ray structure alone (Thien & al., 1985). The scalariform perforation plates of the latter taxon are primitive in having up to 30 bars against less than 20 in the extant *Platanus*.

Large lobed to entire leaves with a coarse pinnate-palmate venation pattern are prominent in the Cretaceous plant localities since the late Albian and are assigned either to *Palatanus* or to the leaf-genera *Platanites, Credneria, Protophyllum, Aspidiophyllum*, etc. (e.g. Vakhrameev, 1952). Epidermal characters were studied in a number of such leaves (Fig. 20), yet none of the Cretaceous species showed the stomatal structures with raised subsidiary cells characteristic of extant *Platanus*. Some morphotypes include the peltate perfoliate leaves with several pairs of infrabasal veins. In the extant *Platanus*, perfoliation is characteristic of stipules rather than of foliage leaves. In addition, the Cretaceous platanoids often show marginal glands that are lacking in the plane leaves. These observations suggest that platanophylls comprise forms convergently similar in leaf shape and venation rather than closely related. Actually, some reproductive structures associated with platanoid leaves are related to the Altingiaceae rather than to the Platanaceae (see below under hamamelids), these two groups being superficially similar in both the leaf and inflorescence morphologies, but not necessarily close phylogenetically.

The earliest platanoid records from the late Albian of Maryland, eastern North America (Crane & al., 1986), represent pentamerous pistillate and staminate flowers, the latter producing pollen that was described as tricolpate, but with an exposed ectocolpus characteristically gaping in the middle (Crane & al., 1986.), as in tricolporate forms, (e.g. *Sarbaya radiata* Krassilov & Shilin, 1995).

Sarbaya is a fruiting axis from the Cenomanian-Turonian of Kazakhstan, bearing several sessile staminate heads (Plates 33, 34; Fig. 21). The flowers are distinct, 4-staminate, with the tepals much shorter than the stamens. The latter show a short filament, a massive connective and the relatively narrow lateral pollen sacs. The pollen grains are semitectate tricolporate with long slit-like ectocolpi slightly gaping over the porous mesoapertures. The endoapertures are lalongate, forming a thin equatorial zone,

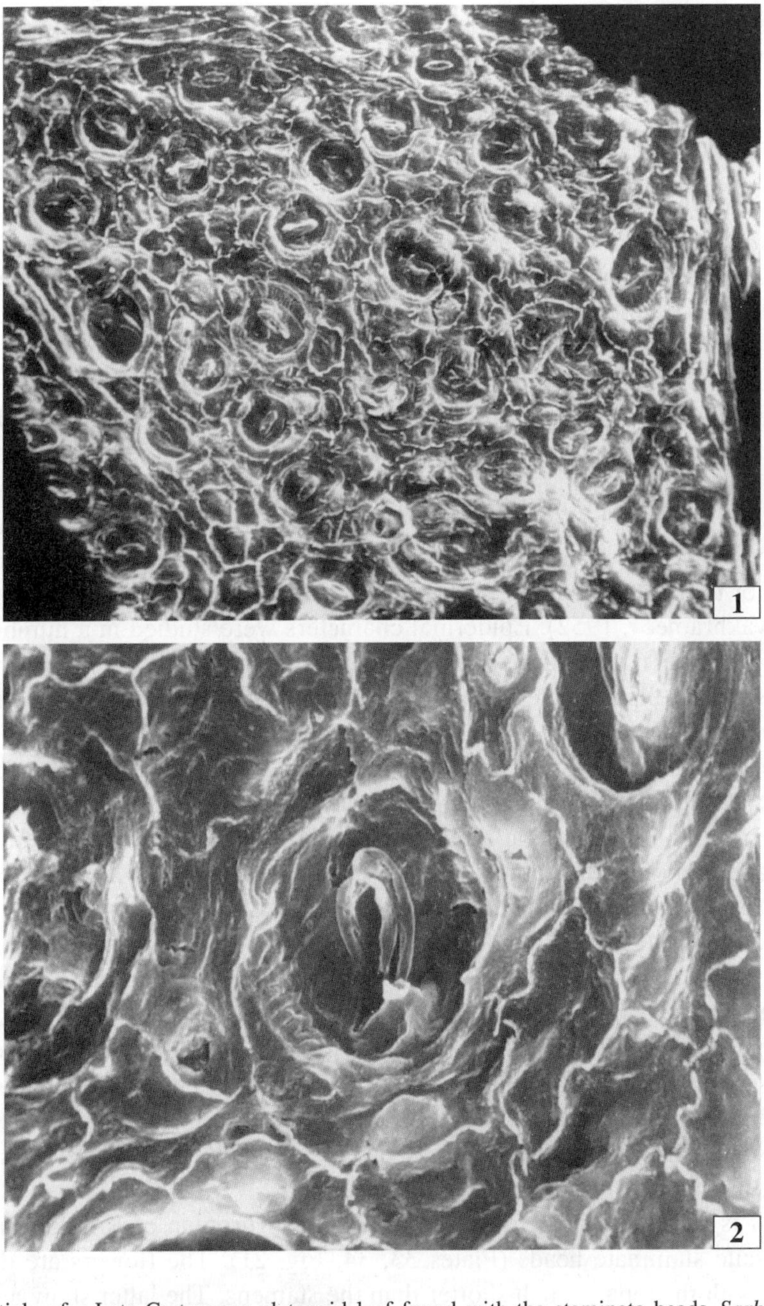

Fig. 20. Cuticle of a Late Cretaceous platanoid leaf found with the staminate heads *Sarbaya radiata* (Krassilov & Shilin, 1995): (1) stomata in the areole of tertiary veins, SEM, x200, (2) stoma showing a radially striate peristomatal flange and narrow subsidiary cells, SEM, x1 400.

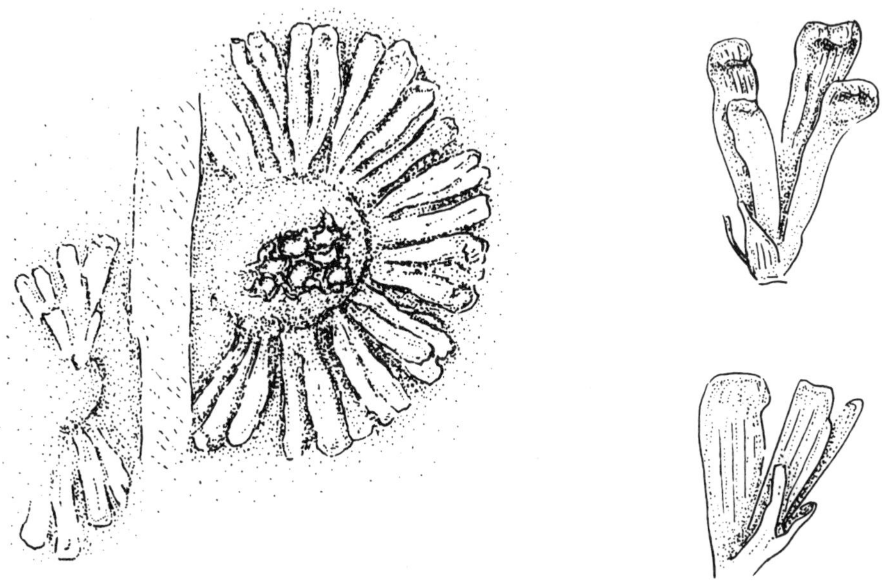

Fig. 21. *Sarbaya radiata* Krassil. & Shilin, from the Late Cretaceous (Cenomanian) of Kazakhstan: staminate heads and individual florets (after Krassilov & Shilin, 1995).

or endocingulum. The associated leaves show a platanoid venation pattern and the prominent marginal glands, with the stomata flanked by a thickly cutinized, transversely striated flange. Of special interest are the pollen grains that appear tricolpate in SEM, but show well-developed porous structures in transmitted light. The endocingulate structures resemble those of some extant rosaceans (Retisma, 1966; Van Leeuwen & al., 1988) indicating possible phylogenetic links of the early platanoids with the rosid group.

Nymphaeids

The leaf morphologies typical of the Nymphaeales and Nelumbonales (*Nelumbites*) appear early in the Cretaceous records, but their nympaheid affinities remain problematic in the abscence of substantiating anatomical or reproductive evidence. The earliest angiosperm pollen morphotypes *Afropollis* and *Asteropollis* are currently related to Magnoliales, although their apertural characters occur also in the Cabombaceae and other families of nymphaealean affinities, while some early tricolpates might have been produced by aquatic plants related to nelumoleans. However the typical nymphaeid seeds appear not earlier than the Senonian (Knobloch & Mai, 1986).

Ranunculids

Fossil ranunculids are represented by the leaf morphotypes characteristic of the present-day Ranunculidae as well as by floral structures borne in cymose or racemose inflorescences, with numerous non-laminar stamens, tricolpate pollen grains, apocarpous or partly syncarpous gynoecia of many laterally compressed, thin-walled, ventricidal follicles that are spiral or whorled on the apically protruding (toroid) receptacles, and with ventribasal, anatropous, typically bitegmic ovules. These floral features are shared with such orders as the Illiciales and Trochodendrales, usually assigned to magnoliids or, the latter, also to hamamelids.

Ranunculids in the above sense are fairly distinct from the typical magnoliids that are characterized by solitary flowers, laminar stamens, monosulcate pollen and thick-walled dorsicidal carpels, as well as from the typical hamamelids having bicarpellate gynoecia. The early *Paeonia*-like forms seem related to ranunculids s.l. (Krassilov & al., 1983).

Prominent in the early angiosperm leaf records are small leaves grading from the compound ternate or palmate to palmately lobed to entire reniform or peltate, with acrodromous or actinodromous venation. Some morphotypes of this group, such as *"Cissites"*, are obviously similar to the leaves of extant *Trollius, Acontium, Anemone* and other ranunculacean genera. They are linked by intermediate leaf morphologies with the compound *Vitiphyllum* on the one hand, and with the indistinctly lobed or entire *Menispermites, Proteaephyllum* and *Populophyllum* on the other. The most complete series of ranunculoid morphotypes (though not recognized as such in the original systematic treatments) came from the Albian Potomac Group of Maryland, eastern North America (Fontaine, 1899; Berry, 1911; Hickey & Doyle, 1977). The stomatal structures described for some of them (Upchurch, 1984) are anomocytic or hemiparacytic, with scarcely specialized subsidiary cells and poorly developed stomatal ledges. A parallel, though somewhat less complete, series in the roughly contemporaneous deposits of western Kazakhstan is represented by *Cissites, Menispermites* and *"Populus" ("Cercidiphyllum") potomacense* (Vakhrameev, 1952). The cuticular structure of the Late Cretaceous *Menispermites katiae* (Krassilov & al., 1988) is much like in the Albian representatives. The tracheary elements of the primary veins show circular to slit-like pits in conjunction with helical thickenings (Fig. 22). Characteristic of the Cretaceous ranunculids was an irregular, "disorganized" venation pattern considered as a primitive feature (Hickey, 1973). However, some extant ranunculids have an even more archaic open venation, as in *Circaeaster*.

Caspiocarpus from the middle Albian of Kazakhstan (Vakhrameev & Krassilov, 1979) is a leafy shoot bearing a terminal paniculate infructescence (Plates 35,36; Fig. 23). The shoot axis is thin, grooved, slightly expanded at the leaf node. The leaves are opposite, petiolate, with the blade palmately lobed, of the *Cissites* morphotype. The panicles arise at a short distance above the leaf node, tripartite, with the central part elongate, cylindrical, tapering toward a terminal loose raceme and the lateral branches

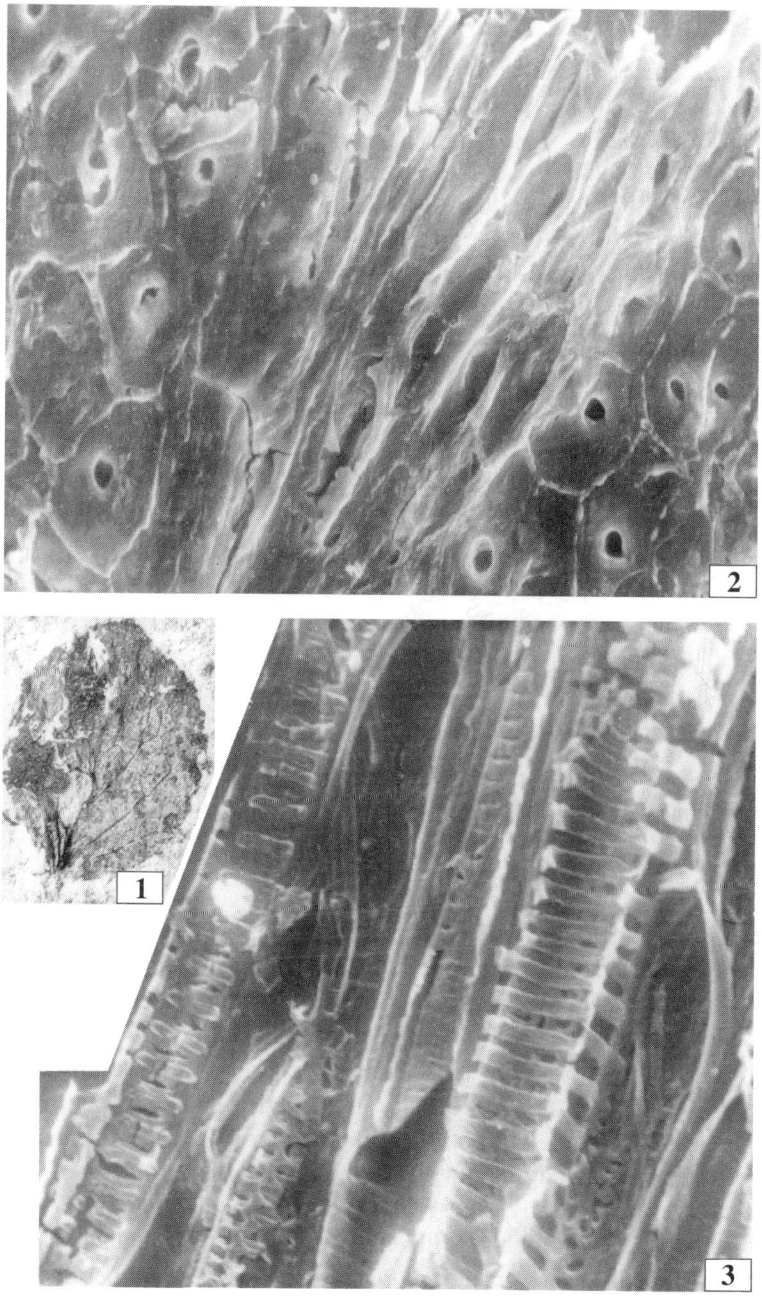

Fig. 22. *Menispermites katiae* Krassil., an example of ranunculoid leaf morphotype from the terminal Cretaceous - lowermost Paleocene of the Lesser Kuril Islands (Krassilov & al., 1987): (1) leaf, x2, (2) differentiation of the upper epidermis into costal and intercostal zones, the latter with papillate cells, SEM, x1 500, (3) tracheary elements of the midrib, SEM, x1 500.

Fig. 23. *Caspiocarpus paniculiger* Krassil. & Vachr., a ranunculoid angiosperm from the Early Cretaceous of Kazakhstan: flowering shoot with leaves, a dehiscent follicle and ovule (after Vakhrameev & Krassilov, 1979; Krassilov, 1984).

much shorter, more strongly condensed, conical or capitate. The inflorescence branches bear secondary paraclades in the form of spirally arranged racemes with 4-5 to 10 follicles each. The latter are spirally arranged, imbricate, elliptical, about 1 mm long, with a short beak, pubescent, ventricidal but also dehiscing a short distance dorsally, occasionally preserved with minute appendages at the base, perhaps representing persistent perianth lobes. The ovules are 1-3 per follicle, small, flattened, anatropous, bitegmic, with the outer integument two cells thick, forming a micropyle, and the inner integument basally three cells thick, wedging out above the midlength of the ovule. The nucellus is relatively massive, free distally, shortly beaked.

In this fossil, the thin ventricidal follicles are typical of the Ranunculaceae, while both ventral and dorsal dehiscences occur in *Glaucidium* and in the Berberidaceae. The ovule morphology is likewise ranunculoid. Of the two integuments, the outer one is typically shorter in the Ranunculaceae, but the reverse relations occur in the extant *Aquilegia, Hydrastis* and in the Berberidaceae. The outer integument is extremely thin in *Ranunculus* and even thinner in *Circaeaster* where it is only two cells thick. The shoot habit suggests a herbaceous life-form. The leaves are comparable with many extant ranunculids as well as with the fossil morphotypes assigned to *Cissites* and *Vitiphyllum*.

A similar shoot from the Early Cretaceous (Aptian) Koonwarra locality, southern Australia (Taylor & Hickey, 1990), bears small, long-petiolate, rhomboid leaves with a pinnate-palmate venation. The leaves are attached below a racemose bracteate inflorescence or infructescence of numerous crowded follicles. A chloranthacean affinity has been suggested on the basis of the axillary bracts or bracteoles, while the leaf comparisons included the Saururaceae and Aristolochiaceae. Both the leaves and the infructescence resemble those of the slightly younger *Caspiocarpus*, but the details of the follicles and ovules are as yet not available for the Australian fossil.

Archaeanthus from the Cenomanian of Kansas, North America (Dilcher & Crane, 1984), is a strobilate gynoecial organ borne apically on a peduncle that shows many small stamen scars and the larger perianth scars. An elongate receptacle bears 100-130 spirally arranged, shortly pedicellate follicles (Fig. 24) that are thin-walled, transversely striate, ventricidal. The ovules are about 100 per follicle, small, anatropous. The bud scales are calyptral. The associated leaves are petiolate, bilobed, of the *Liriophyllum* type.

Though originally assigned to the Magnoliaceae, *Archaeanthus* is in fact closer to ranunculids on account of numerous free follicles (in magnoliids, the follicles are typically connate, or, if free as in the Austrobaileyaceae, then much less numerous) that are thin-walled (rather than thick-walled as in magnoliids) and contain a great number of small ovules. Furthermore, the stamens are numerous, leaving small rounded scars (in the Eupomatiaceae alone among the magnoliids, the stamens are likewise numerous but at least the lower ones are laminar). The leaves resemble those of *Liriodendron*, but differ in the venation with acroscopically laminaless distal branches, as in *Helleborus*. The resin ducts in the parenchymous tissues occur not only in magnoliids, but also in the Menispermaceae.

Caloda delevoryana from the Cenomanian of Kansas, North America (Dilcher & Kovach, 1986), is a catkin-like raceme of dense, spirally arranged, pedicellate flowers showing a whorl of many distinct follicles on a slightly expanded receptacle. Similar structures have been described from the Senonian of Sakhalin (Krassilov, 1979). Such apparently primitive gynoecial morphologies might have evolved into a more compact syncarpous structure, such as *Elsemaria* from the Senonian of Hokkaido, Japan (Nishida, 1994), with ten laterally adnate follicles whorled around the central column and containing up to 40 pendulous anatropous bitegmic ovules.

Protomonimia from the Late Cretaceous (Turonian) of Japan is based on the permineralized sectioned material (Nishida & Nishida, 1988) and is reconstructed as a globous apocarpous gynoecium of many spirally arranged follicles on a woody peduncle supplied with collateral vascular bundles that contain vessels with oblique scalariform perforation plates. The follicles are thick-walled, with anomocytic stomata, and showing adaxial stigmatic crests. The ovules are numerous, marginal, sessile, anatropous bitegmic, with both the outer thick and the inner membranous integuments forming a collar-like extension around the micropyle.

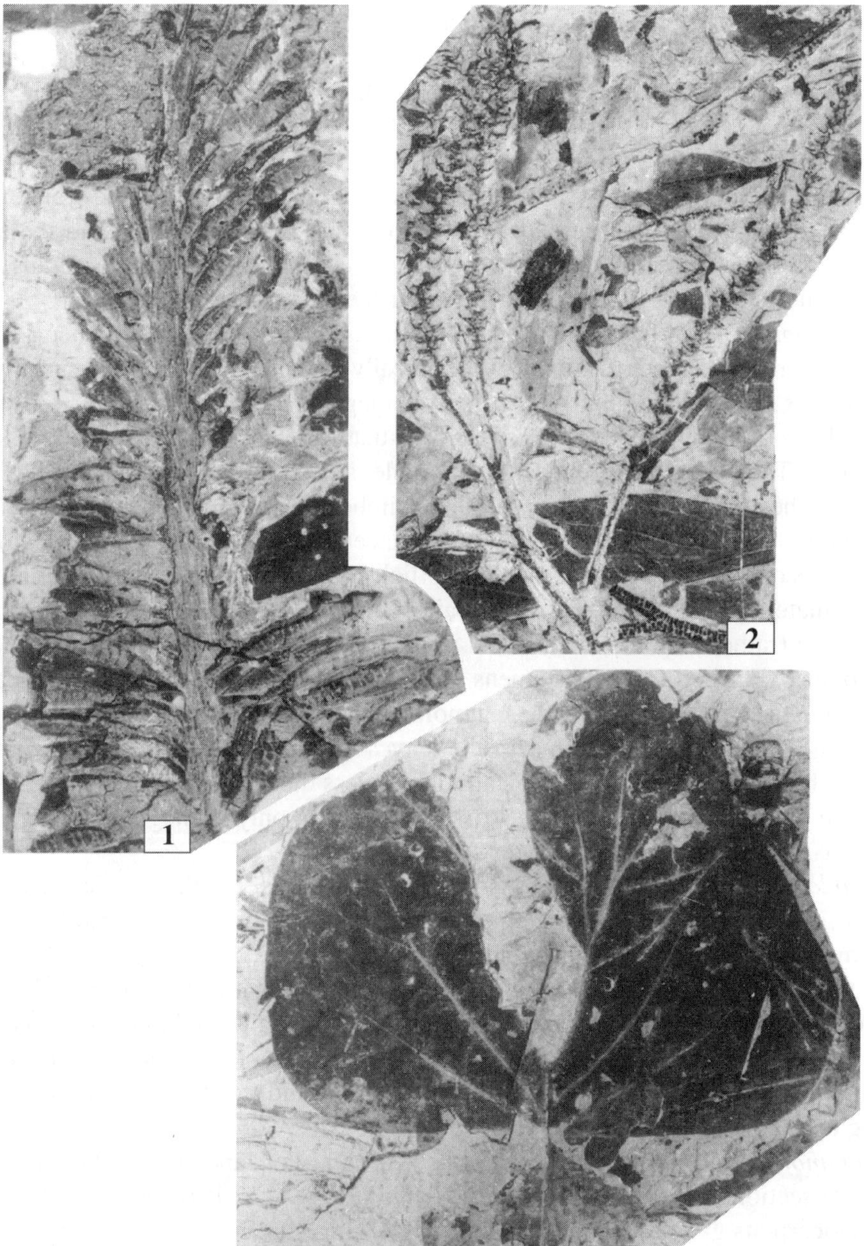

Fig. 24. Angiosperms from the Cenomanian Dakota Formation, Kansas, North America (Dilcher & Crane, 1984; Retallack & Dilcher, 1981c): (1) *Archaeanthus,* pistillate spike, x1, (2) *Prisca,* bracteolate panicle, x1, (3) *Liriophyllum,* supposed leaf of the *Archaeanthus* plant, x1 (courtesy of D.L. Dilcher)

This fossil was compared with the Monimiaceae on the basis of the supposedly concave receptacle, although some cross-sections seem to suggest a sterile central column. The anomocytic stomata is a ranunculid feature that makes a lauralean affinity problematic. The general structure of the gynoecium, as well as the carpel morphology, seem to support a trochodendracean affinity that was indicated by Nishida & Nishida (1988) as an alternative possibility. Similar gynoecial structures are described from the Turonian of Japan by Ohana & Kimura (1987).

Nordenskioldia, widespread in the terminal Cretaceous and Paleocene (Heer, 1970), is based on the fruits that are borne in the long, loosely spicate infructescences that may bear proximal buds and dismembered flower remains, the latter showing thin membranous tepals. The fruits are sessile, alternate or in pairs, compound, consisting of about 20 follicles enclosed in the ribbed bivalved structures interpreted as either exocarps (Manchester & al., 1991) or calyptras (Krassilov & Fotyanova, 1995). These structures spread disk-like and remain on the axis from which the fruits are shed (Fig. 25). The follicles are radially disposed around the central column, or torus, laterally flattened and adpressed, proximally connate but separable when ripe, with shortly reflexed free styles, and containing a solitary seed. The associated leaves are simple, entire, elliptical to reniform, palmatinerve, with two strong acrodromous laterals that branch on the outside.

Nordenskioldia has been compared with *Trochodendron* on account of the similarly arranged follicles (Kryshtofovich, 1958; Crane & al., 1990; Manchester & al., 1991). However, in *Trochodendron* the leaves are pinnatinerve, the inflorescence is cymose rather than spicate, the flower lacks both perianth and calyptra, and the fruit is a septicidal capsule rather than a follicetum. The more similar leaf and fruit structures, the latter sometimes calyptrate and typically consisting of numerous whorled apocarpous follicles, can be found in the extant Menispermaceae, while the one-seeded follicles spreading radially on toroid receptacles are typical of the Illiciaceae. *Nordenskioldia* thus appears to belong in the ranunculid circle, but represents an extinct family.

Hyrcantha from the Middle Albian of Kazakhstan (Krassilov & al., 1983) is a loose paniculate or thyrsoid inflorescence, bracteate, with the branches racemose, axillar to small bracts, the lower ones forked. The flowers are pedicellate, monoclinous, with a calyx of small tepals, persistent and pendent at fruit (Plates 37,38; Fig. 26). The stamens are numerous, with slender filaments, shorter than carpels. The gynoecium is apocarpous of 3-5 follicles, the latter are flask-shaped, thick-walled, ventricidal, crowned by a broad sessile stigma. The associated leaves are compound ternate (Fig. 27), with the leaflets shortly petiolulate, elliptical or oblanceolate, microserrulate, showing the scattered anomocytic stomata.

Similar bracteate inflorescences bearing monoclinous flowers with numerous stamens and tricarpellate apocarpous gynoecia occur in *Thalictrum*, an isolated genus of ranunculid alliance. However, in combination with the persistent calyx and sessile stigma, these characters suggest a more close affinity with *Paeonia*, formerly assigned to the Ranunculales but presently more often considered as representing a

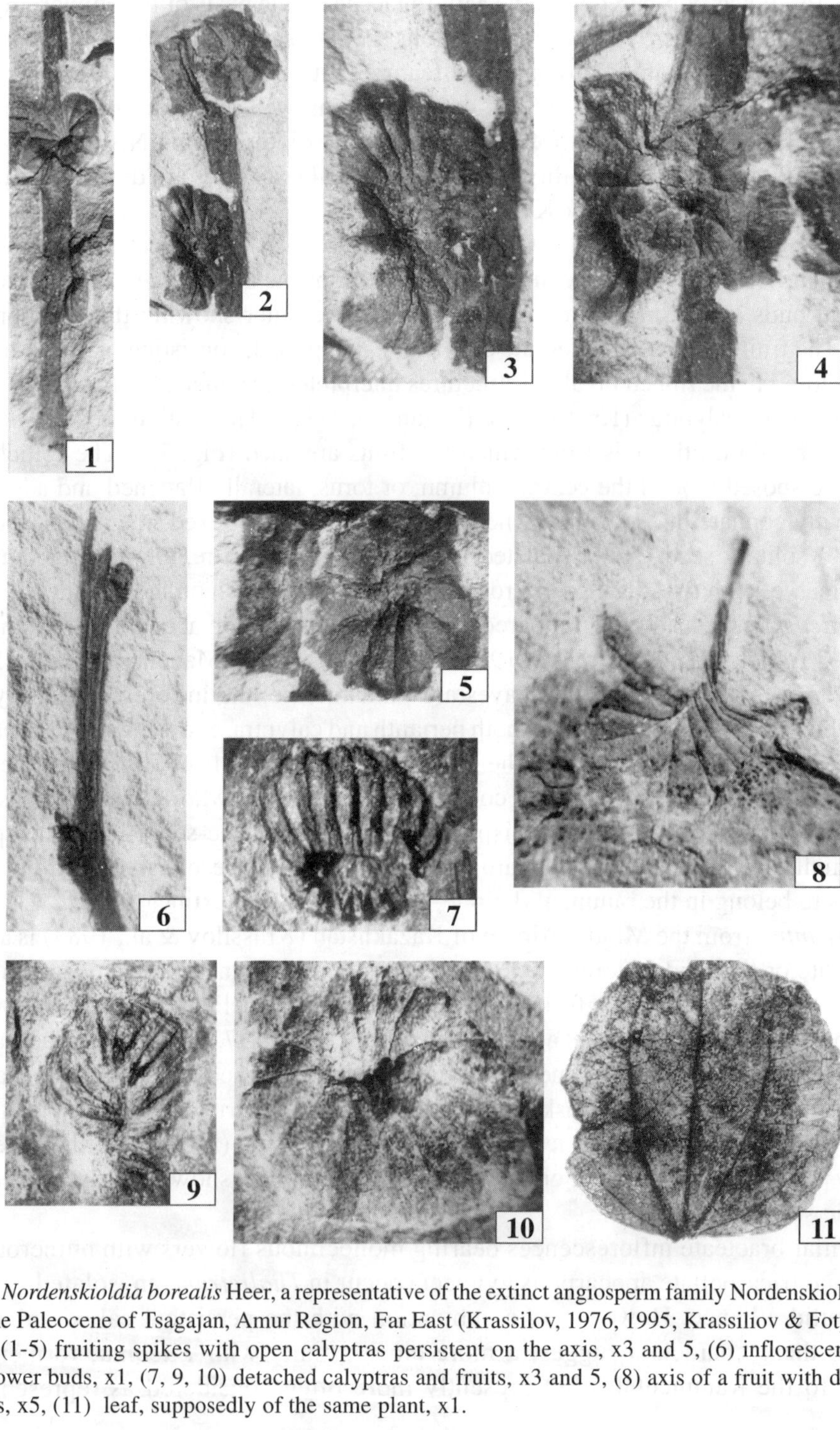

Fig.25. *Nordenskioldia borealis* Heer, a representative of the extinct angiosperm family Nordenskioldiaceae from the Paleocene of Tsagajan, Amur Region, Far East (Krassilov, 1976, 1995; Krassiliov & Fotyanova, 1995): (1-5) fruiting spikes with open calyptras persistent on the axis, x3 and 5, (6) inflorescence axis with flower buds, x1, (7, 9, 10) detached calyptras and fruits, x3 and 5, (8) axis of a fruit with detached fruitlets, x5, (11) leaf, supposedly of the same plant, x1.

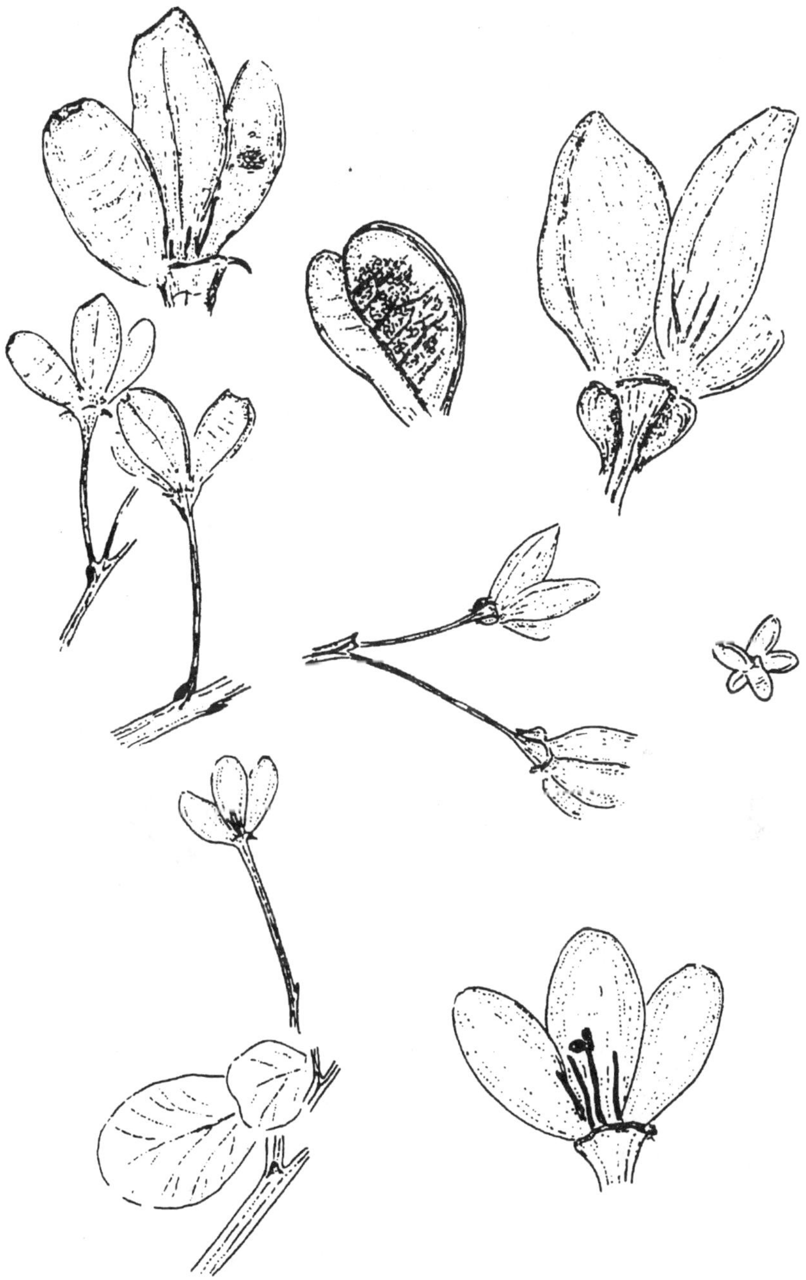

Fig. 26. *Hyrcantha karatscheensis* (Vachr.) Krassil., a paeonialean inflorescence from the Early Cretaceous of Kazakhstan and individual flowers (after Vakhrameev, 1952; Krassilov & al., 1983).

Fig. 27. *"Leguminosites" karatscheensis* Vachr., ternate leaves found with *Hyrcantha karatscheensis* (Vachr.) Krassil., a paeonialean inflorescence from the Early Cretaceous of Kazakhstan, in comparison with leaves of the extant *Paeonia caucasica* (bottom left), after Krassilov & al. (1983).

monotypic order related to the Dilleniales mainly on account of the dendroid androecial traces (Cronquist, 1981). The associated leaves have their morphological analogues in the extant *Paeonia caucasica*.

Appomattoxia is a spiny fruit from the Potomac Group resembling *Circaeaster* (Friis & al., 1995). It is unilocular, with a solitary ovule that is described as pendulous orthotropous. Monosulcate pollen grains found in the stigmatic area may suggest a piperoid-monocotylid, rather than ranunculid, affinity (Crane & al., 1994). Superficially at least, these fruits resemble *Tyrmocarpus,* the spiny capsules from the Neocomian of the Tyrma Basin, eastern Asia (Krassilov, 1977a, 1989).

Hamamelids

As suggested above under "platanoids", a number of the plane-like leaf species, or platanophylls, might actually belong to hamamelids that are closer to the Hamamelidaceae or the Altingiaceae than to the Platanaceae. Among them, *"Platanus" reynoldsii*, a leaf species widespread in the terminal Cretaceous and in the Paleocene, associated with *Steinhauera*, the altingioid pistillate heads, and with *Tricolpopollianthus*, the staminate heads of indistinct flowers producing small tricolpate pollen of a hamamelidacean rather than platanacean aspect (Krassilov, 1979). *Evacarpa*, another Paleocene example bearing on the affinities of early platanoids (Fig. 28), is a definitely hamamelidacean carpellate head associating with platanoid leaves (Maslova & Krassilov, 1997). The following fossil genera here tentatively assigned to hamamelids have been differently interpreted in their original descriptions.

Lesqueria from the Cenomanian of North America (Crane & Dilcher, 1984) is a pedunculate pistillate head with a massive convex receptacle bearing the helically arranged, laminar appendages in the peripheral zones and about 250 crowded fruitlets in the central zone. The fruitlets are elongate, keeled dorsally and containing about 20 ovules disposed in two rows. Though described as follicles, the fruitlets actually show two stylodes, are distally split in the median plane, and leave rhomboid areoles when shed. They are thus comparable with bicarpellate hamamelid gynoecia immersed - less deeply than in the geologically younger forms - in the hypanthial areoles, while the whole structure resembles the pistillate heads of the Paleocene *Steinhauera* (Kirchheimer, 1957; Knobloch, 1960; Krassilov, 1976) and *Evacarpa* (Maslova & Krassilov, in press), as well as of the extant *Altingia*.

Nyssidium, Jenkinsella, Joffrea and *Trochodendrocarpus* constitute a group of closely related fruit morphotypes appearing in the Late Cretaceous and widespread in the Early Tertiary (Heer, 1869; Knowlton, 1930; Reid & Chandler, 1933; Kryshtofovich, 1958; Krassilov, 1976; Crane & Stockey, 1986). They are large racemose to paniculate infructescences of loose, spirally arranged, paired, adaxially dehiscent fruits. The latter contain numerous winged seeds. The accompanying *Trochodendroides*-type leaves are longly petiolate, elliptical to reniform, dentate to crenulate to lobed, with a palmate actinodromous to palindromous venation pattern.

Some variants of these polymorphic leaves resemble those of the extant *Cercidiphyllum*, prompting a systematic affinity with the latter. The reproductive structures are profoundly different, however. *Cercidiphyllum*, although sometimes placed in hamamelids, is a fairly isolated genus having pseudanthia of four whorled, basically decussate, bracteate follicles that are abaxially dehiscent and produce wingless seeds of irregular form, while the fossil genera differ in the racemose infructescence structure, paired fruits, a dehiscence mode, and the seed morphology, all suggesting a hamamelidalean affinity.

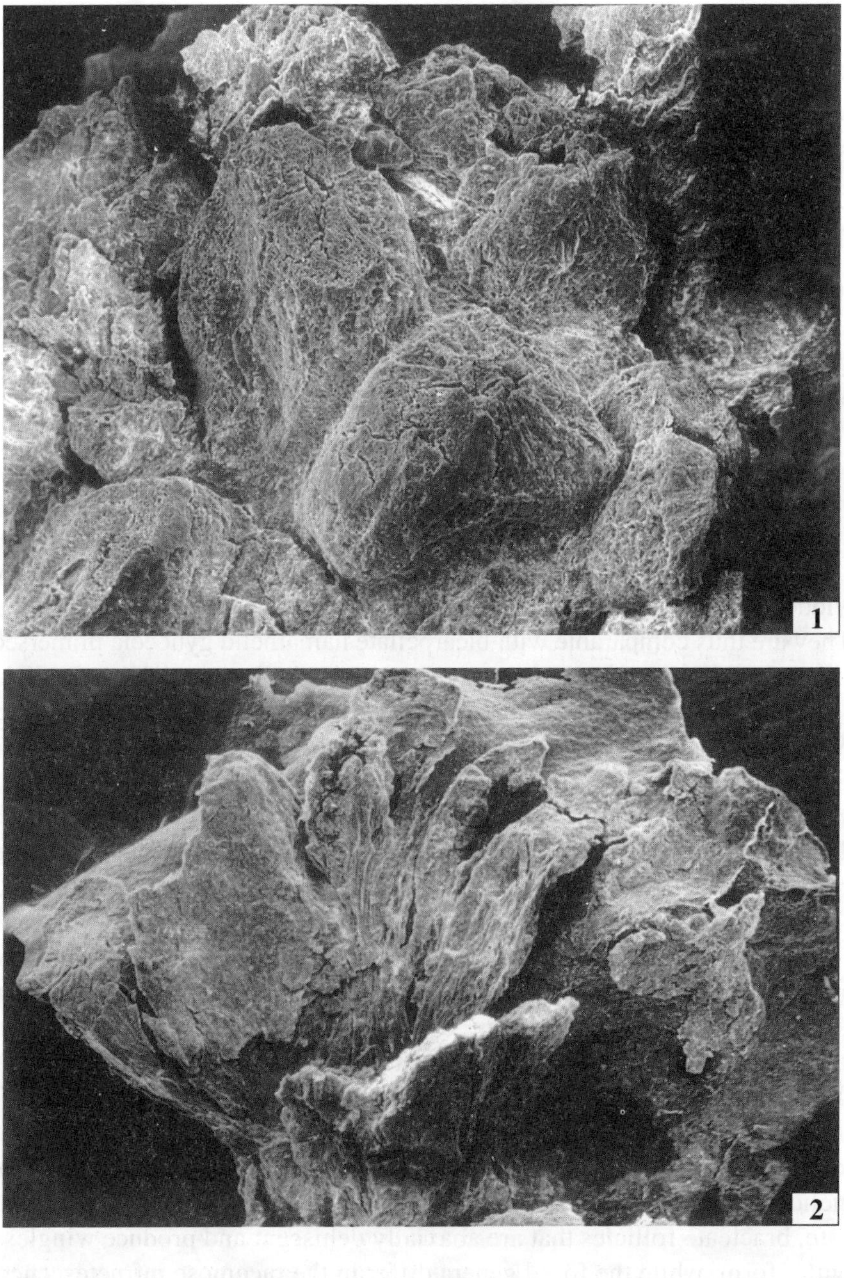

Fig. 28. *Evacarpa polysperma* N. Maslova & Krassil., a liquidambroid pistillate head from the Paleocene of Kamchatka, North-East Russia (Maslova & Krassilov, 1997), SEM: (1) part of the head showing flowers of paired carpels surrounded by intrafloral phyllomes; styles are shed leaving elliptical scars, x50; (2) flower with opened carpels, x30.

Rosoids

Cretaceous fossils here included in the rosoid group are characterized by the prevailingly monoclinous dichlamideous flowers that are solitary or borne in cymose inflorescences. The perianth is mostly 5-membered, with the calyx persistent at fruit. The androecium is haplostemonous, rarely diplostemonous. The gynoecium is syncarpous 2-5-carpellate, with free styles and with two to several anatropous ovules per locule.

Asterocelastrus Velenovsky & Viniklar, 1926, emend. Krassilov & Pacltová, 1988, from the Cenomanian of Bohemia, is a pentamerous actinomorphous flower with a coriaceous stellate calyx, persistent at fruit (Plate 39). The calyx lobes are distally free for about half their length, proximally connate and adnate to the ovary. The latter is syncarpous with free styles, forming a 5-locular, radially dehiscent capsule. The ovules are basal anatropous, two per locule. In the characters of syncarpy and the calyx to ovary fusion, it is closer to Rosaceae than to Celastraceae, a formerly suggested alliance. Among the extant Rosaceae, the Maloidea is similar in having the epigynous flowers with various extents of syncarpy and synsepaly, as well as in the basal ovules. In the Maloidea, the ovaries are occasionally 5-locular (*Stenvaesia:* Sterling, 1965). At the same time, capsulate fruits are more common in the extant members of the Saxifragales that differ from the fossil form in the greater number of ovules per locule and the insignificant perianth to gynoecium fusion. Remarkably, in *Asterocelastrus*, fusion of the floral parts, an allegedly advanced feature, is more extensive than in the geologically younger *Paleorosa* (Basinger, 1976), as well as in many extant forms.

Scandianthus Friis & Scarby, 1982, from the Late Cretaceous (Senonian) of Sweden, is similar to the above form in the pentamerous calyx and capsulate fruit, but the latter is dicarpellate 1-locular with many pendulous seeds. This form shows a peculiar combination of characters occurring separately in different families of the extant Saxifragales, such as free sepals (Crassulaceae), a dimerous gynoecium (Saxifragaceae) and the apical pendulous placentae (Vahliaceae). *Actinocalyx*, another floral form from the same locality (Friis, 1985), also has a pentamerous perianth with distinct sepals but with fused petals forming a floral tube. The androecium is haplostemonous, with stamens basally adnate to the floral tube and shorter than the latter. The gynoecium is syncarpous, 3-locular, with many anatropous ovules and developing into a loculicidal capsule. It is compared with the Diapensiaceae and Ericaceae, but is different in the free styles and the basal, rather than axial, ovules. The latter characters may indicate a rosoid, rather than dillenioid, affinity.

Sarysua Krassilov & Shilin in Krassilov & al., 1983, from the Senonian of Kazakhstan, is a cymoid inflorescence of many monoclinous dichlamideous flowers (Figs. 29, 30; Plate 40) that are actinomorphic, with the pedicel shorter than the gynoecium. The receptacle is massive, bearing clavate glands below the stamens. The calyx is persistent at fruit, with the sepals lanceolate, caudate, reflexed. The androecium is haplostemonous, of about ten stamens and staminodes, the latter hamate. The gynoecium is syncarpous, with five free styles of varying length. In the terminal flowers, the gynoecia are shorter but with longer styles. The ovary is 5-lobed, pubescent. The styles are glabrous, solid, with the epidermis and subepidermal tissue similar to those of the stamen filaments. The stigma is terminal, capitate.

Combined, the above characters indicate affinities with the Geraniales among which *Linum* is similar in the dichasial inflorescence, actinomorphic flower with persistent calyx, nectaries in the staminate disk, the 5-lobed syncarpous ovary and at least partly

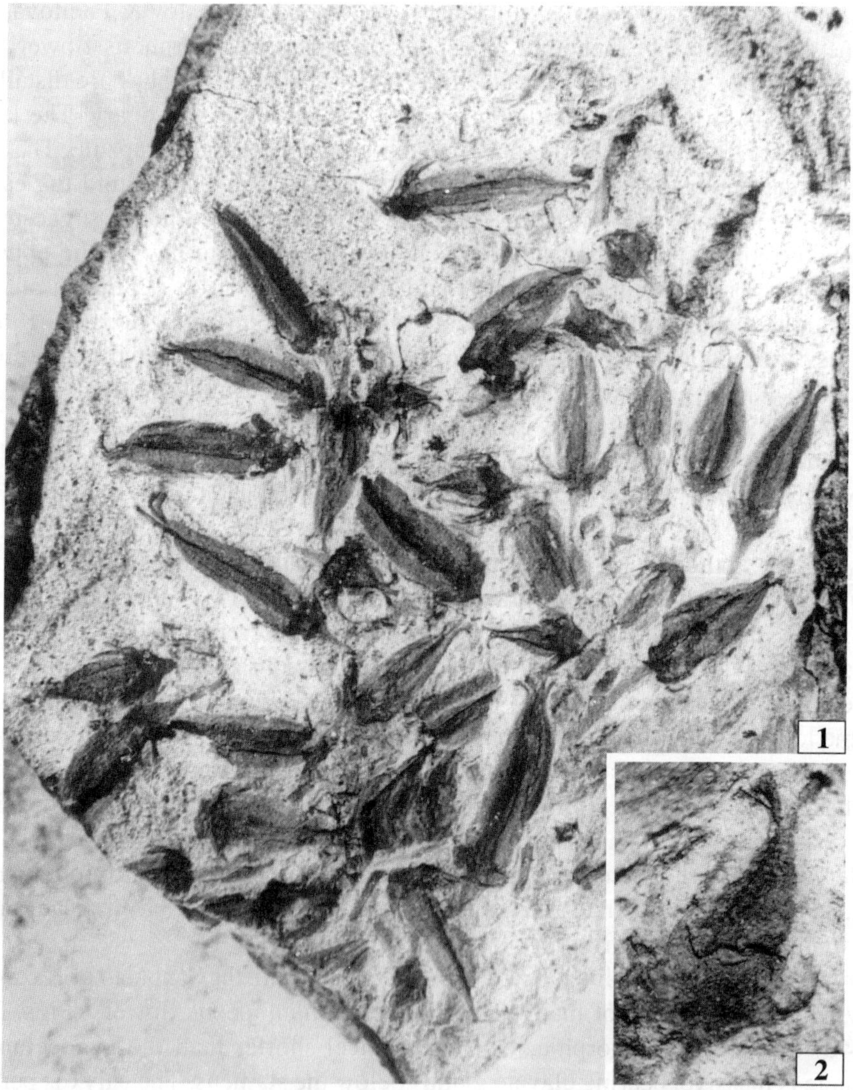

Fig. 29. *Sarysua pomona* Krassil. & Shilin, flowers from the Late Cretaceous of Kazakhstan (Krassilov & al., 1983): (1) flowers crowded in a manner suggesting a single cymose inflorescence, with shorter flowers in the central part, x3, (2) one of the shorter flowers, x10.

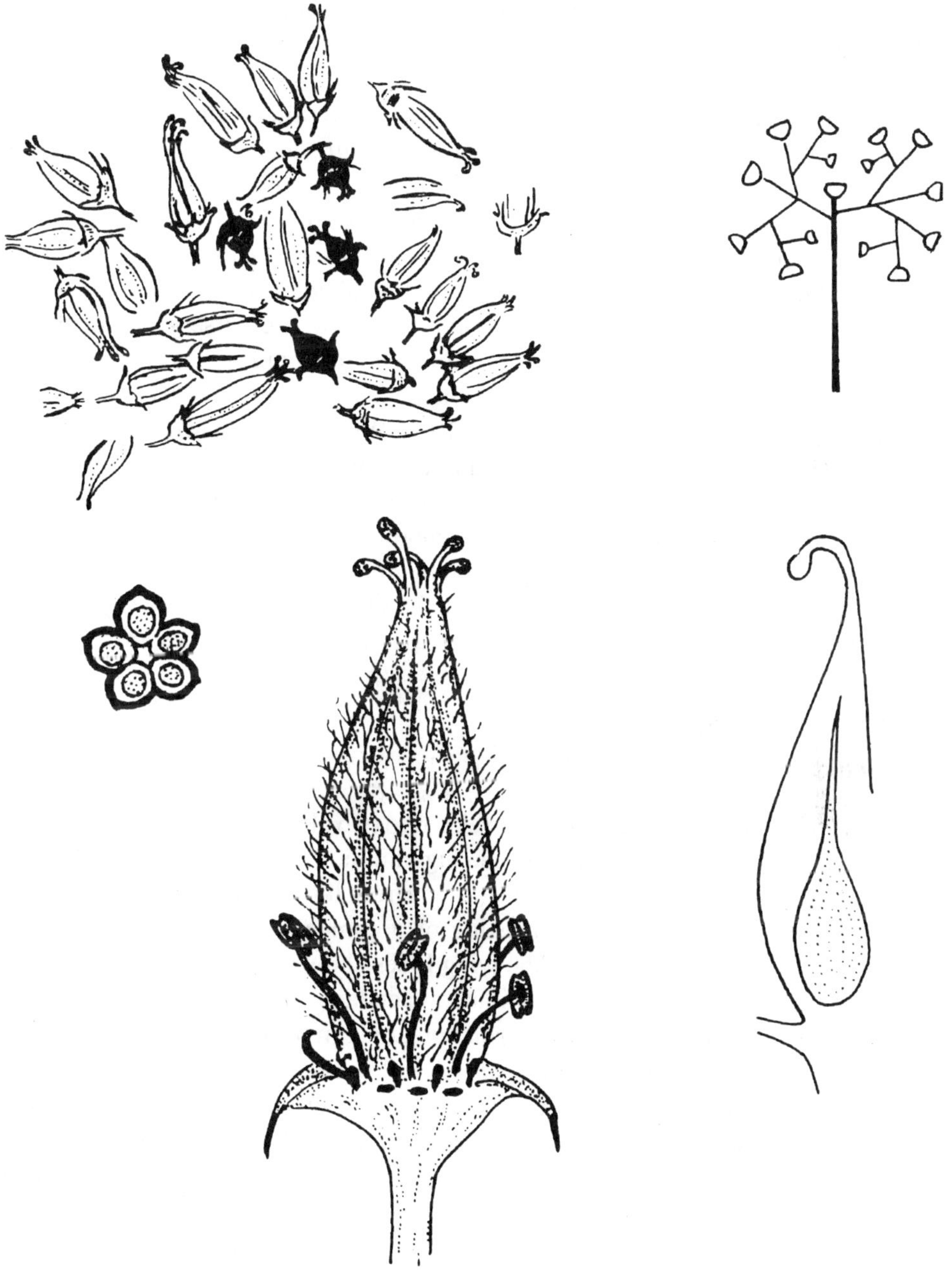

Fig. 30. *Sarysua pomona* Krassil. & Shilin from the Late Cretaceous of Kazakhstan: disintegrated cymose inflorescence (short flowers blackened) and reconstruction, individual flower and carpel with locule stippled (after Krassilov & al., 1983).

free styles. In *Oxalis*, the styles are free and often glabrous, bearing capitate stigmas, while in *Geranium* the stamens are scarcely connate at base and some of them are reduced to staminodes, the ovary is prominently 5-lobed and often pubescent, with the locules distally contracted and sterile.

Sarysua represents an early syncarpous form with solid styles and an incipient heterostyly. In the geranialean genera *Linum* and *Oxalis*, heterostyly is involved in a highly elaborate breeding system (Weller, 1976). Remarkably, style lengths in *Linum flavium* are typically 6-8 mm and 8-11 mm, almost like the two style length modes in *Sarysua*. Another interesting observation is a striking resemblance between the stamen filaments and the styles in diameters and the epidermal characters.

Among the Cretaceous pollen types, *Aquilapollenites*, a diverse group of the Late Cretaceous dispersed pollen grains dominant in the temperate zone, is characterized by the short colpate apertures raised on special appendages. Superficially similar pollen morphologies occur in the Santalales (Jarzen, 1977; Wiggins, 1982; Muller, 1984), while the associated *Proteacidites, Cranwellia* and *Woodehousia* morphotypes were compared with pollen grains of the Proteaceae, Loranthaceae and Geraniales respectively (Khlonova, 1961; Muller, 1974, 1984). Such palynolomorphological similarities reflect scarcely more than remote affinities. For instance, *Proteacidites* with annulate pores is comparable not only with the Proteaceae, but also with the Onagraceae, conceivably representing a progenitorial myrtalean stock. Taken together, these pollen types are evidence of basic radiation of the rosoid complex.

Myricoids

This group includes early angiosperms related to two closely allied orders, Myricales and Juglandales. An early appearance of the juglandioid fruit morphology is evidenced by *Ramonicarya* Krassilov & Dobruskina (1995) from the Aptian of Israel (Plate 41), a shortly pedicellate samaroid fruit consisting of a comparatively large bilobed wing, abaxially attached to a small nutlet. The wing lobes show an irregular reticulate venation. On the adaxial face there is a spatulate appendage (winglet) onlapping a stylar part of the nutlet, with its two stigmatic arms spreading in the plane parallel to the wing. This fruit superficially resembles *Hooleya,* a fossil genus of Platicaryeae that has two diverging wings with irregular reticulate venation. In the Platicaryeae, however, the wings are lateral and their primary veins are flabellate. The Engelhardieae is closer in having involucrate fruits the involucres of which consist of a large wing-like bract and a much smaller adaxial appendage, or winglet, formed of bracteoles and perianth lobes. The nutlet is typically pedicellate, with the extended style bearing a biarmed stigma (*Engelhardia*), but the bract-wing is trilobed rather than bilobed. The lateral veins of the wing lobes in extant *Oreomunnea* and related fossil forms are parallel to the midrib, but in *Alfaropsis* they diverge to the margins as in the Aptian fruit. *Alfaropsis* is distinct, however, in a much reduced winglet.

Gurvanella and *Erenia* from the Early Cretaceous of Mongolia (Krassilov, 1982b) are bilocular winged fruits with persistent funnel-shaped or biarmed stigmas. The wing is reticulate or, in the latter genus, smooth, slightly oblique and avoiding the pedicel. The latter feature is characteristic of some superficially similar juglandioid fruits, such as *Pterocaryopsis*. Their attribution to juglandioids remains as yet problematic.

Common in the Cretaceous and Paleogene are the *Debeya-Myricaephyllum-Dryophyllum* leaf morphotypes grading from the compound palmate to simple, with the leaflets or, in the latter case, leaves elongate, serrate with craspedodromous or semicamptodromous lateral veins. The venation pattern in *Dryophyllum* indicates a juglandalean rather than the previously postulated fagalean affinity (Jones & Dilcher, 1990). The cuticle, studied in three *Debeya* species from the Late Cretaceous of Sakhalin and Kuril Islands (Krassilov, 1973c, 1979; Krassilov & al., 1988), is thick showing sparse stomata and several types of hairs and glands. The stomata are amphicyclic, with both guard cells and subsidiary cells sunken below the level of the encircling cells that form a stomatal pit, and typically with a thick peristomatal ridge as well as the prominent guard cell ledges. These leaf morphotypes occur in the *Normapolles* palynological provinces and within its ecotone with the *Aquilapollenites* province (Herngreen & Chlonova, 1981). Some of them might belong to the *Normapolles*-producing plants.

The *Normapolles*-producing flowers are of several types (Friis, 1983, 1985), having in common the bracteate perianths, inferior ovaries and basal orthotropous ovules. *Caryanthus* among them is a small zygomorphous epigynous flower with a short bracteate wing at fruit and with two bracteoles adnate to the ovary. The calyx consists of two heteromorphous - a narrow median and a broad lateral - pairs of tepals. The stamens are 6-8, opposite the lateral tepals. The gynoecium shows two styles persistent in fruit. In *Manningia*, the flowers are actinomorphous, with 3-carpellate gynoecia. *Antiquocarya* is a fruit with persistent calyx of 6 tepals and with three stylodes.

Normapolles is a diverse group of brevaxonate iso- or heteropolar tri(col)porate forms with prominent apical apertures. The group first appeared in the Cenomanian, initially as the relatively primitive probrevaxonate forms with simple apertures rapidly evolving into more elaborate types (Pacltová, 1971, 1977; Kedves, 1982, 1983; Kedves & Párdutz, 1983b). Their infrastructure grades from granular to columellar, and they vary also in the development of endexine which is lacking in *Atlanthopollis*, but is apertural and lamellate in *Complexipollis*. In *Oculopollis*, a tetrahedral nexine geometry has been described by Médus (1983). Zavada & Dilcher (1986) have traced different *Normapolles* subtypes of Kedves' classification up to various groups of amentifers.

Graminoids

Cretaceous plants of graminoid aspect may actually include progenitorial forms of not only the Poales but also the Restionales or Cyperales or even Juncales that are not always distinguishable as fossils. The graminoid leaf morphologies appear in the Early

Cretaceous as parallelinerve leaves supposedly belonging to the gnetalean proangiosperms (see above). On the other hand, such Late Cretaceous leaf morphotypes as *Dammarophyllum* or *Krannera* might actually belong to graminoids. The distinctive pollen grains of restionaceous affinities are reported from the Maastrichtian (Hochuli, 1979). However, some early pollen morphotypes, despite their entirely different current assignments, might actually belong to graminoid monocots. Thus *Walkeripollis* tetrads occurring in the Albian and even earlier, late Barremian to Aptian, pollen assemblages, are usually related to the

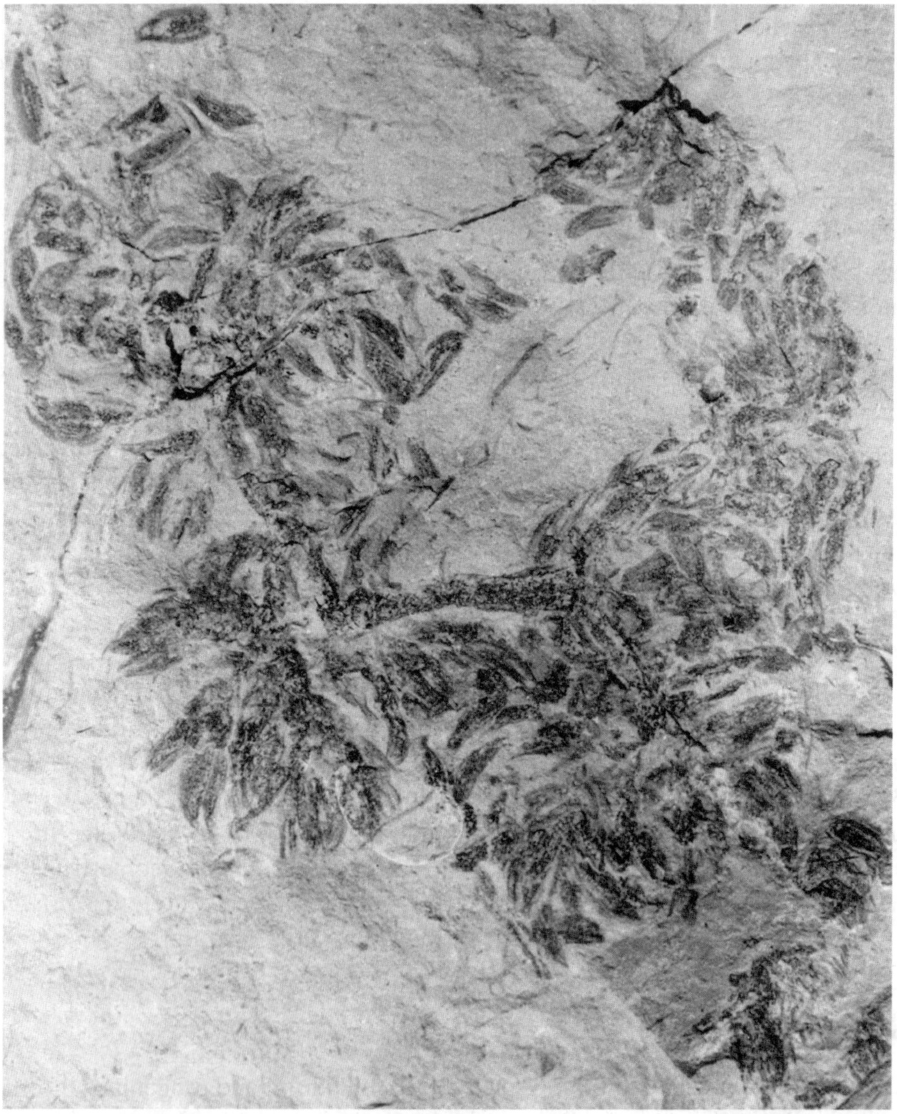

Fig. 31. *Gerofitia*, a cyperaceous corymbiform inflorescence from the Late Cretaceous (Turonian) of Israel (Krassilov & Dobruskina, 1997), x3.

Magnoliales, although they are calymmate, developing a common ectexine, a step in the direction of cryptotetrad condition characteristic of the present day Cyperaceae. The following examples may support a rather early appearance of graminoids.

Gerofitia Krassilov & Dobruskina (1997), from the early Turonian of Israel, is a loosely paniculate corymbiform infructescense with subopposite ascending or (proximal) pendent branches subtended by the drooping inflorescence bracts and bearing the fascicles of densely crowded, but individually discernible, spikelets (Fig. 31). The spikelets con-

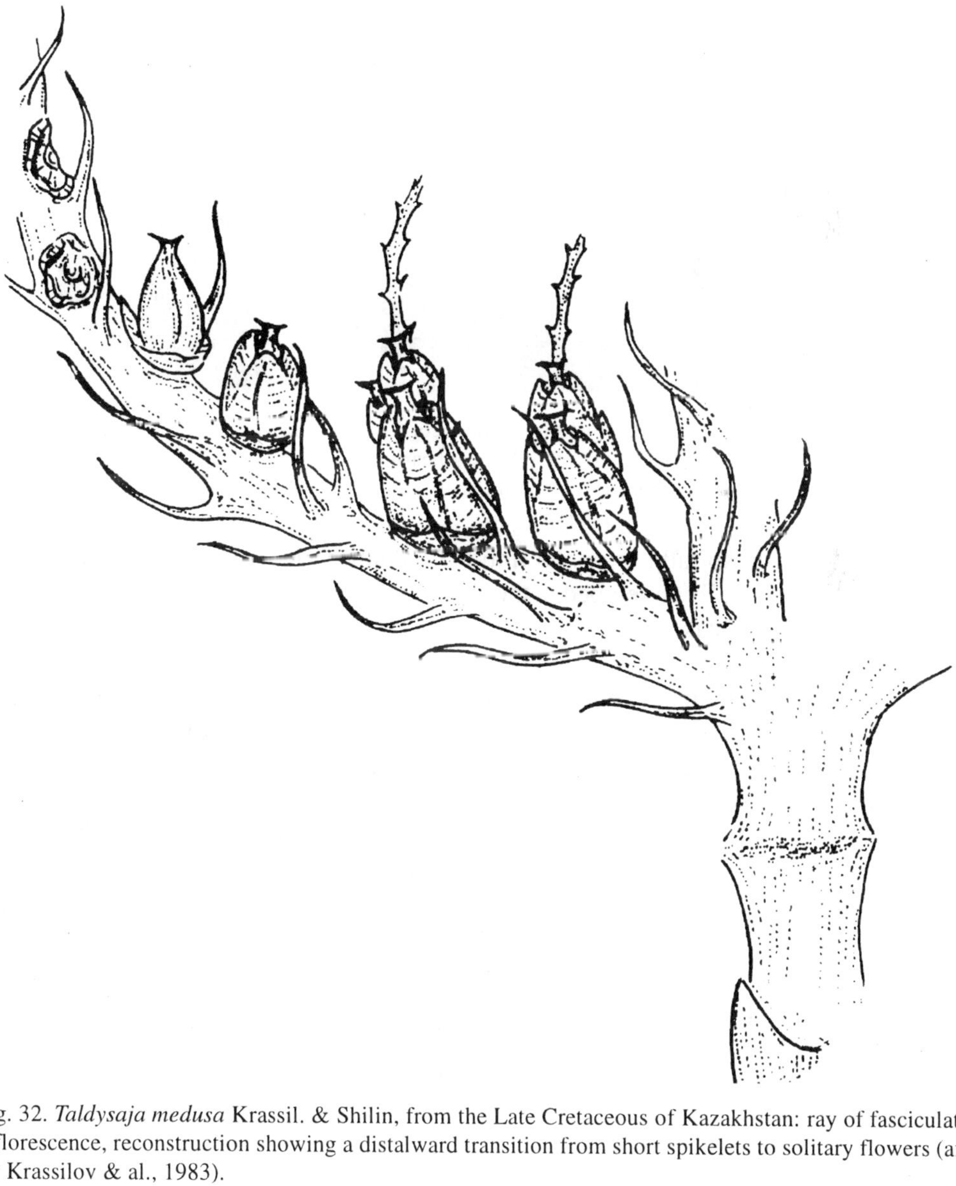

Fig. 32. *Taldysaja medusa* Krassil. & Shilin, from the Late Cretaceous of Kazakhstan: ray of fasciculate inflorescence, reconstruction showing a distalward transition from short spikelets to solitary flowers (after Krassilov & al., 1983).

Fig. 33. *Taldysaja medusa* Krassil. & Shilin, a monocot inflorescence from the Late Cretaceous of Kazakhstan (Krassilov & al., 1983): (1) fascicle rays bearing spiny bracts and three spikelets, with a rachilla distinct in the distal spikelet, x10, (2) bifid peduncular bract or prophyll, x10.

sist of five to seven spirally arranged imbricate glumes, the distal ones axillating the achenes that are adaxially enveloped in the shorter scales (prophylls) forming the utricles with pincers-shaped beaks.

Taldysaja Krassilov & Shilin in Krassilov & al., 1983, from the Senonian of Kazakhstan, is a fasciculate pedunculate inflorescence (Plate 42; Figs. 32, 33) with a long peduncle bearing a bifid prophyll and paired bracts. The radially spreading rays of the fascicle are spiny, apically sterile, proximally bearing 1-2-flowered spikelets axillary to the spines. The distal spikelets are reduced to solitary flowers (anthoids) leaving the raised areole-like elliptical scars when detached. The rachillae of the spikelets are spinulate, with longly protruding sterile apices. The flowers show imbricate perianth scales and a flask-shaped gynoecium. The associated leaves are taeniate parallelinerve, of the *"Dammarophyllum"* type (Shilin, 1986).

Taldysaja is notable as an early graminoid with the spines of the rays and the spinules of the rachillae interpreted as glumes and glumellae respectively. The bifid flask-shaped pistils are cyperoid, while the sterile tips of the rachillae projecting beyond the anthoids occur in many sedge species. At the same time, the prophyll and inflorescence bract morphologies, as well as the areolae of the rays are rather of a poalean aspect.

Phylogenetic relationships of Cretaceous angiosperms

The macrodiversity of Cretaceous angiosperms increased from about four orders in the pre-Albian records to about 21 in the Senonian records, that is, at least three-fold in about 20 million years. In gymnosperms, a comparable rise of the ordinal diversity took about 70 million years.

Actually, however, a larger part of the ordinal diversity is due to the late Albian-Cenomanian burst of adaptive radiation when angiosperms first emerged as a numerically important group. Forms of the opposite morphocline poles, such as hypogynous and epigynous flowers or paracarpous and syncarpous gynoecia, are recognizable at that stage already, while, according to the traditional criteria of primitiveness, some Cenomanian forms (e.g. *Asterocelastrus* among the Rosales, see above) are even more advanced than the extant members of the group. These data suggest that floral morphoclines are of little phylogenetic significance.

In the preceding 15 million years, angiosperms were rare and the tempo of their evolution was less impressive, with much the same groups represented in the Neocomian to Aptian and early Albian records. This initial diversity might have resulted either from a burst at the yet unrecorded low-density monophyletic starting point or from polyphyletic origins. However, the former possibility is unlikely because of the apparently negative correlation between rarity and evolution rates. Furthermore, the monophyly view assumes that angiosperm groups having long geological records should be closer to each other back in time, which does not seem to be the case.

Among the common features of the first appearing angiosperms are herbaceous or small woody growth forms, caulomic leaves, racemose or capitate inflorescences of many inconspicuous flowers, and small seeds. These are obviously grade characters related to either developmental acceleration or dispersal environments, both discussed in the following chapters. Similarities scarcely extend beyond these features, though.

Prominent among the early records are morphotypes that are related to taxonomically isolated extant genera constituting monotypic or oligotypic families, such as the Platanaceae, Chloranthaceae, Ceratophyllaceae, Paeoniaceae, etc. Their higher rank assignments are, for the most part, ambiguous, for they do not fit in any of the well-defined angiosperm orders. Thus, the Chloranthaceae has been assigned to either Laurales or Piperales, the latter perhaps closer but still remotely related (Dahlgren, 1983). The Ceratophyllaceae are likewise remotely related to Nymphaeales, but are similar also to aquatic monocots. On evidence of the fossil record, they were fairly distinct morphologically from the start, never combining the diagnostic features of each other or of any larger order. In contrast, the ranunculids were the most diversified group of the Early Cretaceous angiosperms, with polymorphic foliar and floral characters evidencing their great morphogenic potentials. On the other hand, their affinities to extant taxa are less obvious than in the case of the monotypic groups.

Sinse the fossil data are as yet insufficient for intergroup comparisons, the latter involve also characters of the allied extant taxa. The Platanaceae are usually assigned to hamamelids, although sharing with the latter the general aspects of leaves and floral heads alone while differing in the tricolporate pollen grains, hypogynous flowers, apocarpous gynoecia, orthotropous ovules and achenocarpous fruits. The orthotropous ovule morphology occurs in most of the monotypic groups, including the early appearing *Ceratophyllum*-like forms (Dilcher & al., 1996) and the piperoid line. The ovules arising from the floral apex, as the well as the cupular gynoecia developing from annular primordia encircling the apex, as in *Piperomia* (Tucker, 1986), occur also in the graminoid monocots and myricoid dicots, including the Juglandales that produce catkin-like inflorescences, but are otherwise different from the hamamelid amentifers (Thorne, 1973).

The fossil platanoids are recognized by the palmato-pinnate leaf venation, the characteristic stomatal structures with raised subsidiary cells, the likewise peculiar stamens with longly decurrent pollen sacs and protruding connectives, as well as by the one-seeded carpels and orthotropous ovules. Characteristic of the Cretaceous platanoids are the peltate perfoliate leaf morphotypes that are shared with the Nymphaeales and Piperales (*Piperomia*) and are recorded among the first appearing angiosperms (*"Nelumbites"*: Berry, 1911). The tracheary elements with both scalariform of many bars and porous perforations are typical of both platanoids and piperoids, while the latter approach monocots (Burger, 1977) and nymphaeids in the scattered vascular bundles and inconsistent development of vessels (in *Sarcandra* occurring in the roots: Carlquist, 1987). Similarities between the piperoids and numphaeids include the heteromorphous cotyledons forming a massive haustorium, and other embryological features some of which are shared also with monocots. Thus, typical of piperoids are tetrasporic embyo sacs, while in nymphaeids the coleoptyle usually appears before the root, as in the graminoid monocots (Philomen & Shah, 1985).

These comparisons may evidence a certain degree of relatedness between the platanoids, piperoids (chloranthoids) and nymphaeids (ceratophylls) that might have arisen as paraphyletic lineages from a non-angiosperm ancestral group. They could be marginally related to the graminoid monocots, while the mid-Cretaceous burst of platanoids might have given rise to the ancestral rosoids sharing with them the pentamerous floral structures and certain palynological characters (see above under *Sarbaja*). At the same time, their fossil record gives no evidence of their either progenitorial or fraternal relations with magnoliids. They also seem fairly distant from their contemporaneous early ranunculids.

The Cretaceous ranunculids had simple or compound leaves, a basically dichopodial venation pattern, the typically anomocytic stomata, toroid flowers, numerous apeltate stamens, tricolpate pollen, free to partly connate ascidiform carpels, and the ventribasal anatropous ovules. Presently ranunculids include several satellite genera that fall apart from the major orders and are sometimes separated as monotypic orders, such as the

Glaucidiales (Takhtajan, 1966), while *Kingdonia* and *Circaeaster* are sometimes considered as the most archaic of the living angiosperms (Foster & Gifford, 1974).

As noted above, several taxa traditionally assigned to magnoliids or hamamelids are essentially ranunculoid in their floral morphologies. Thus, the Illiciales share with the typical ranunculids the toroid flowers, apeltate stamens, tricolpate pollen, and the cyclically disposed, laterally compressed carpels. In *Trochodendron*, an archaic genus variously placed in the Magnoliales, Hamamelidales, Cercidiphyllales or Trochodendrales (Takhtajan, 1966; Dahlgren, 1983; Endress, 1986a), the stamens are typically ranunculoid, numerous, with distinct filaments and lacking prominent apophyses. The floral axis is slightly raised above the bases of the likewise ranunculoid carpels that are cyclically arranged, laterally connate, with free reflexed styles, and are dorsiventricidal, as in *Glaucidium*. These genera are here considered as representing a woody branch of the ranunculid cluster. *Paeonia* constitutes a presently monotypic order that could be more diverse in the past, including such extinct forms as *Hyrcantha* that also resembled *Glaucidium* and *Thalictrum*, thus strengthening the alliance of the Paeoniales with ranunculids (Krassilov & al., 1983). On the other hand, the Paeoniales are related to dilleniids, primarily on account of their androecial similarities. Ranunculids and dilleniids also share phylogenetically significant palynological characters (Nowicke & Skvarla, 1979). Caryophyllids may belong in the same cluster but their fossil records are even scantier than in the latter group.

With the exclusion of such artificial links as *Trochodendron, Cercidiphyllum*, etc., the magnoliids and hamamelids are left well separated from each other, as well as from the ranunculids. Their shared features are primarily related to their pseudanthial flowers with inner phyllomes. Their geological histories are also totally different. The hamamelids start in the Cenomanian (*Lesqueria*) and rise to dominance in the terminal Cretaceous to early Paleocene, while magnoliids, after exclusion of ambiguous pollen records (see above), are scarcely known before the Turonian and are rare until the Eocene. Their relationships are further discussed in the concluding chapters.

Environments

The Cretaceous global environments were determined by extensive riffting and subsequent submergence of cratonic areas. Inasmuch as land biomass is many times larger than marine biomass, any sea expansion causes a roughly proportionate loss not only of the terrestrial, but also of the total biomass. In the Late Cretaceous, altogether about 60 million sq. km, or about 40% of the present-day land area, were covered by epicontinental seas. A marine transgression of such an extent would result in a drastic biomass reduction. Since biomass is a major sink for atmospheric CO_2, the latter's concentration would considerably increase, in turn causing a greenhouse warming. On evidence of isotopic palaeotemperatures and biotic indicators, notably a more than 1,500 km shift of zonal boundaries, the mid-Cretaceous mean annual temperatures might have been about 5-6°C higher than at present.

A fairy distinct phytogeographical boundary dividing fossil plant localities with abundant bennettites, brachyphyllous conifers or gnetaleans ("brachyphylls") to the south and those dominated by ginkgoaleans and czekanowskialeans to the north occurred at about 50°N in Asia, ascending to about 60°N in Europe (Fig. 34). Key genera of these provinces are the pachycaul *Cycadeoidea* and the heteroblastic ribbon-leaved *Phoenicopsis* respectively (Krassilov, 1972a). Studies in eastern and central Asia have shown that not only the dominants, but also the subordinate components were restricted to one

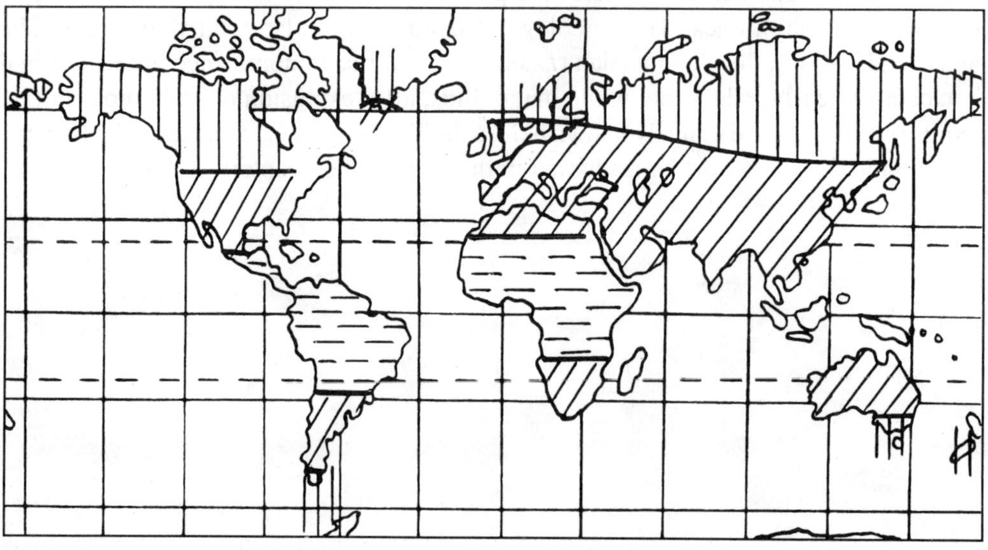

Fig. 34. Cretaceous climatic and vegetational zones (after Krassilov, 1965): vertical lines - temperate; oblique lines - subtropical with xeromorphic vegetation; horizontal dashes - equatorial with impoverished xeromorphic vegetation.

or another province or they underwent certain changes across their boundary. Incidentally, the giant horsetails more than 5 cm in diameter occurred in the southern province alone. Among ferns, the osmundacean *Todites* and the cyathacean *Coniopteris* with aphlebial basal pinnules were replaced to the north by *Osmunda* and *Dicksonia* respectively. Nilssonias with deciduous dwarf shoots were subdominant in many northern and ecotonal fossil floras, becoming rare to the south. Among conifers, the araucarias and podocarps were mostly confined to the southern province, whereas the dwarf-shoot shedding pinaceans with *Pityophyllum*-type needle leaves and *Pityospermum*-type winged seeds were more common in the north. Deciduousness of the northern province dominants (and subdominants, including the pinaceans and nilssonias, the latter also shedding dwarf-shoots) is amply documented by their forming leaf-mats, while their association with coal beds suggests a seasonal summer-wet climate. In contrast, leaf-mats are rather uncommon in the southern zone localities. If formed of the *Ptilophyllum-Otozamites* type bennettite leaves, they usually occur in littoral facies. Thus, the *Cycadeoidea*/*Phoenicopsis* province boundary might have corresponded to the summer-green/ evergreen or winter-green zonal boundary. Its sharpness could hardly be explained by the temperature differences alone, for the Cretaceous temperature gradient was much lower than at present, with equatorial means nearly the same as at present against about 0°C at the poles (Sellwood & al., 1994). Scaly leaves, thick cuticles, sunken stomata, pubescence, and other xeromorphic features of the southern province dominants may suggest precipitations, rather than temperature, being a major differentiation factor across the boundary.

A latitudinal differentiation within these two broad zones was scarcely discernible, with the equatorial zone apparently much drier than at present. The present day low latitude humid zone is due to a rapid cooling of the ascending tropical air masses shedding rain over the equator. In the warmer climate, these air masses might retain water vapour until they spread to a considerable distance from the equator that, consequently, received a lower precipitation than the extratropical zones. In addition, the low-latitude anticyclones spreading from over the latitudinal Tethys seas (like those, but much more extensive than, over the present-day Mediterranean), made tropical rain forests unlikely. In effect, relatively more biomass was concentrated in high latitudes than in low latitudes, as evidenced by the latitudinal distribution of the Cretaceous coal localities (see Krassilov, 1985), with the largest reserves in northern Siberia and Alaska.

Among the various feedbacks of the build-up of atmospheric CO_2, the increased productivity of terrestrial vegetation is of certain interest as a factor of early angiosperm evolution. Presently, a 10% higher crop yield is expected as a potential economically positive effect of the doubling atmospheric CO_2. The greenhouse climate would also give advantage to high growth rates over drought resistance because of a considerably decreasing water uptake. Hence, angiosperms seem to have appeared in an environmental situation that encouraged high growth rates that became their major advantage over the typically slow-growing gymnosperms (Favre-Duchrtre, 1979).

In the Cretaceous, large phytomass production is attested by an exceptionally high phytophage diversity, including both invertebrate and vertebrate plant-eaters, with such first appearing specialized forms as semenivores and frugivores. Giant herbivores, vanishing at the Cretaceos/Tertiary boundary and reappearing in the early Eocene, could be indicative of high growth rates in the cropped plant communities. In its turn, the increased phytophage pressure might have instigated a further developmental acceleration and prolific reproduction in the cropped plant populations, as well as reciprocal interactions, including, on the plant part, specialized entomophilous and zoochorous adaptations.

Angiosperm cradle plant communities

There are conflicting theories of tropical versus temperate, as well as upland versus lowland, or even aquatic, angiosperm origins that are based on palaeontological, as well as on actualistic, evidence (see Kräusel, 1956; Axelrod, 1959, 1872; Retallack & Dilcher, 1981b; Tiffney, 1984, 1985). Tropical rain-forest is often given preference as a cradle of angiosperms, although this plant formation is relatively recent, with no Cretaceous analogues yet recorded. Actually, the extant angiosperms that are considered primitive are concentrated in the warm temperate zonal vegetation belts, such as the montane forests of Yunnan, China, comprising diverse magnoliacean genera, such as *Liriodendron, Manglietia, Manglietiastrum, Paramichelia, Parakmeria*, etc., accompanied by *Tetracentron, Euptelea*, and *Cercidiphyllum*, each representing a relict family, as well as by *Semiliquidambar* and other lower hamamelids, *Davidia* among the Nyssaceae, and the archaic open-veined ranunculid genera *Circaeaster* and *Kingdonia* (Fu Li-kuo, 1992). These forests are the remnants of a cosmopolitan Miocene Arcto-Tertiary flora having its mixed conifer - broad leaved precursors in the Paleogene, but having little in common with any of the recorded Cretaceous assemblages (except in the terminal Cretaceous: Krassilov, 1976, 1979; Golovneva, 1995). Hence, an actualistic approach fails in giving any idea of the angiosperm cradle plant community.

In palaeobotany, plant community reconstructions are based primarily on the quantitative relations of fossil plant species and the relevant taphonomic data (Krassilov, 1972c). Dominance is used as a major criterion, and the ensuing classification is, by necessity, far less detailed than one achievable by the modern phytosociological methods. The input of plant material is controlled by sedimentation processes that, in turn, are governed by tectonic, geomorphological and climatic conditions in each particular sedimentary basin. Plant fossils from a single sedimentary basin constitute a local fossil flora (taphoflora) reflecting vegetation of both the deposition site, or autochthonous, and the borders, or allochthonous. For a plant palaeoecologist, it is essential not to lump all material as a "fossil flora", but to deal discretely with each discernible fossil plant assemblage, the latter conceived of as an episodic vegetation sample. Such samples usually contain both autochthonous (hypoautochthonous) and allochthonous components discernible by their numerical representation and taphonomic evidence of in situ burial versus long-distance transport, assisted by intrabasinal and interbasinal comparisons.

A bedding-plane assemblage is usually dominated by a single or a few species that might be dominant in the source community or just locally abundant or depositionally amassed. Their actual status is further clarified by their frequencies in other roughly contemporaneous plant beds. Regional dominants may not be the most locally abundant, but are distinguished by their high frequencies all over the basin. For instance, *Czekanowskia* is locally abundant in a number of Siberian Mesozoic basins representing a widespread but facially restricted plant community type, the Czekanowskietum. Also common in the same basins are fern-beds, *Phoenicopsis*-beds and *Ginkgoites*-beds. Their lithological facies allow some inferences of the vegetational structure. Most commonly, the fern-beds occur in coal-bearing deposits where they intermingle with the Czekanowskietum in a pattern suggestive of a flood-plain or coastal swamp-marsh mosaic, whereas the Phoenicopsietum is rare in the coastal facies, being most frequent in all facies of the alluvial domain as a dominant lowland plant community. *Ginkgoites* is mostly confined to drifted plant beds, supposedly representing a slope or upland community. Such orderly changing occurrences along a facies gradient are indicative of a vegetational catena.

The catenic, as well as the frequently coincident seral successions can be inferred from the vertical sequences of fossil plant assemblages that replace each other following a retreating coastal line. For example, in the Upper Jurassic of the Bureya Basin, Russian Far East, the lower plant beds overlying the Callovian ammonite facies still contain marine phytoplankton and limulid remains alongside with terrestrial plant macrofossils, the latter representing fern marshes and the Czekanowskietum as the pioneer plant communities of a recently emerging land (Krassilov, 1972b,c). Upsection, the extinct ginkgoalean genus *Pseudotorellia*, the cycadophyte *Nilssonia*, and the pinaceous *Pityophyllum* dwarf shoots appear in the Czekanowskietum signalling an enrichment of the seral communities. Further on, they are replaced by a brachyphyllous conifer, *Elatides*, in association with *Sphenobaiera*, the bennettitalean leaf genus *Pterophyllum*, and occasional *Phoenicopsis*, the latter then raised to dominance in the overlying alluvial facies.

Certainly, some of such sequences reflect lateral, along-shore, facies migrations rather than sere - a community build-up succession. Their interpretations have to be corroborated by parallel sequences in the same or other contemporaneous basins. In the case of the Jurassic successions, a remarkable parallelism has been observed between the above-mentioned Bureyan sequence and those following the retreating seas in the adequately studied sections of southern Primorye (= Maritime Province, Russian Far East), Ukraine and other areas (Krassilov, 1972c). At the same time, the brachyphyll and bennettite stages were much more diverse taxonomically in the European sequences than in the temperate Siberian province.

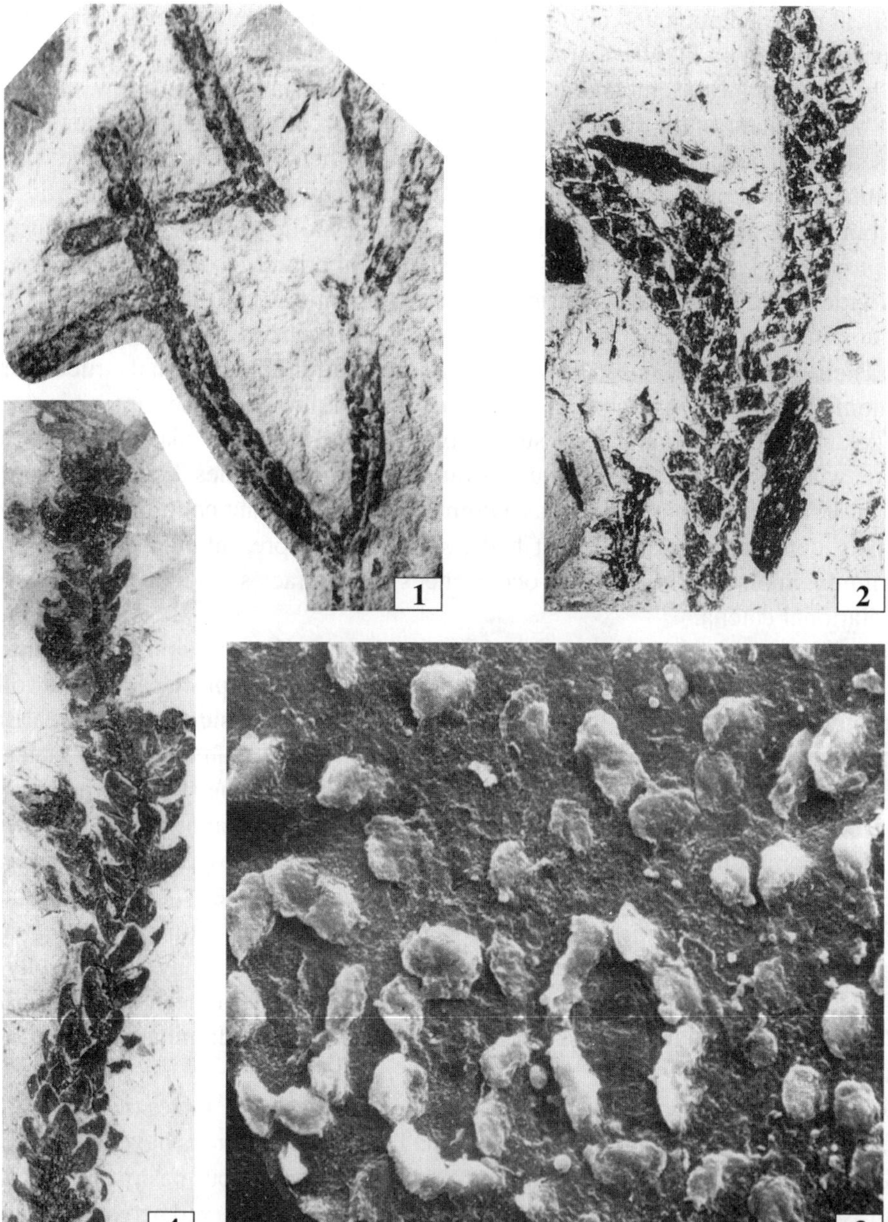

Fig. 35. Brachyphylls of xeromorphic Early Cretaceous plant assemblages of Mongolia (Krassilov, 1982): (1-3) *Brachyphyllum densiramosum* Krassil., variable shoot morphologies and leaf cuticle with stomatal pits marked by papillae, (1) x1, (2) x2, (3) SEM, x800; (4) *Araucaria mongolica* Krassil., spariously branching shoot with thick scaly leaves, x2.

Much the same sequence of the build-up stages have persisted in the Early Cretaceous. However, in the central Asiatic regions, including the vast sedimentary basins of Mongolia and Transbaikalia, fern marshes were drastically reduced and sometimes lost as a distinct member of the catenic and seral successions (Krassilov, 1982b). They were replaced by the communities of the later seral (and catenic) stages, mostly by the *Classopollis*-producing hirmerellaceans and other brachyphylls (the term refers to the not always distinguishable shoots of conifers and gnetaleans with helical or decussate, adpressed, thickly cutinized, amphistomatic or epistomatic scale-leaves, see Fig. 35), as well as by bennettites of the *Ptilophyllum-Otozamites* group. Numerically, representation of the two latter groups in the coastal mud flat - flood plain plant bed assemblages increased considerably, while their coriaceous, thickly cutinized leaves gave these assemblages a xeromorphic aspect. In effect, their containing floras have been conceived of as xerophytic. Consequently, the floristic changes at the Jurassic/Cretaceous boundary were ascribed to aridization, either regional or global (Vakhrameev, 1981).

Similar but less radical vegetational changes, with the *Weichselia*-dominated coastal fern-marsh facies in the Wealden, took place simultaneously in the European and Middle Eastern - North African sectors of the southern zone, whereas in the temperate zone the corresponding events were far less prominent, with dicksoniaceous fern marshes surviving well in the Early Cretaceous. Although global climatic changes might have played certain roles in the abovementioned vegetational changes, the xeromorphic brachyphyll-bennettite assemblages could scarcely be taken as evidence of that. They are constantly associated with littoral facies, including the coal-bearing paralic, as well as the lacustrine-lagoonal calcareous deposits. Thick pinnules of *Otozamites lacustris*, a characteristic species of the Lower Cretaceous lacustrine facies in Mongolia and Transbaikalia (Krassilov, 1982b), had xeromorphic leaf pinnae with revolute margins and with stomata sunken in thick aerenchymatous spongy tissue suggesting a periodic submergence (Fig. 36). In this case, xeromorphism pertains to a helophyte, rather than xerophyte, habitat. Upright stems of *Classopollis*-producing plants have been found in situ in littoral deposits (Francis, 1983), with their pollen grains common both in the marginal and marine facies. In addition, their pollen grains and scale leaf cuticles have been found in the gut compressions of the exceptionally well-preserved katydids and other large insects from the calcareous deposits of the Late Jurassic Karatau Lake (Krassilov & al., 1997a; Fig.11). Being neither long-distance fliers, nor durable enough for withstanding even a moderate-distance transportation intact, those insects conceivably inhabited the coastal vegetation dominated by xeromorphic brachyphylls.

As is well known, xeromorphism is a characteristic feature of the present-day littoral plant communities, and perhaps it has remained such through the geological history. At the same time, the climatic control of the brachyphyll-bennettite communities is evident in their being far more prominent, in terms of taxonomic diversity as well as frequency of respective plant-bed assemblages, in the southern (*Cycadeoidea*) phytogeographical province extending into the ecotonal zone 50-60°N, but scarcely further in

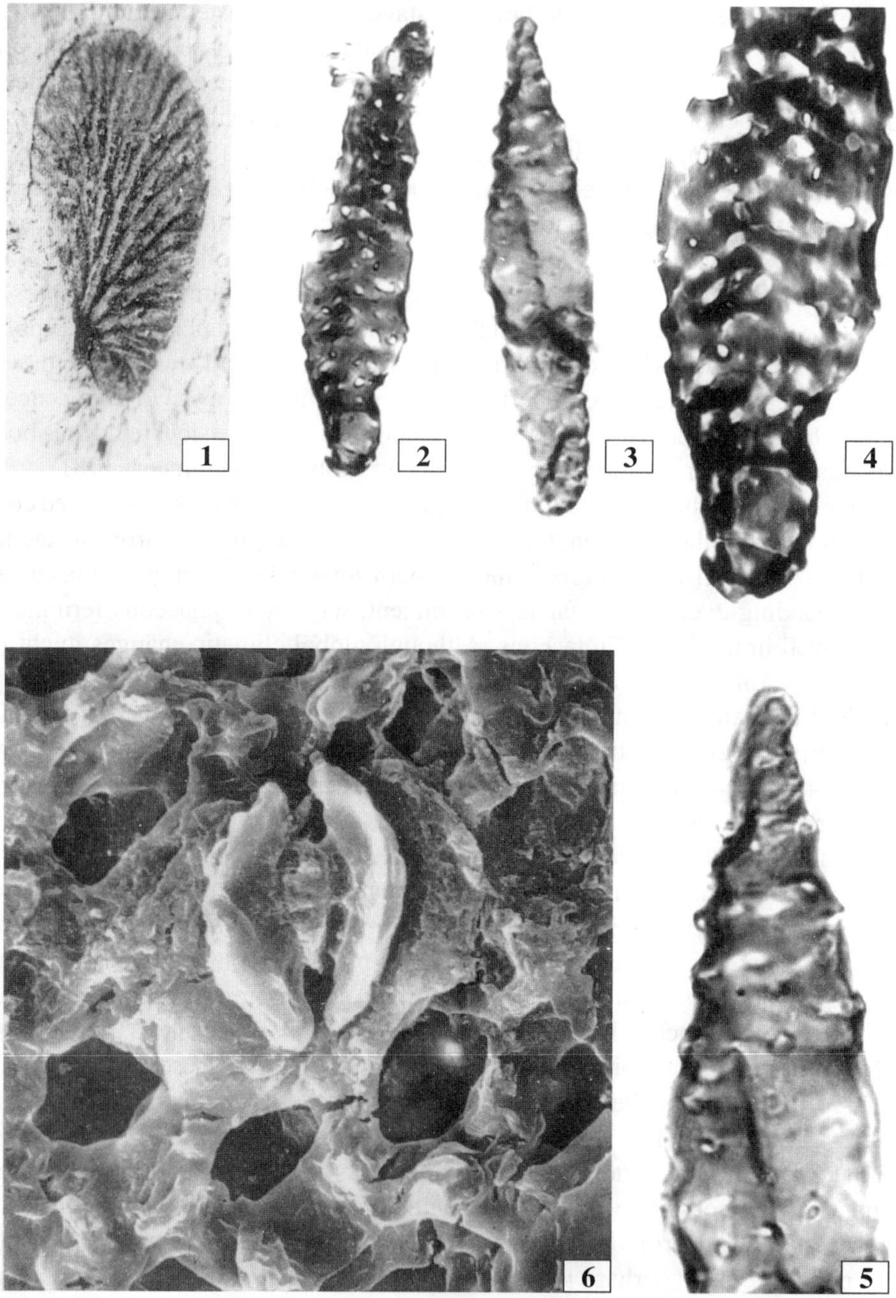

Fig. 36. *Otozamites lacustris* Krassil. a bennettite from the Early Cretaceous of Mongolia (Krassilov, 1982b): (1) pinnule with ribbed veins, x7, (2-5) vessel-like tracheary elements of the midrib showing a helical or reticulate radial wall thickenings and a porous perforation (2, 4) or an oblique perforation plate at the lower end (3), x300 and (5), x600, (6) stoma sunken in aerenchymous mesophyll, SEM, x1 000.

the northern (*Phoenicopsis*) zone. The impression of a southern zone xeromorphism, as well as the sharpness of the zonal boundary (see above), might be, to a considerable extent, due to a climatic threshold controlling littoral vegetation that was poorly developed in the northern zone in comparison with the southern zone.

For us the significance of the brachyphyll-bennettite assemblages lies in their comprising early angiosperm pollen and macrofossil records that were present neither in the contemporaneous fern-marshes nor in the Phoenicopsietum and Ginkgoetum associations of the inland basins and slopes. This conclusion is based primarily on the study of the Lower Cretaceous localities in Transbaikalia, Mongolia and Kazakhstan in central Asia, as well as in the Primorye region, the Far East, where a search for early angiosperms went alongside with palaeoecological analysis (Krassilov, 1967, 1973, 1982b, 1986; Krassilov & Bugdaeva, 1982, 1988; Krassilov & al., 1983, Vakhrameev & Krassilov, 1979; Krassilov & Makulbekov, 1995). For other areas, palaeoecological data are more sketchy. However, even floristic descriptions reveal a leading role of brachyphylls and bennettites in the Wealden-type floras of western Europe and Atlantic coasts (Seward, 1904-1905; Berry, 1911; Teixeira, 1948; Harris, 1953; Watson, 1982, Watson & Sincock, 1992). Remarkably, the brachyphyll-bennettite plant beds have yielded not only the first angiosperms and angiosperm-like fossils, but also the remains of advanced proangiosperms of bennettitalean gnetalean affinities. In the Baisa locality on the Vitim River, late Neocomian of Transbaikalia, the angiosperm leaves *"Dicotylophyllum" pusillum* and the pollen grains *Asteropollis* and *Tricolpites* (Vakhrameev & Kotova, 1977) are accompanied by the abundant cotton-grass-like disseminules of *Baisia* (Krassilov & Bugdaeva, 1982). *Eoantha*, a bracteate preflower with a four-lobed gynoecium and with *Ephedripites* pollen grains in the pollen chambers of orthotropous ovules came from the same plant-bed (Krassilov, 1986). A recently found attached flower has made it possible to assign the associated graminoid leaves to the *Eoantha* plant. Additional proangiosperm species of gnetalean affinities are represented by the *Baisianthus* androclades bearing bracteate nodal whorls of cupulate sporangiophores, and the *Vitimantha* preflowers with leafy reticulinerve perianth bracts (see under "Selected fossil links" above).

No less diverse are the angiospermoid and proangiosperm fossils in the roughly contemporaneous localities Manlay, Gurvan-Eren and Bon-Tsagan in the western Gobi, Mongolia (Krassilov, 1982b). The angiospermoid fruits *Gurvanella* and *Erenia* are accompanied there by the monocot-like *Cyperacites*, *Graminophyllum* and unassigned *Sparganium*-like and *Potomageton*-like fruiting axes (Figs 37, 38), as well as by the pappose reed-mace-like *Typhaera*. Their preservation is unfavourable for detailed morphological studies, thereby making their angiospermous or proangiospermous status uncertain. However, comparisons with the better preserved Baisa material are rather in favour of the latter, at least in the case of *Cyperacites* and *Graminophyllum* that resemble the inflorescences of *Baisianthus* and the leaves of *Eoantha* plant respectively.

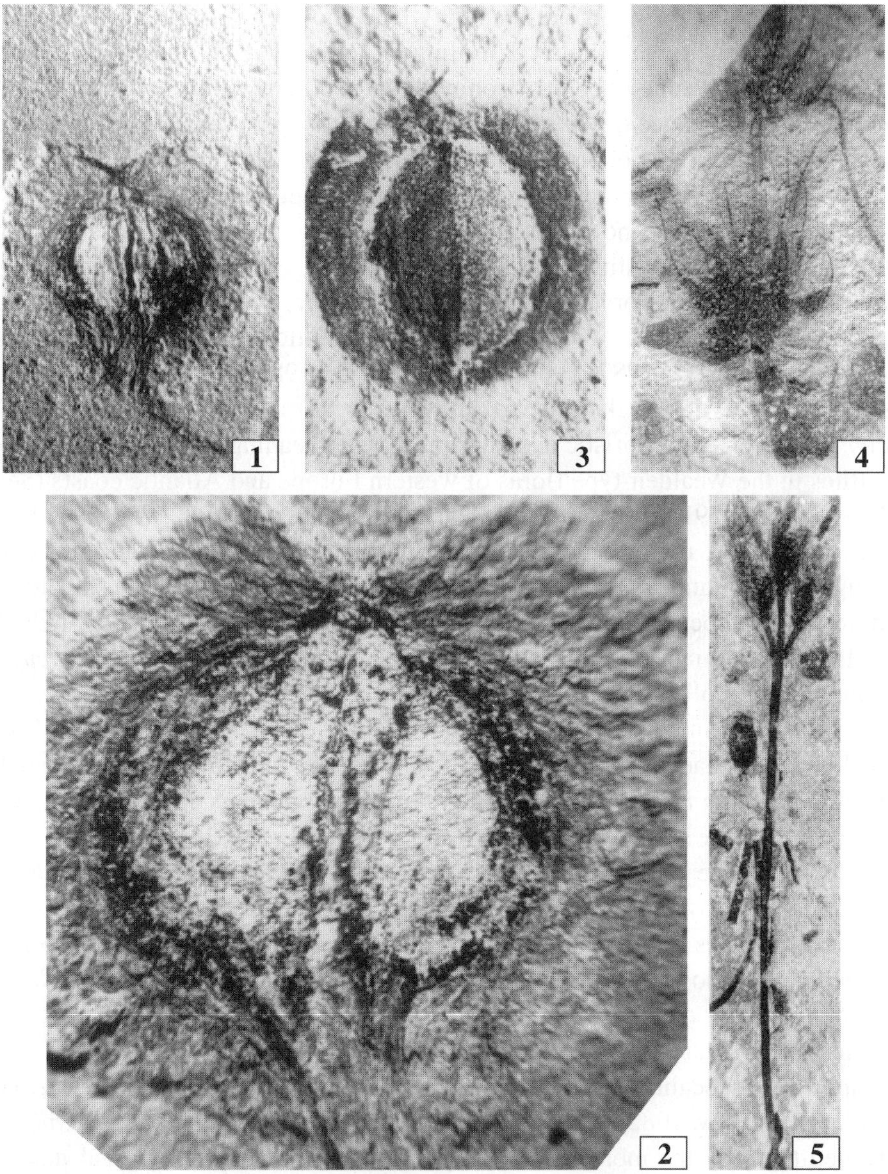

Fig. 37. Angiosperm and angiosperm-like reproductive structures from the Early Cretaceous of Mongolia (Krassilov, 1982): (1, 2) *Gurvanella dictyoptera* Krassil., winged fruit with reticulate wings, Gurvan-Eren Assemblage, x3 and 10, (3) *Erenia stenoptera* Krassil., another type of winged fruit with smooth wings and a bifid persistent stigma, same asemblage, x10, (4) *Sparganium*-like axis bearing bracteate heads of longly awned fruits or cupules, Manlaj Assemblage, x3, (5) *"Cyperacites"*, a fasciculate inflorescence with filiform bracts, same assemblage, x2.

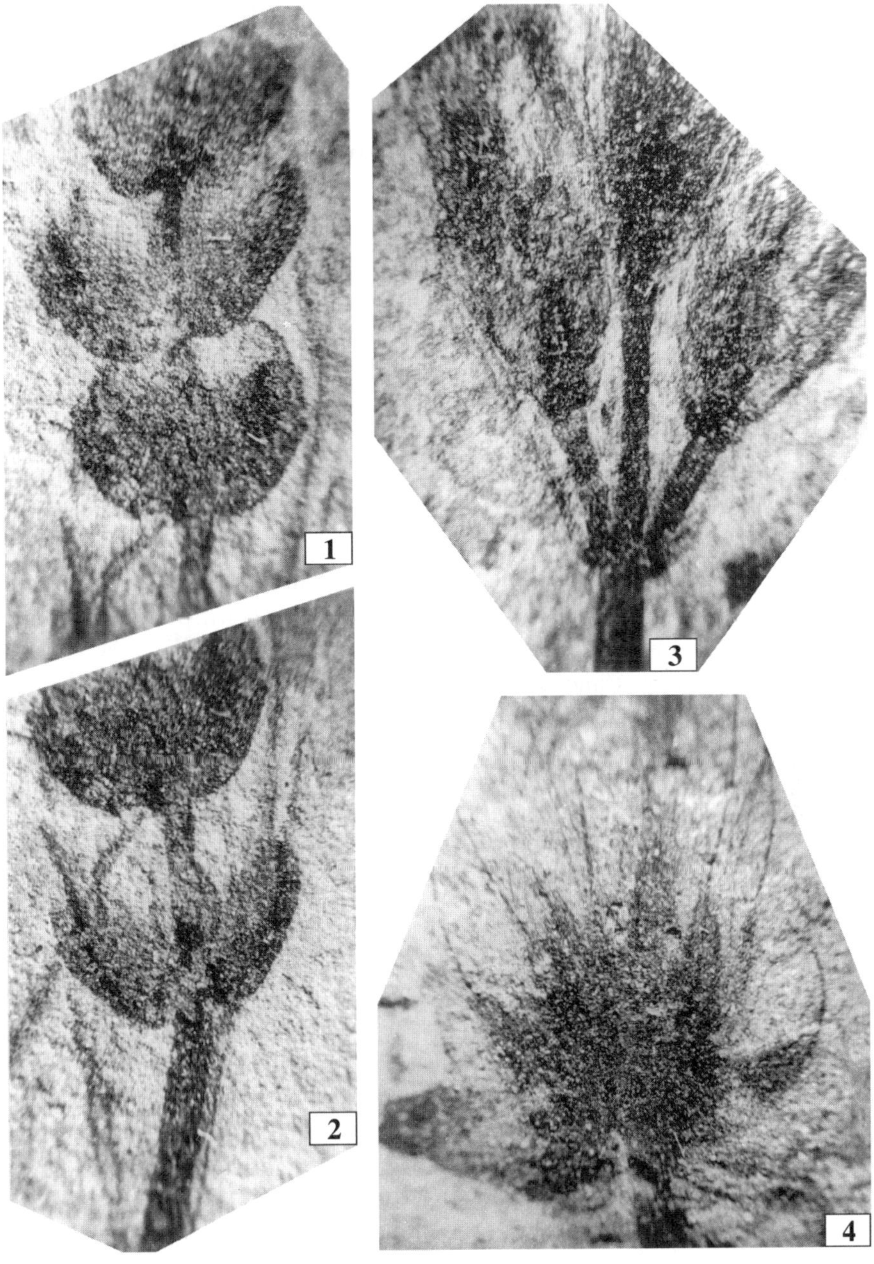

Fig. 38. Angiosperm-like reproductive structures from the Early Cretaceous of Mongolia (Krassilov, 1982): (1, 2) *Potamogeton*-like spike of awned fruits or cupules from the Manlaj Assemblage, x9, (3) "*Cyperacites*", details of spikes and filiform bracts, same locality, x10, (4) *Sparganium*-like bracteate head of longly awned fruits or cupules, Manlaj Assemblage, x7.

Similar situations are described in other parts of the world. In the Potomac flora of the Atlantic coast, early angiosperms appeared in association with *Drewria*, a herbaceous gnetalean plant (Crane & Upchurch, 1987). In the English Wealden, the entry of angiosperms in the pollen record was paralleled by the rise in the gnetoid pollen frequencies (Hughes & McDougal, 1987). In the Koonwarra Fossil Bed, southern Australia, angiospermous ceratophyllacean fruits (Dilcher & al., 1996) and racemose inflorescences of follicular carpels (Taylor & Hickey, 1990) are joined by the ephedroid *Leongathia* (Douglas, 1969; Krassilov & al., 1996) and perhaps some other gnetaleans represented by ovulate bracts and bracteate pollen cones (Drinnan & Chambers, 1986). In a small collection from the Aptian "amphibian bed" of Makhtesh-Ramon, Israel, angiospermoid fruits are found together with *Sagenopteris*-type leaves of caytonialean proangiosperms (Krassilov & Dobruskina, 1995).

Palynological studies in northern Africa have revealed the intercorrelated frequencies of the earliest angiospermoid pollen of *Afropollis, Schrankipollis,* and *Brenneripollis* types and the proangiospermous *Classopollis, Ephedripites*, etc., through the Barremian to Cenomanian sequences (Penny, 1991; Schrank, 1992; Ibrahim, 1996), with *Classopollis* coming to dominate over the *Afropollis* group after the initial increase of the latter in the late Barremian to early Aptian. The diverse gnetoid pollen and the recently found macrofossils appear with the early angiosperm pollen records in the Aripe Basin, Brasil (Pons & al., 1992).

To summarize, the appearance of angiosperms was not an occasional breakthrough against a static background, pushing other plants aside, but was rather a collective breakthrough involving a number of species growing side by side in the expanding brachyphyll community, their parallel, although not uniformly successful, advance channeled by this community evolution.

Centres of origin

As argued in the preceding chapter, the present-day supposedly primitive angiosperms have survived as relicts of the Arcto-Tertiary flora having little in common with the Cretaceous floras that comprised early angiosperms. Therefore, concentrations of such relicts are scarcely indicative of a centre of angiosperm origin. The latter cannot be deduced from a phytogeographical analysis of extant vegetation. One can hardly rely also on the currently earliest records of solitary angiosperm-like fossils for, in the first place, most of such records are not entirely convincing and, secondly, they depend on chances of preservation and collecting.

Angiosperm origins are here conceived of as a process with neither strictly definable starting points nor deadlines. A semitectate pollen grain signals a step in the morphological evolution related to the sporophytic self-incompatibility syndrome rather than the appearance of angiosperms. Such records go back in time at least to the Late Triassic (Cornet, 1979). More convincing are the combined records of several angiosperm traits. Such records first appeared, after a period of single-trait signals, in the late Neocomian to Aptian, about 120-115 Ma. Close to these dates are the associate records of authentical angiosperms and diverse proangiosperms, notably the advanced gnetaleans, in central Asia, Middle East, Atlantic coasts, Australia, etc. (see above). They not only testify to the early angiosperm appearances in these areas, but are also evidence of the ongoing process of angiospermization in the cradle plant communities that are crucial for recognition of geographical centres of angiosperm origin.

In both Northern and Southern hemispheres, the critical localities occur in the ecotonal zones of temperate summer-green and subtropical evergreen to winter-green vegetation, about 50ºN and 40ºS respectively. Angiospermization might have been conceivably going on all along the extent of these zones. However, most of the actual records are confined to downfaulted grabens and semigrabens of the Early Cretaceous rift systems conceivably constituting multiple angiospermization centres. Thus, the Transbaikalian and Mongolian basins are linear depressions of the extensive rift system striking northeast from Mongolia to the Sea of Okhotsk. This system has been initiated by a mid-Paleozoic rifting, while some of its segments, notably the Lake Baikal Rift, are still active. A massive volcanic event close to the Jurassic/Cretaceous boundary produced one of the world's largest basaltic provinces extending from western Gobi to the Amur River Basin. In the Cretaceous, the rift zones were marked by tectonic instability reflected in their sedimentary facies (Krassilov, 1985).

The lower Cretaceous lacustrine facies of the Mongolo-Okhotskian rift zone are typical of stratified lakes, with the thick, finely laminated black shale sequences intervened by psammitic and carbonate interbeds, the latter abounding in fossil fish and aquatic insect larvae remains. Large dragon fly, may-fly and beetle (coptoclavid) larvae pile up on the bedding planes suggesting mass mortalities. Since all these organisms are sensitive to pH, their mass deaths might have been caused by abrupt pH fluctuations inflicted by volcanogenic acid rains affecting also the surrounding wetlands and terrestrial biota. These taphonomic data suggest heavily impacted environments of the rift-bound angiospermization centres.

Rise of angiosperms as a biocoenotic process

The clandestine period of angiosperm evolution came to an end at about the Albian/Cenomanian boundary above which, or locally slightly below which, they became fairly common. Before the relatively recent discoveries of the earlier angiosperms, their Late Cretaceous mass appearances, as if from nowhere, seemed mysterious. One of the for the time being popular hypotheses holds that angiosperms could have descended from highlands where they grew for million years, perhaps even since the Paleozoic, without leaving fossil traces. The highland hypothesis accords with the Lyellian-Darwinian doctrine of imperfection of the fossil record. It is true that plant communities proximal to the site of deposition tend to be overrepresented in the fossil record at the expense of the more distant plants. However, while speaking of million-year time spans, we have to consider the contribution of long-distance transport in the form of wind-borne pollen and seed rain constituting a minor, but still recognizable, component of fossil plant assemblages (Hughes, 1961, 1976). Moreover, climatic fluctuations would repeatedly inflict upland-lowland shifts of altitudinal belts bringing distant vegetation closer to the deposition sites. Yet angiosperms did not appear during the major mid-Jurassic cooling event (Krassilov, 1973a and elsewhere). These general considerations are supported by the late Albian and Cenomanian angiosperms being far more numerous and diverse in the marginal epicontinental sea deposits (Berry, 1916; Vakhrameev, 1952) than in the contemporaneous deposits of intermontane depressions (Krassilov, 1989; Krassilov & Shorokhova, 1989). Such evidence is in favour of the coastal hypothesis of angiosperm dispersal (Retallack & Dilcher, 1981b) that implies some trust in the fossil record. The mid-Cretaceous time is marked by widespread tectonic and volcanic events of the Austroalpine phase that brought about frequent sea-level fluctuations and concomitant climatic changes. Of the latter, the late Albian to early Cenomanian cooling (Krassilov, 1973d, 1975b) could affect, in the first place, the thermophilous brachyphylls and bennettites of angiosperm cradle communities.

The mid-Cretaceous rise of angiosperms was paralleled by a concurrent rise to dominance of such morphologically modern taxodiacean and pinacean conifers as *Sequoia, Parataxodium* (Arnold & Lowther, 1955), *Pseudolarix* (Krassilov, 1982b), etc. These events were accompanied by the appearance of *Cycas* and other modern cycads (Krassilov, 1979; Krassilov & al., 1990). The situation was thus more complicated than a mere competition-induced replacement of gymnosperms by angiosperms. We can rather speak of a wholesale floristic innovation and the appearance of new plant communities comprising the angiosperms as their dominant or subordinate components alongside with the modernized gymnosperms.

Vegetation types can persist for a long time as stable systems of climax and successional communities. Their spatial ranges depend on the tolerance of their climax species, the successional species often having broader tolerances and potentially wider ranges. Insofar as successional species are poor competitors, the respective seral phases are open to new species entries by immigration or innovation. The early angiosperm appearances in the brachyphyll-bennettite communities might have been such entries. However, seral developments are equifinal in the sense that their end-products but little depend on the starting point. A climax phase can persist irrespective of the changes occurring in the pioneer and successional phases, while the radical restructurings are inflicted by environmental impacts that truncate the seral sequences halting their development at a successional stage. If such impacts are of a vegetation zone or global extent, they drive the climax species to extinction, with the successional species then exploring their vacant niches. Their rise to dominance is, thus, conjoined with functional and morphological innovations. These are postulates of the climax cut-off model (Krassilov, 1992, 1995) as opposed to the widely held species competitive replacement model.

The Mesozoic plant assemblage records suggest a leading role of the community replacement and community restructuring, rather than intracommunity competition, in the extinctions and origins of plant higher taxa as distinct adaptive types. All plant communities were simultaneously, although differently, affected by the Jurassic/Cretaceous boundary and, later, the mid-Cretaceous events. The respective vegetational events can be summarized as follows:

(1) The coastal fern marshes, world-wide in extent in the Jurassic, became spatially restricted at about the Jurassic/Cretaceous boundary but persisted to the end of Cretaceous, although their dominant tree fern component was reduced and partly replaced by the osmundacean and gleicheniacean genera. Judging by their numerical representation in the respective plant-beds (Krassilov, 1979), angiosperms scarcely penetrated the fern marshes even in the Late Cretaceous.

(2) The mesophytic forests with ginkgoaleans, czekanowskialeans and archaic conifers remained the least affected at the Jurassic/Cretaceous boundary but declined in the mid-Cretaceous surviving as relict communities with no or an insignificant angiosperm representation. Until the end of the Late Cretaceous epoch, their replacing communities of the modernized taxodiacean and pinacean conifers contained angiosperms as a subordinate component represented by laurophylls as well as by the small-leaved *Trochodendroides, Araliaephyllum, Menispermites, Liriophyllum* and a few other morphotypes perhaps representing a successional vegetation, some of them climbers.

In the flood-plain facies domain, the taxodiacean assemblages comprised the platanoid *Credneria-Protophyllum-Aspidiophyllum* morphotypes as their codominant component. The platanoid leaves, accompanied by the staminate and carpellate heads as well as by the tricolpate-tricolporate dispersed pollen grain assemblages, often formed leaf-mats in the coarse-grained laminated psammitic levee deposits as well as in the clayey oxbow lenses. These riparian mixed broadleaved platanophyll-conifer communities were widespread in both intracontinental and marginal basins of western Europe, Kazakhstan, West Siberia, Sakhalin, mid-western North America, etc.

(3) The xeromorphic brachyphyll-bennettite communities emerged as a dominant coastal to littoral vegetation type at about the Jurassic/Cretaceous boundary, replacing fern marshes in central Asia and, locally, in other sectors of the warm summer-dry zone and its ecotone with the temperate zone. The widespread xeromorphism might have been partly related to greenhouse climate. Notably, in both gymnosperms and the later appearing angiosperms, xeromorphic epidermal features correlate with a low stomatal index. Periodic penetration of xeromorphic vegetation into the temperate zone coincided with transgressions of the Tethys seas over the northern land masses.

Angiosperms appeared as minor components of these communities concomitantly with their expansion and proceeded in this role for about 15 million years. The long period of rarity suggests that early angiosperms had been competitively inferior to the community dominants until the climax phase was undermined by cooling and other environmental impacts related to the global tectono-volcanic event at about the Albian/Cenomanian boundary. Both hirmerellacean brachyphylls and bennettites survived through the Cretaceous and the latter even increased their frequencies in the Maastrichtian (Krassilov, 1979). However, since the late Cenomanian, they lost their dominant status, thereby leaving the respective communities undersaturated and opening new perspectives for the early angiosperms. The brachyphyll-bennettitalean climax cut-off was not an isolated vegetational event, but co-occurred with mass extinctions and replacements of the then dominant groups in other plant communities as well as in marine biotas.

Xeromorphic angiosperm assemblages comprise the small to medium-sized compound to lobed leaf morphotypes of the *Debeya-Dryophyllum* group as well as the narrow-leaved laurophylls of the *Proteoides* morphotype, and occasional palm leaves. The dicotylid leaf types are xeromorphic, often coriaceous, with thick cuticles, sunken stomata and a low stomatal index (Figs. 39, 40). They associate with fossil wood of shrubs or small trees (Stopes & Fujii, 1911, Shimakura, 1037; Sitholey, 1954; Prakash, 1965; Page, 1981) and with dispersed pollen of the *Normapolles* group. The non-angiosperm components are represented by brachyphylls and cycadophytes, infrequent on the basinal scale, but common locally. Taxodiacean conifer remains are less common. A typical example studied by the present author (Krassilov, 1984b) comes from the middle Cenomanian shallow marine carbonate deposits of Sel-Buchra Mountain, Crimea, with similar occurrences in western Europe, Israel, North America and elsewhere within the subtropical and warm-temperate ecotonal zones.

On evidence of the leaf morphologies and associated fossil wood, the xeromorphic angiosperm communities of the Debeyo-Dryophylletum type are conceived of as coastal shrublands or low woodlands, their brachyphyll component suggesting autochthonous origins from the Early Cretaceous brachyphyll-bennettite community. In the alluvial facies domain, the xeromorphic plant assemblages have merged with the platanophyll-dominated riparian ones, the latter most prominent in the northern temperate realm. At about the Cretaceous/Tertiary boundary, the Debeyo-Dryophylletum, although still prominent in the Paleogene, have receded before the advancing conifer-broadleaved forests.

Reproductive biology

It has long been suspected that plant-animal interactions might have played certain, perhaps even leading, roles in the origin of angiospermy. Many extant angiosperms considered primitive are insect-pollinated. Among them the beetle-pollination, or can-

Fig. 39. Typical angiosperms of the Late Cretaceous plant assemblages of Sakhalin (Krassilov, 1979): (1) *Debeya tikhonovichii* (Krysht.) Krassil., compound leaf of juglandacean-myricacean affinities, x1, (2) *Araliaephyllum polevoi* (Krysht.), trilobed lauracean leaf, x1.

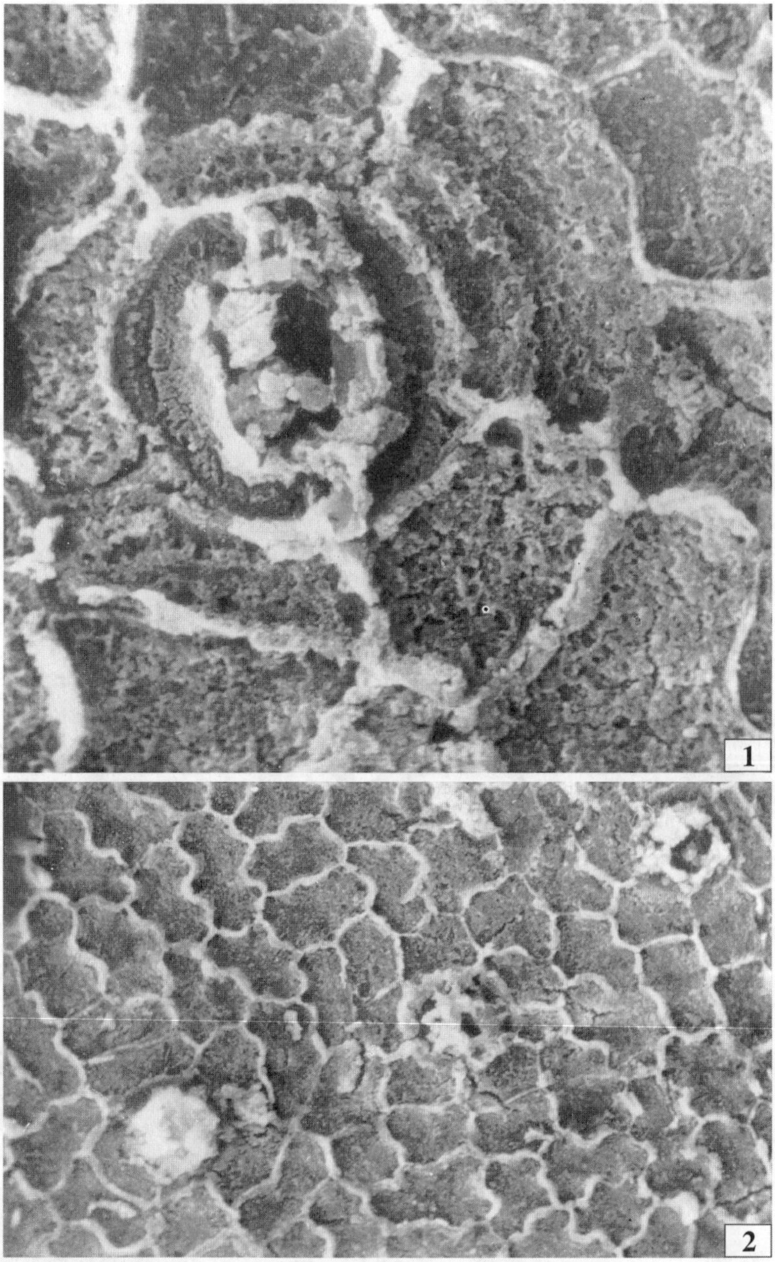

Fig. 40. Cuticle of a xeromorphic angiosperm *Debeya* from the terminal Cretaceous - lowermost Paleocene turbiditic island arc foredeep deposits of the Lesser Kuril Islands, Far East (Krassilov & al., 1987): (1) stoma with sunken guard cells and a prominent peristomatal ridge, SEM, x1 000, (2) well-spaced stomata of the lower epidermis with a low stomatal index, x250.

tharophily, is characteristic of the Magnoliaceae and is often considered as primary for all angiosperms. However, most early Cretaceous angiosperms had inflorescences of many minute inconspicuous flowers, as in *Caspiocarpus* (Vakhrameev & Krassilov, 1979), that looked anything but cantharophilous.

Extant cycads and gnetaleans are known to have insect pollinators, some of them perhaps inherited from their Mesozoic precursors. The morphology of bennettite pre-flowers, bowl-shaped with glands and perhaps filled with nectar (Harris, 1969), suggests insect visits for which there is some direct evidence (Delevoryas, 1965; Crepet, 1974). The caytonialean proangiosperms had fleshy cupules, while their pollen was found in coprolites (Harris, 1964).

A direct evidence of pollinivory has been obtained by maceration of the compressed gut contents preserved in fossil insects of different geological ages. Material came from the extensive entomological collections deposited in the Paleontological Institute, Moscow and was studied by the present author jointly with Prof. A.P. Rasnitsyn and his collaborators in the Laboratory of Arthropods of the same institute. The coaly gut compressions of the Early Cretaceous xyelid insects from the Baisa locality, Transbaikalia, remarkable for the diversity of proangiospermous plants (see above), have hitherto yielded only bisaccate pollen with laterally connected sacci, as in caytonialeans and some peltasperms (Krassilov & Rasnitsyn, 1982). While these studies are in progress, some interesting data have come from other localities.

Striate pollen grains were found in the gut compressions of extinct hypoperlid and grylloblattid insects (Fig. 41) from the Permian of the Urals (Rasnitsyn & Krassilov, 1996). These data have some bearing on parallelisms of pollen evolution in gymnosperms and, by implication, in angiosperms. Striate pollen grains constitute a dominant element of the Permian spore-pollen assemblages and are characterized by the external layer of their exine being divided, or only proximally so, into parallel or subparallel bands, or taeniae. Their number and pattern are diagnostic of the numerous form-genera differing from each other also in the presence of sacci and their configuration, with the asaccate morphotypes occasionally also showing rudimentary sacci. Typically, they differ from the modern saccate pollen of pinaceous conifers in a protosaccate, rather than eusaccate, infrastructure, with the inner partitions filling the sacci and attached to the corpus.

Striate pollen types were produced by at least three different groups of gymnosperms: glossopterids, peltasperms, and conifers (Clement-Westerhof, 1984; Meyen, 1978; Zavada, 1991), the former dominant in Gondwanaland and the two latter common to dominant in the Euramerican, Subanagaridan and Cathaysian provinces of Laurasia. Thus, the global spread of striate pollen was due to parallel developments. Meyen (1978) has suggested parallel successions of the non-striate to striate protosaccate to striate asaccate morphotypes in both the northern and southern provinces.

Aridity is generally considered as a leading factor of floristic evolution in the Permian. In particular, the scale-leaved conifers as well as peltasperms with thick leaf cuticles are

Fig 41. Taeniate pollen in the gut contents of pollinivorous insects from the Early Permian of the Urals (Rasnitsyn & Krassilov, 1996): (1) *Idelopsocus diradiatus* Rasn., a hypoperlid insect with pollen clump preserved at the end of the hindgut, x6.5, (2) pollen clump with some partly digested grains striped of the tectum and showing an infratectal reticulum, SEM, x800. (3) well-preserved *Lunatisporites*-type pollen grain from the clump shown in (2), SEM, x1 500, (4) *Vittatina*-type pollen grain from the gut of another pollinivore, *Sojanidelia floralis* Rasn., from the same locality, x1 200.

conceived of as representing a xeric type of Permian vegetation distinguishable palynologically by the prevalence of striate pollen grains. However, their frequent association with coal-bearing facies makes this interpretation improbable. Glossoprterids, or at least some of them, produced the *Vertebraria*-type aerenchymatous underground organs evidencing a peat bog or mangrove swamp habitat. Since the helophyte, rather than xerophyte, adaptations prevailed among the producers of striate pollen grains, aridity seems to have never been a leading factor in their morphological evolution.

Pollinivory provides an alterative explanation of the remarkable parallelism of palynological characters in the Permian gymnosperms. Both the protosacci (as functionally different from the bladders of anemophilous conifers) and taeniae might have had a volume-regulating (harmomegathy) function, at the same time provoking pollinivory by increasing the pollen mass. Harmomegathy functions could have been even more important for pollen transferred by large slow-flying insects than for any wind-borne grains.

In addition, insects are known to transfer various gene-transducing microorganisms that can confer parallel genetic changes in the target plants. This mechanism is particularly important for pollen grains that are haploid organs with the mainly single gene-controlled exinal characters.

The effects of the plant-insect interactions could scarcely have been restricted to pollen organs alone. The larger the pollen grains, the fewer of them could compete for room in the target pollen chamber, thus creating a selection pressure in favour of the larger or, alternatively, the densely aggregated ovules, as well as stimulating adaptive modifications of their receptive structures. These processes might have brought about the burst of seed plant diversity in the Permian, as well as the subsequent burst in the Mesozoic.

This Paleozoic story is here narrated in some detail because of its relevance to the Mesozoic story. The Mesozoic xeromorphic brachyphyll-bennettite communities were, in a syngenetic sense, direct descendants of the Permian conifer-peltasperm communities, even their respective coniferid elements could have been related phylogenetically (Clement-Westerhof & van Konijnenburg-van Cittert, 1991). The recent discovery of angiospermoid *Classopollis* pollen grains in the intestines of Jurassic grasshopper-like katydid insects (Krassilov & al., 1997a: Fig. 42) adds to the ecological similarity of the Permian and Mesozoic xeromorphic communities in several ways. In the first place, they provide evidence of pollinivory as a mode of plant-animal interactions involving the dominant plants of both communities. Furthermore, both the Permian and Mesozoic pollinivores were large slow-flying insects that could hardly have been transported intact over a considerable distance, thus suggesting proximity of their habitats to the deposition sites. They also testify to the most advanced pollen types of each of the epoches being related to the plant-insect interactions, with possible implications for plant reproductive biology.

The first appearing angiosperm pollen grains are recognized as such by the features of the self-incompatibility syndrome (Zavada, 1984; Zavada & Taylor, 1986a)

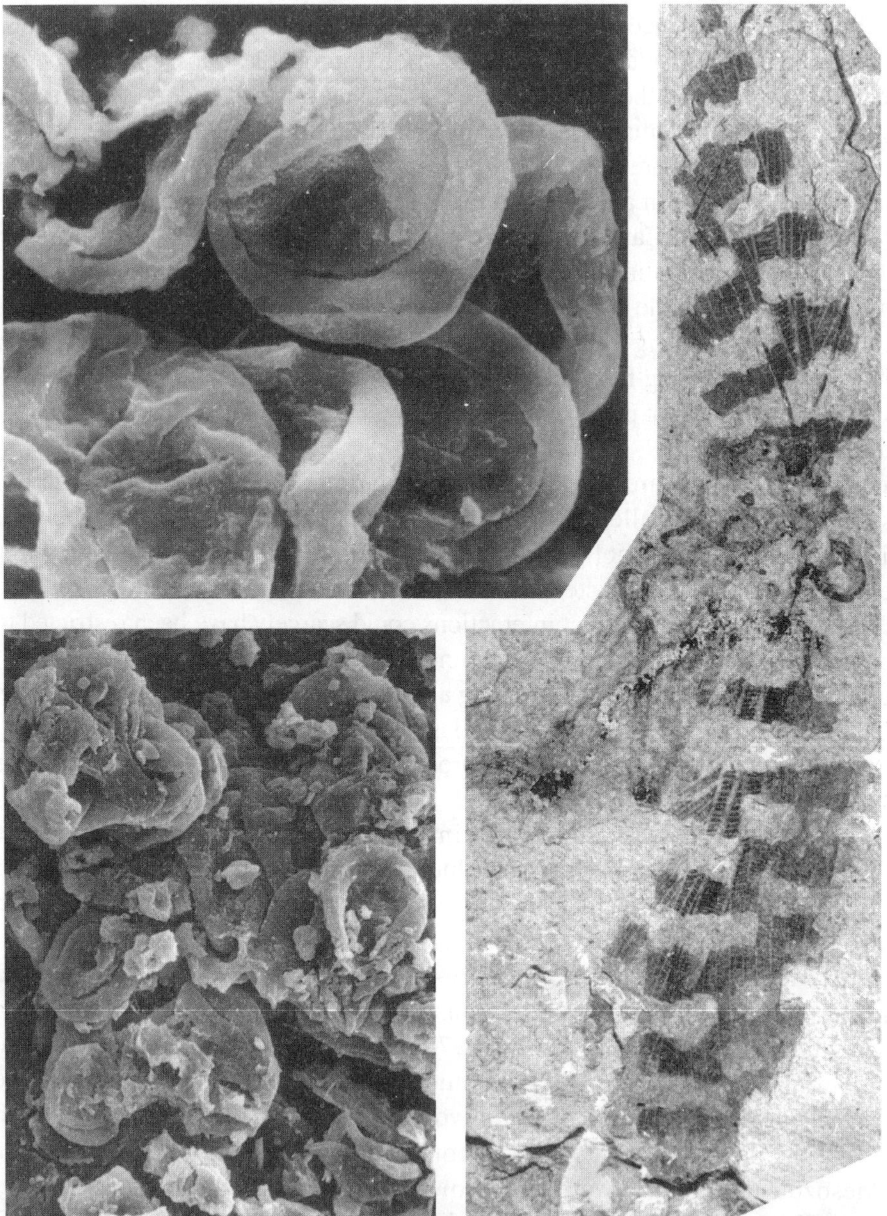

Fig. 42. Proangiosperm pollen in the guts of pollinivorous insects (Krassilov & al., 1997): *Aboilus amplus* Gorochov, a large katydid from the Late Jurassic of Karatau, Kazakhstan, x1, with a dark intestine compression filled with clumped *Classopollis* pollen grains, SEM, x830 and 1 700.

that is conceived of as a form of sexual selection distinguishing angiosperms from gymnosperms. The sporophytic self-incompatibility mediated, by the pollen-stigma interaction, might have appeared before the gametophytic self-incompatibility resulting from the pollen-stylar conducting tissue interactions, the latter mode correlating with solid, rather than open, styles recorded not earlier than the Senonian (Krassilov & al., 1983). However, a less advanced form of pollen selection by micropylar structures, such as the stigmatic (Narben) micropyles (Kügler, 1970), occurs also in gymnosperms: pollen grains are captured in the slime excreted by the micropyle and are then sucked in by a biochemically different pollination drop excreted by the nucellus (Owens & Blake, 1984). This mechanism anticipates the two successive secretory stages of angiosperm pollination. Moreover, in the extant *Pseudolarix*, pollen sometimes germinates in the micropylar canal (Allen, 1943), while in *Caytonia,* I have observed pollen in the tubes connecting the exostome with the ovules (Krassilov, 1977b). The infratectal structures serving for storage of recognition substances might evolve on the basis of the honeycomb infrastructures of protosaccate gymnosperm pollen grains in relation to the micropylar pollen selection function. In angiosperms, this function might have developed primarily for preventing the abortion of gynoecia inflicted by mispollination. Judging by the frequent occurrences of aborted ovules and cupules, notably those of *Caytonia*, this phenomenon actually took place in Mesozoic seed plants. In extant angiosperms, mispollination sometimes causes a seed yield reduction, as in *Ipomea aggregata* (Kohn & Waser, 1985). As successional plants, the early angiosperms tended to increase their seed yield. Thus, gynoecial abortions had to be prevented.

Advances in the pollen selection function made angiosperms preadapted to the development of monoclinous ("bisexual") reproductive organs that occasionally appear also in conifers (Tosh & Powell, 1986) but are unstable in them because of their imperfect self-incompatibility systems. Even in bennettites and gnetophytes, the morphological monocliny usually digresses to the functional dicliny, with the androecial (in *Williamsonia*) or gynoecial (in the pollen preflowers of *Welwitschia*) organs transformed into secretory structures. In the insect-pollinated diclinous gymnosperms, the insects are often attracted to the ovules by sweet pollination drops. For angiosperms lacking the pollination drop mechanism, stamens are the major attraction. Monocliny is therefore essential as a component of the accomplished angiospermous entomophily syndrome including also specialized pollen morphologies, as in *Stellatipollis* and related forms.

The katydid example shows that proangiospermous and, by implication, their succeeding early angiospermous plants could use unconventional pollinators, while beetles and other then existing anthophilous insects were feeding in the more conspicuous bennettite preflowers (for insects they were flowers irrespective of what plant morphologists might think of them). Some extant pollinators of the "primitive" Winteraceae, such as *Sabatinca* (Micropterigidae), are closely related to the fossil forms (*Para-*

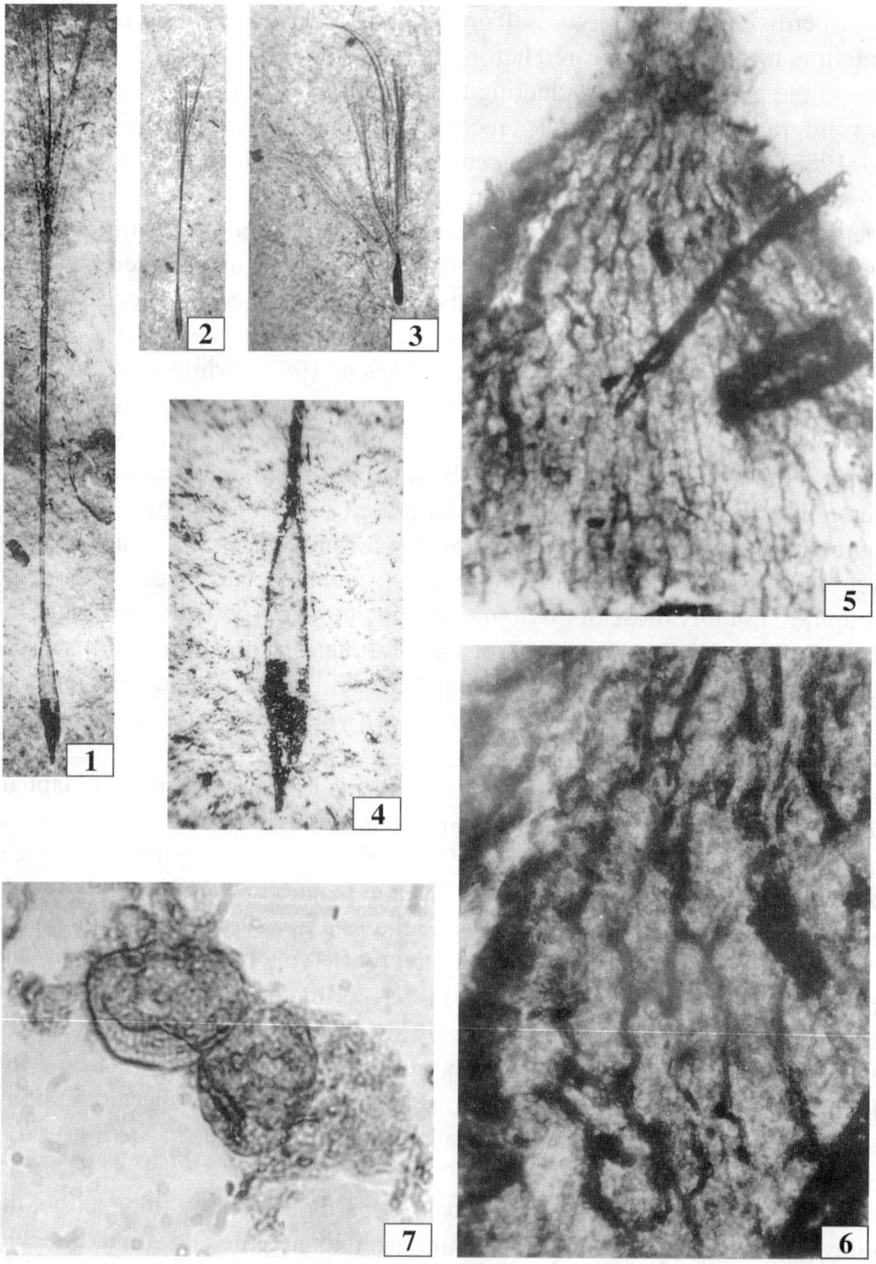

Fig. 43. *Problematospermum ovale* Turutanova-Ketova, presumably bennettitalean pappose seeds from the Late Jurassic of Karatau, Kazakhstan (Krassilov, 1973b), x1 (2, 3), 5 (1) and 7 (4); sinuous cell walls of the seed coat, x58 and 146 (5, 6) and pollen grains stuck to the micropylar tube, x562 (7).

sabatinca) appearing in the early Neocomian (Thien & al., 1985) prior to the advent of angiosperms. With a decline of bennettites in the mid-Cretaceous, their insect symbionts passed over to their replacing angiosperms. This was a turning point in the history of terrestrial ecosystems marked by the first appearances of the insect groups that formed a core of the Cenozoic entomofaunas (Zherikhin, 1979). The angiosperms exploring the temporarily vacant, after the mass extinction of bennettites, cantharophilous niche, have developed large solitary flowers of many parts that are characteristic of cantharophily and are often considered primitive (e.g. Takhtajan, 1966).

As it was suggested elsewhere (Krassilov, 1981), the mid-Cretaceous extinction of giant sauropods and their replacement by the newly appearing dinosaur herbivores with crashing dental batteries could have been related to the bennettite by angiosperm replacement that, at the life-form level, meant a substitution of the sparsely branched pachycaul forms with large leaves by the copiously branched leptocaul forms with small leaves characteristic of early angiosperms. The ensuing increase in shoot mass at the expense of the drastically reduced leaf mass required adequate dietary and morphological innovations in the larger herbivores. Their indiscriminate shoot browsing (Wing & Tiffney, 1987), perhaps involving reproductive shoots, might have promoted endozoochory to a leading plant dispersal strategy. At the same time, a comparative analysis of the dental morphologies and digestion systems indicates semenivory as a primary phytophagous adaptation in mammals (Agadjanian, 1996). Semenivory was obviously related to the appearance, in *Caytonia* already, of fleshy cupules containing many small seeds. The overall significance of this feeding habit could have increased with the advance of the Cretaceous multituberculates, some of them arboreal (Lillegraven & al., 1979), coming to dominance in the small vertebrate herbivore niche. There was a concomitant increase in the multituberculate teeth and the small *Spermatites* - *Costatheca* type seed frequencies in the fossil assemblages. The simultaneous rise of angiosperms and mammals might thus have been due to coevolution of semenivoruos (frugivory) and endozoochorous strategies.

Epizoochory is another dispersal strategy related to the early angiosperm-mammal interaction. Pubescence is a characteristic feature of bennettitalean reproductive structures (Harris, 1969) that, in the derived forms, produced the pappose disseminules, as in *Problematospermum* (Fig. 43), *Typhaera* or *Baisia* (Krassilov, 1973b, 1982b; Krassilov & Bugdaeva, 1982). These structures, perhaps primarily anemochorous, were preadapted to dispersal by sticking to furred animals. The glochid-like appendages, as in the ceratophyllacean fruits from the Aptian of Australia (Douglas, 1969; Dilcher & al., 1996;) and the contemporaneous fruit-like structures from eastern Asia (Krassilov, 1967, 1977; Plate 43) point to epizoochory as an early appearing dispersal mechanism.

One more aspect of the plant-phytophage interaction is related to the biochemical defense evolution in angiosperms. In their progenitorial stocks, accelerartion might involve not only the morphogenic, but also the biochemical processes, resulting in abbreviation of the shikimate pathway in the biosynthesis of detergent alkaloids (Kubitzki & Gottlieb, 1984b) and the loss of tannins and lignins. Early angiosperms might have thereby been especially vulnerable to phytophage pressures that had kept them rare until they developed their own arsenal of detergents, some of which had functionally reversed to attractants in the course of subsequent biochemical evolution (Pellmyr & Thien, 1986).

A general scheme of ecological evolution

It follows from the preceding discussion that angiosperm origins were an integral part of the Mesozoic plant community evolution. Angiosperms entered the fossil record after an expansion of the xeromorphic brachyphyll-bennettite community that had replaced coastal fern marshes in the Mesozoic catenic successions. Typically, the angiosperms appeared in clusters of a few species representing different life-forms, some of them aquatic, and were accompanied by the diverse gnetophytes and other proangiospermous plants that evolved in the same direction, acquiring certain traits of the angiospermy syndrome in parallel with the conventional flowering plants.

The palaeontological evidence of a "collective breakthrough" accomplished by the proangiosperms growing side by side in the "cradle communities" sheds new light on parallel evolution as a result of biotic interactions, involving insects, such as pollinivorous katydids, that might have transferred gene-transducing microorganisms, thus spreading mutations across the community. The first appearances are confined to the broad ecotone of the warm evergreen or summer-green and the temperate summer-green zones at about 50°N and 40°S, with the major dispersal centres in the active rift zones. These circumstances suggest geological impacts as a factor impelling a developmental acceleration and condensation, the prevailing modes of morphological evolution in early angiosperms.

Among the Mesozoic vegetation types, the xeromorphic brachyphyll-bennettite assemblages of the helophyte to littoral facies domains comprise most of the earliest pre-Albian angiosperm records. Expansion of xeromorphic assemblages under greenhouse conditions gave impulse to ecological and morphological innovations, angiospermy one of them.

In the early angiosperm evolution, there was a "clandestine period", about 120-100 Ma, of occasional records suggestive of a subordinate role the angiosperms played in the brachyphyll-bennettite communities. At this stage, the herbaceous forms are somewhat better documented than woody forms, with the fossil wood records exceedingly rare and perhaps not always authentical. Typical for this stage were small entire or lobed leaf blades, an irregular venation network, the inflorescences of inconspicuous diclinous flowers, and small to medium-sized follicles and achenes.

An important feature of the early angiosperm reproductive biology was an advanced pollen selection system as evidenced by the exinal characters of the self-incompatibility syndrome. Initially, the development of the latter might have been related to an increased seed yield by preventing the ovule (carpel, flower) abortion inflicted by mispollination. Angiosperms might have been predominantly diclinous at this stage. With elaboration, self-incompatibility enabled angiosperms to acquire monocliny (unstable in gymnosperms), that, in turn, promoted entomophily to a leading pollination mode.

Rarity during the clandestine period might have been related to the developmental acceleration involving not only morphological but also biochemical traits and resulting in the loss of certain defense compounds (notably lignins and, in the magnolian line, tannins) rendering early angiosperms an easy prey to various phytophages.

The bennettite and brachyphyll mass extinctions and the compensatory rise of angiosperms at about 100 Ma marked the end of the clandestine period. Insofar as both these extinct groups were relatively thermophilous, the late Albian-early Cenomanian cooling might have been a major factor of their decline. The appearance at that time of a huge cirum-Pacific volcanic belt could have contributed to extinctions either indirectly, through its climatic effects, or directly, as a source of acid rains that affected both plants and their insect symbionts. Vast marine transgressions in cratonic areas have caused a reductuion of terrestrial biomass that seves as a major sink of atmospheric carbon dioxide. The ensuing greenhouse conditions have promoted high growth rates giving the early angiosperms an advantage over the slow growing gymnosperms.

In effect, the climax phase of the brachyphyll-bennettite communities was undermined and was replaced by the likewise xeromorphic Debeyo-Dryophylletum association with shrubby dicotylids and palms, the relict brachyphylls still locally abundant. The simultaneous replacements of the mesophytic arboreal communities gave rise to the taxodiacean-laurophyll and taxodiacean-platanophyll forest types, the latter mostly riparian. The mid-Cretaceous rise of angiosperms was paralleled by the rise of modern conifers and the appearance of modern cycads. It was, thus, not an isolated event that could be ascribed to a competitive replacement of gymnosperms by angiosperms, but was instead related to the wholesale restructuring of dominant plant communities.

The replacing plants converged on the replaced. The mid-Cretaceous angiosperms inherited the symbionts and consumers of the replaced bennettites and brachyphylls thereby acquiring the bennettite-like cantharophilous flowers currently considered primitive in the extant flowering plants.

Since most of the early angiosperms had smaller leaves than bennettites, the replacement resulted in a lower leaf mass to shoot mass ratio. The larger vertebrate herbivores had to cope with the leaf mass reduction by acquiring new nutritional habits, such as shoot browsing involving reproductive shoots and promoting endozoochory. No less important for endozoochorous dispersal and the development of the respective fruit morphologies, to a certain extent anticipated by *Caytonia* among proangiosperms, was semenivory, a primary phytophagous adaptation in multituberculates and other early mammals. These developments coincided with the appearance of abundant small seeds in the Late Cretaceous microfossil plant assemblages. Epizoochorous adaptations, arising on the basis of pappose wind-borne disseminules widespread among the proangiosperms, constituted another aspect of the early angiosperm - early mammal coevolution that led both groups to dominance in their respective domains.

The next stage of angiosperm evolution is related to the Cretaceous/Tertiary boundary events. The most clearly marked among them were the Laramid tectonism affecting both the circum-Pacific and Alpine belts, the concomitant retreat of epicontinental seas and the global cooling inflicting a complete restructuring of the dominant plant communities. The fern marsh communities disappeared, and there was a reduction of xeromorphic shrublands, the cradle of angiosperms that came to dominance there in the Late Cretaceous. These highly productive and resilient vegetation types supported, through the Mesozoic era, large dinosaur herbivores that perished with them.

In the advancing deciduous forests, there was a replacement of *Sequoia*-like conifers and laurophyll angiosperms, dominant in the Late Cretaceous, by *Metasequoia* and the broadleaved hamamelid genera that had previously constituted a riparian component. Their emergence as the dominants of deciduous forests gave the latter a modern broadleaved aspect, but in fact these Paleocene forests were dominated by extinct genera, such as *Trochodendroides, Steinhauera* (Plate 44) and/or *Nordenskioldia*. They soon receded before the then appearing modern amentifers. Yet these chapters of angiosperm history are beyond the scope of this book.

Phylogeny

Phylogeny is interesting not so as a guide to who bore whom as a reflection of evolutionary processes. The angiosperm phylogeny incorporates the processes of seed plant evolution that started long before the appearance of angiosperms. As outlined in the preceding chapters, the seed plant evolution is divisible into chronological stages, each characterized by the prevalence of morphological characters that define successive evolutionary grades. Angiosperms are successors of the hydrosperm, pteridosperm, cycadophyte and proangiosperm grades. Their precursory character states have been accumulating from grade to grade until their almost complete set appeared at the proangiosperm level.

Monophyly versus polyphyly of angiosperms is a much debated, yet little-studied, problem. As Hughes (1961) noticed, decisions are often made in advance of the study. Indeed, the problem has some non-scientific aspects, such as the Biblical Noah legend of each species arising from a single pair, or an aesthetic preference of tree topography over bush topography, or, more importantly, a subconscious urge for unity impedes satisfaction until all items are brought to one end. In effect, monophyletic systems are much more popular than polyphyletic systems, of which even the most thoroughly worked out ones (e.g. Kuznetsov, 1936) have had little chance of survival.

The cladistic techniques provide a seemingly objective approach to the problem by numerical estimates of intergroup versus outgroup similarities. It is assumed that, for each group, there is a common ancestor, the problem is only how deep in the tree it can be found. The problem thus boils down to the closest common ancestor being or not being a member of the group concerned. Correspondingly, the group is either holophyletic, neatly interconnected by the alternative character states arising from each other, or paraphyletic, with the character states arising independently in the clades that diverged before the appearance of the group. But the conclusions are, to a considerable extant, predetermined by the basic assumptions.

A phylogeny of even a single genus can be biased by the assumption of monophyly, while an unprejudiced analysis indicates polyphyly (Lanyon, 1994). In the case of such a diverse and long studied group as angiosperms, the problem is further complicated by the ambiguous character cline polarities, pseudohomologies, semantic inconsistencies, etc. If the typical characters are primitive, then the common ancestor will certainly be an angiosperm. If, on the other hand, the aberrant characters are primitive, then it may not be an angiosperm. Whatever the outgroup, the unity of angiosperms can be easily corroborated on the basis of charismatic characters, such as flower, fruit, double fertilization, etc. For example:

	Gnetophytes	Dicotyledons	Monocotyledons
Flower	-	+	+
Stamen	-	+	+
Carpel	-	+	+
Cupule	+	-	-

Before taking this table seriously, we ought to look into definitions and possible homologies of the characters used. Otherwise their presence or absence depends on what we mean by them.

Another problem is consistency of intragroup divisions. If, in the above example, some taxa nominally assigned to monocots are, on the bulk of the evidence, closer to dicots or vice versa, then the morphological distance between the two would appear smaller than it actually may be. But this is just what is constantly happening to the higher-rank angiosperm divisions. The magnoliids, for example, are fairly distinct from the ranunculids on the basis of their gynoecial, as well as their stamen and pollen morphologies. But, with a questionable inclusion of the Illiciaceae and Schizandraceae in the former group, the distinctions fade out. Angiosperm divisions seem not so naturally interlinked as intermixed by misassignments.

Morphology

The following compilation of typical characters may serve as a morphological diagnosis of angiosperms:

Cretaceous - extant; life-form arboreal, leptocaul or pachycaul or brachycormic, herbaceous, aquatic, geophytic, epiphytic, etc. Roots typically tetrarch, lateral roots often diarch or triarch, in monocots often polyarch. Apical meristem stratified, with tunica-corpus differentiation, but the latter less distinct or lacking in monocots; secondary growth by apical, in monocots more typically by lateral, meristem. Primary vascular system of free or interconnected symposia, with vascular bundles collateral or - more often in monocots - concentric, large and small alternating. Wood, with a few exceptions, porous, with vessels; sieve elements with companion cells and with S-plasmids. Nodes 1-lacunar with two leaf traces, less often 3- or multilacunar. Shoots often heteromorphic, sometimes heteroblastic. Leaves spiral or decussate, less commonly verticillate, expanded laterally by marginal and plate meristems, simple or compound, sometimes sheathing, some peltate or ascidiform, bifacial or, in monocots, unifacial, caulomic or phyllodic; stipules common; venation reticulate, pinnate or palmate, with 3-5 orders of areoles, the ultimate ones often filled with blind veinlets, or, more often in monocots, primary veins parallel or acrodromous, with commissural cross-veins, with alternating large and small bundles and with interbundle sclerenchymous strands; stomata anomocytic, cyclocytic, paracytic or tetracytic, with subsidiary cells mesogenous, perigenous or mesoperigenous, other types (about 20) less common.

Reproductive shoots with apical floral meristem typically producing cyclic aggregates of androecial and/or gynoecial structures (flowers) sometimes intermingled with intrafloral phyllomes; flowers solitary or gregarious in inflorescences of many types, diclinous or monoclinous, with perianth components heteromorphic or homomorphic, their members free or connate, or sometimes perianth lacking. Pollen organs, or stamens, arranged in 1-2, rarely more cycles, solitary or in fascicles, supplied with solitary or dendroid interconnecting traces respectively, rarely forked or branched or laminar and supplied with 1-3 traces, developmentally centripetal or centrifugal, typically filamentous, with anthers terminal or lateral, in laminar stamens abaxial, rarely adaxial or marginal; anther consisting of a sterile connective, this sometimes protruding (produced) apically, and a bilobed synangium, 4-sporangiate, rarely of 2 or many sporangia.

Pollen grains dispersed singly, less commonly as monads or persistent tetrads, polyads or pollinia, bi- or trinucleate, these conditions correlating with dry or wet stigma types respectively, ana-, cata-, trichotomo-, or zonosulcate, variously colpate, porate or colporate, rarely inaperturate, coated with intine and exine, in aquatic forms sometimes with intine alone, asaccate but sometimes cavate, with airspace between sexine and nexine, rarely with residual sacci; covered with trifine or pollenkit, tectate or semitectate, with tectum reticulate or perforate (these structures correlating with sporophytic or gametophytic self-incompatibility), rarely atectate; pollen wall infrastructure columellate or granulate; lamination of nexine lacking (with exceptions) or lamellae present under apertures alone.

Gynoecium of solitary to many closed or partly open compound ovuliferous structures, or pistils, exposed or partly or totally sunken below and connate with perianth and/or androecium, or sunken in receptacle, provided with pollen receptive structures, or stigmas, these raised on hollow or solid styles or sessile, sometimes formed by surrounding structures, as in receptacular hyperstigma (*Tambourissa*); pistils consisting of one or two to many ovuliphores, or carpels, these distinct, connate or partly so, with ovules produced basally, apically, marginally, on sutures or proximal to them or scattered on walls or arising from floral axis axillary to carpels or acarpellate; ovules orthotropous or anatropous or of transitional polarities, uni- or bitegmic, rarely with 3-4 integuments, with micropyle formed by inner or outer integument or both, sometimes zigzagged; none or seldom one of integuments, rarely both, vascularized, some cutinized and stomatiferous; nucelli sometimes beaked, with or without parietal tissue and with archesporium more or less reduced.

Pollen tubes siphonogamous, reaching the embryo sacs by various routes, often discharged into synergid; megasporocyte single, rarely many (*Paeonia*); megaspore tetrads of five basic types, often linear or T-shaped; embryo sac mono-, bi- or tetrasporic, seldom with haustoria or protruding into style, of different developmental types, more commonly 8-nucleate with micropylar and chalazal groups of nuclei or, in tetrasporic embryo sacs, also with lateral groups or with chalazal nuclei lost (*Oenothera*) or sometimes multiplied (*Sparganium*); syngamy involving two or more nuclei, seldom wanting, endosperm nucleus triploid or polyploid, seldom diploid; seeds shed at different stages of embryogenesis, dispersed separately and sometimes arillate or dispersed with fruits, these dehiscent, septicidal or loculicidal, or indehiscent, dry or fleshy of many types, formed of gynoecia or their parts or involving other parts of flowers or even inflorescences. Embryogenesis cellular, seldom initially free nuclear (*Paeonia*), endosperm development free nuclear, less commonly cellular or first cellular then free nuclear. Proembryo cell pattern of several types, differing in contribution of the basal cell, rarely irregular (*Degeneria*). Cotyledons one, terminal or pseudoterminal and haustorial, or two, seldom more, sometimes heteromorphic and one haustorial (*Piperomia*) or cotyledons unidentified.

Character sharing

Angiosperms are usually thought of as sharing more characters with each other than with any of non-angiospermous seed plants. This may be due to the functionally intercorrelated angiospermy syndrome comprising a great number of discrete characters related to production, pollination and dispersion of enclosed ovules, and constituting the bulk of the commonly analyzed characters. With the functionally not or weekly correlated character sets the situation might have been be entirely different. A morphological analysis of the present-day angiosperms shows the following:

(1) The typical or diagnostic characters are those shared by a considerable number of angiosperm species. None of such characters is shared by all species currently assigned to angiosperms. There are alleged angiosperms lacking a distinct stratification of the apical meristem, with an anomalous secondary growth, homoxylic (the Winteraceae, Trochodendraceae, Tetracentraceae, Nymphaeaceae, Ceratophyllaceae and a few Monimiaceae; *Sarcandra* (Chloranthaceae) has been considered as vesselless, but actually it has metaxylem vessels: Carlquist, 1987; Takahashi, 1988), lacking typical sieve element companion cells (*Austrobaileya*: Behnke, 1986), with open venation (*Kingdonia, Circaeaster*: Foster & Gifford, 1974), with the flowers "reduced" to a single sporangiophore (*Hedyosmum*) or ovuliphore (*Cercidiphyllum*), with sacci (*Lactoris*: Zavada & Taylor, 1986b), with lamellate endexine (*Austrobaileya*: Zavada, 1985; *Degeneria*, some Liliaceae, etc.) or the endexine lacking (*Hedycarya*, some monocots), with the pistils open (*Drimys, Platanus, Sassafras*, etc.), with intracarpellate pollen germination (some Annonaceae, Alismataceae, Liliaceae: Vasil & Johri, 1964; Johri, 1984), acarpellate

(Sattler, 1974a), with both integuments vascularized (Euphorbiaceae), with a massive perisperm (Piperaceae), with many chalazal embryo sac nuclei (*Sparaganium*), lacking a typical double fertilization (Onagraceae), with embryogenesis nuclear up to 64-128 or even 256-nucleate stages (*Paeonia*: Yakovlev, 1973), etc. Combined, all these "aberrant" states would give a plant profoundly different from typical angiosperms and hardly assignable to angiosperms at all.

(2) Most of the "aberrant" angiosperm character states are typical for some other seed plant groups. And, on the contrary, virtually all the typical angiosperm features occur as aberrant in other seed plant groups, e.g. the apical tunica-corpus differentiation (Araucariaceae, gnetaleans), vessels (in gnetaleans, previously thought of as being non-homologous to those of angiosperms, but actually perhaps homologous: Muhammad & Sattler, 1982) and in bennettites (Krassilov, 1982b), unifacial leaves (*Acmopyle, Dacridium*), stipules (cycads), a hierarchical reticulate venation (gigantopterids, *Gnetum*), mesogenous stomata (bennettites: Krassilov, 1978a; gnetaleans: Maheshwari & Vasil, 1961; some pteridosperms and conifers: Johnson & Riding, 1981; Reihman & Schabilon, 1985), a flower-like arrangement of the tepals as well as of the androecial and gynoecial organs (bennettites), monocliny (hydrospermaleans, bennettites, gnetaleans, occasionally conifers: Tosh & Powell, 1986), persistent tetrads (*Classopollis*), multiaperturate pollen (*Welwitschia, Classopollis*), a lack of endexine lamination or this discernible at the early developmental stages alone (*Crossotheca* among pteridosperms, the extant ginkgo: Rohr, 1977), closed ovuliphores, sometimes with stigmatic crests (*Leptostrobus*), bitegmic ovules (*Caytonia*: Krassilov, 1984a), extraovular pollen germination (Araucariaceae, *Saxegothea, Tsuga*, some Cupressaceae, probably Hirmerellaceae), archegonialess embryo sacs (*Gnetum, Welwitschia*), fusion of the embryo sac nuclei (gnetaleans, some conifers: Konar & Oberoi, 1969), functioning of both sperms in copulation (regularly in podocarps, gnetaleans, etc.), double fertilization involving the ventral canal cell nucleus (*Ephedra*), cellular embryogenesis (*Sequoia*), fruit-like dispersal units involving periovular structures (pteridosperms, caytonialeans, *Araucaria, Welwitschia, Hirmerella*), etc.

Thus, the morphological boundary between angiosperms and other groups of seed plants is not absolute. Notably, angiosperms are more distinct from the present-day gymnosperms than from the fossil non-angiospermous groups having their ovules enclosed in the pistil-like cupules and with the androecial and gynoecial parts arranged cyclically as in the typical flowers. A possible reason for the disappearance of these angiosperm-like plants might have been that they have evolved into the full-fledged angiosperms. One dealing with the Mesozoic examples alone would hardly draw a major demarcation line between the then existing angiosperms, on the one hand, and all other seed plants, on the other, and would perhaps cluster caytonias and bennettites with angiosperms rather than with conifers.

(3) Not a single character considered typical of angiosperms is uniform within the group. There are at least two types of vascular sympodia, open and reticulate, the latter more common in herbs; four basic types of nodal anatomy; two basic leaf types, one occurring in dicots, the other, supposedly phyllodic (Kaplan, 1970), in monocots, but the "dicotyledonous" type occurring in such monocot families as Smilacaceae, Philesiaceae, Dioscoreaceae, Stenomeridaceae, Triliaceae and Taccaceae, mostly climbers (Conover, 1982); about 20 stomatographic types, also non-randomly distributed between dicots and monocots, the latter being more uniform in this character (Tomlinson, 1974; Rasmussen, 1983; Inamdar & al., 1986); many distinct flower types distinguished on the basis of their having the full ("perfect") or only a partial complement of the floral organs, a single set or two sets or no perianth organs, as well as being monoclinous or diclinous, in the relative positions of the perianth, androecium and gynoecium on the floral receptacle, in their morphologically recognizable or supposed pseudanthial or euanthial origins, etc.

To this can be added two types of androecia, with solitary or fasciculate stamens; two developmental types of pollen grains and about 16 apertural types, of which the anasulcate one is more common in monocots; six major placentation types; two basic types of ovule polarity with modifications and two integumental types, also with modifications; 11 embryo sac types, five types of proembryo cell patterns; and, eventually, two cotyledon types, formally diagnostic of dicots and monocots, although at least 11 dicot species have a single cotyledon. The caryological and biochemical characters (not included in the above description, because only an insignificant portion of angiosperm species have been hitherto appropriately studied) are likewise separable into distinct types differing in the chromosome numbers (with at least three distinct modal levels of n=5-8, 12-14 and 19-21), the differentiation of sex chromosomes, etc. (Stebbins, 1974; Grant, 1975; Gottlieb, 1984; Kubitzki & Gottlieb, 1984a, b; Antonov, 1995). Remarkably, the Magnoliaceae is isolated biochemically by lacking tannins and some other compounds that are common in the rest of angiosperms as well as in non-angiospermous groups (Blagoveshchenskiy, 1975; Cronquist, 1977).

(4) The character states assignable to a certain morphological type are not necessarily homologous in all group members and, in fact, are rarely so. For example, two leaf traces of a 1-lacunar node can arise form a single or, as in *Austrobaileya*, from different axial sympodia; the compound leaves can result from a localized activity of the marginal meristem or the leaflet primordia are differentiated before the marginal meristem (Periasamy & Muraganathan, 1985); both the paracytic and cyclocytic stomata can be developmentally mesogenous, perigenous or mesoperigenous; the spiral arrangement of the flower parts is either primary (Endress, 1983) or derived from a cyclic arrangement (Tucker, 1960); the developmentally fasciculate stamens arising from a common primordial knob may appear solitary in the ripe androecia (Pauzé & Sattler, 1978); they can arise either in centrifugal or centripetal sequences; the "typical" follicles often thought

of as the leafy ovuliferous structures can be neither leafy in their shape and vascularization - that is, vascularized otherwise than by the canonical single dorsal and two ventral traces, incidentally receiving five traces, as in the Ranunculaceae and Berberidaceae (Chute, 1930; Terabayashi, 1985), or two dorsal traces, as in *Austrobaileya*, or the ventral traces departing from the cortical, rather than axial, bundles - nor ovuliferous (with the ovules arising from the floral axis), or these structures are actually pseudomonomerous (Vink, 1978); some apparently orthotropous ovules are pseudoorthotropous, while the superficially crassinucellar ovules can be pseudocrassinucellar (Davis, 1966; Sporne, 1974); the morphologically similar integuments can be developmentally dermal or subdermal or derived from more than one histological zone, while a unitegmic condition can result from fusion or reduction or shifting (De Boer & Bouman, 1972; Bouman & Calis, 1977), whereas a bitegmic condition can ensue from splitting of a single integument or the peripheral nucellar tissues in respect to the pollen-conducting function (Heslop-Harrison & al., 1985). There are different developmental relations between the primary coats of the ovule and the testa/tegmen histological differentiation of the ripe spermoderm (Boumann, 1984; Schmid, 1986). A single cotyledon can result from fusion (syncotyly) or an asymmetrical development and reduction (heterocotyly), while an apical cotyledon can be developmentally lateral, as in palms (Guignard, 1984) or, if actually terminal as in *Lemna*, then perhaps non-homologous to other cotyledons (Arber, 1925). The caryotypes are likewise liable to pseudohomology due to intricate combinations of polyploidy and aneuploidy (Grant, 1975), whereas some characteristic biochemical compounds actually arise as the end-products of dissimilar biosynthetic pathways (Kubitzki, 1973). Pseudohomology in angiosperms is a prolific source of phylogenetic misconceptions.

Character polarities

For many characters, the extreme states can be linked by intermediate states in a more or less continuous sequences, or morphoclines. When based on contemporaneous forms, the morphoclines can be read in either direction. Historically, two competing schools of angiosperm morphology differed primarily in their giving priority to either a simple to complex or the opposite directions, in particular, a progression from the inconspicuous gregarious flowers of few parts to the conspicuous solitary flowers of many parts, or vice versa (Wettstein, 1907; Bessey, 1897, 1915; Eames, 1961). The underlying reasoning has had nothing to do with the actual chronological trends but has been rooted instead in philosophical doctrines, such as the essentialistic Urpflanze concept holding that any progression is a reduction from the most perfect archetype to its less perfect reflections in the perceptible plant diversity.

A new approach introduced into plant morphology by the theory of evolution was the outgroup comparisons, that is, a character state evaluation by comparison with a close, but distinct group, the pre-existing philosophical differences now affecting the choice of a particular outgroup. While *Ephedra* was used as an outgroup by Wettstein (1907), the then recently discovered bennettites played this role for the Hallierian school. Although with time the relevance of both *Ephedra* and bennettites as outgroups was put to doubt, the polarity controversies remained.

If angiosperms were monophyletic, then all the alternative morphotypes should be derivable, directly or sequentially, from one of them. The comparative plant morphology is governed by this creed and is, thereby, focused on sequencing the morphotypes from the primitive (primary) to the advanced (derived), yet the criteria of advancement are less advanced than might be expected. The following are commonly used:

(a) The statistical, or "the common is primary". Since the growth forms with the nearly continuous subapical primary and lateral secondary thickening meristems occur in not so numerous woody monocots (Chakroun, 1983; Stevenson, 1980a, b; Diggle & De Mason, 1983) and a few dicots, such as the Piperales, it is conceived of as "anomalous" and, by implication, derived; the anatropous and the bitegmic ovules are more widespread (the latter occurring in 152 of 341 dicot families and in 52 of 69 monocot families: Bouman & Calis, 1977) and are, by virtue of it, primary in respect to the orthotropous and the unitegmic ovules; the *"Polygonum"* type of embryo sac development is more common than the others and is, therefore, primary, etc. This approach tends to dismiss rare character states as of no phylogenetic significance. It is inherited from the typological Urpflanze concept (see above) in which all the essential characters occur in the archetype, whereas any departures from the norm are secondary.

(b) The conceptual, determined by a phylogenetic paradigm. This concept is also typological, but the primary versus secondary character states are determined on the basis of a preconceived idea of primitiveness rather than of any statistical considerations. For instance, since all floral organs are modified sporophylls, the laminar stamens, although relatively rare, should be primitive (despite the fact that they are highly specialized, sometimes endowed with a capilliform connective penetrating between the carpels and holding the stamen in a fixed position during pollination, see Canright, 1952, while their vascular anatomy is highly diverse and, in *Victoria* at least, rather caulomic than phyllomic, see Heinsbroek & Heel, 1969, or superficially phyllomic with three traces, but the latter departing from the cortical rather than axial bundles, as in *Degeneria*). For the same reason, a follicular carpel has been proclaimed primitive whatever the interpretation that varied in response to the involute, conduplicate, and conduplicate-peltate ascidiform models of megasporophyll transformation successively replacing each other (Sinnott & Bailey, 1914; Troll, 1932; Bailey & Swamy, 1951; Baum, 1949; Lienfellner, 1950; Rohweder, 1967; Tucker & Gifford, 1974).

These models imply that a primitive placentation should be marginal or submarginal or localized in the ventral Querzone respectively. That such features of the follicular carpels as their tubular growth from annular primordia (Rhoeweder, 1967), characteristic of compound structures, or their vascularization by the double dorsal and sometimes also double ventral traces, as in *Drimys* sect. *Tasmannia* (Tucker, 1975) are scarcely phyllomic, passed unnoticed until some typical follicles have been reinterpreted as pseudomonomerous (Leroy, 1977; Vink, 1978), that is, fairly advanced.

With a change of the paradigm, a totally different morphotype can be selected as primary or primitive. Thus, the appearance of the telome concept replacing the phyllome concept of a basic morphological unit has made the branched stamens, as in the Euphorbiaceae or Monimiaceae, conceivably more primitive than the laminar stamens (Wilson, 1937). Characters are often labelled primary if they occur in a paradigmatic primitive group, such as the Magnoliaceae. This is an insecure approach because, with the primitive status of the Magnoliaceae reconsidered (Bate-Smith, 1972; Boulter & al., 1972; Martin & Dowd, 1984, 1986), all the polarities are to be reversed.

(c) The technological, holding that of two character states the primary is that of which the other can be most readily manufactured. Thus, inaperturate pollen grains are primary because they provide a carte blanche upon which any apertural type can be developed (Muller, 1970; Walker, 1974a). The monoclinal flowers are primary because the diclinal flowers can be produced from them by the loss of either androecium or gynoecium, and a loss is certainly easier than an acquisition. Similarly, all the perianthless flowers are "reduced", even if developmental evidence of the reduction is totally lacking, as in the Saururaceae (Tucker, 1975, 1981). Technologically, inferior ovaries might arise by fusion with the receptacular and/or appendicular tissues, but, in some cases at least, developmental evidence is ambiguous (Douglas, 1944; Komar, 1972; Sattler, 1974a; Leins & Erbar, 1985). On the same ground, a succession from complex to simple is commonly preferred to the opposite. Phytochemical evolution is often assessed technologically, with a precursory compound of a biosynthetic pathway considered as phylogenetically primitive. This approach, though fairly popular, is based on an unwarranted analogy between organic evolution and human technological experience.

(d) The correlational, holding that primitive characters are those correlated with other primitive characters. The logic of such deductions is not entirely convincing, for one can argue that, among contemporaneous species, primitiveness of some characters should be compensated by advancement in some others. In practice, the correlational assessments are often ambiguous. Thus, the 3-lacunar nodal anatomy correlates with stipules that are primary according to Cronquist (1968), but derived according to Takhtajan (1970) and Stebbins (1974), while the 1-lacunar nodes correlate with simple entire leaves and occur in the homoxylic species that are commonly, but not unanimously, considered as primitive. At the same time, they correlate with decussate leaves that are thought of as derived by many morphologists (Bailey, 1956; Carlquist, 1985), but not by Stebbins (1974). The fasciculate stamens might be secondary, for their development is, with a few exceptions, centrifugal, an allegedly derived pattern, but there is a substantial developmental evidence to the contrary (Stebbins, 1974; Sporne, 1974).

In the correlation approach, it is difficult to avoid circularity. Simple entire leaves are primary because they are characteristic of magnoliaceans that are primitive by virtue of having, among other supposedly primary characters, simple entire leaves. This approach does not take into consideration differential rates of character evolution resulting in mosaic morphology. The notion of magnoliaceans as the most primitive living angiosperms, justified or not, does not mean that all their characters, including multilacunar nodal anatomy and the lack of tannins, are primary.

(e) The outgroup-directed polarizations in which a character state is hold primitive if occurring in a taxon marginally related, but not assigned in the group and, on the whole, more primitive. The homoxylic condition, for instance, is hold primary in angiosperms for it is widespread in gymnosperms. This would mean, according to the monophyly concept, that gymnosperms having vessels acquired them in parallel with angiosperms and not as phylogenetic precursors of the latter. However, the homoxylic woods are not uniform in respect to the structure of tracheary elements and they may be primary in some families but secondary in the others, thus requiring a selective outgroup comparison. I have already mentioned that competing phylogenetic concepts are based on the selection of either *Ephedra* or bennettites or *Caytonia* as the outgroup. But these outgroups may not necessarily be mutually exclusive. They and some others could jointly have contributed to the angiosperm morphological diversity. This idea is, however, incompatible with a methodology assuming monophyly as the starting point of a morphological analysis and inevitably leading to monophyletic arrangements as its end-product, thus making the monophyletic concept self-sustainable.

As follows from the above analysis, the typical angiosperm characters are not exclusively confined to angiosperms but occur, though less consistently, in the seed plants that are not formally recognized as angiospermous. On the other hand, the atypical character states are often widespread in non-angiospermous groups. The angiosperm concept is based on the more common character states, with a tendency to dismiss the less common states as aberrations. However, "aberrant" characters tend to occur as typical in geologically older seed plant groups lending support to their interpretation as primary (or, if secondary, then arising by reversion to the ancestral morphology). Combined, the aberrant characters may give an idea of a primordial angiosperm that may be closer to an ancestral form than is the paradigmatic angiosperm. Such form may not be conventionally classified as an angiosperm, thus making angiosperms cladistically paraphyletic, with the typical angiosperm characters, not yet occurring in the common ancestor, appearing in parallel in the descendant lines and, therefore, not uniquely derived.

However, in the widely held opinion, the uniqueness of angiosperms is due, in the first place, to their having flowers, stamens and carpels as well as double fertilization, the "charismatic" structures and processes wanting, at least nominally, in any other seed plants and, therefore, exempt from outgroup comparisons. Yet what is flower except that it is an organ of flowering plants? We have to inquire how far phylogenetic concepts are predetermined by semantics, and not only in the case of paradigmatic angiosperm structures, but also in relation to leaves, pollen and ovules that are nominally recognized both in angiosperms and non-angiosperms.

Structural homologies

Plant organs belonging to different structural categories are weakly if at all functionally correlated and are developmentally controlled by different genomic domains with insignificant interdomain pleiotropic effects, thus acting as independent evolutionary units. In this chapter, the morphological information on the organ homologies, given in the preceding chapters, is briefly reviewed so as to arrive at sustainable semophyletic models.

Leaf

Of the various angiosperm leaf morphologies, the two contrasting types, i.e. the dorsiventrally flattened bifacial with a several-rank areolate venation and with xylem adaxial in the vascular bundles, and the bilaterally flattened unifacial with parallel veins and inverted bundles, are characteristic of dicots and monocots respectively (yet the Smilacaceae, Dioscoreaaceae, Philesiaceae, Taccaceae and a few other monocot families, mostly climbers, have the dicot-type leaves: Conover, 1982). The latter type is more caulomic, although the caulomic features are displayed also in the compound leaves with leaflets developing as outgrowths of the axis (instead of developing as lobes by the localized marginal meristem activities, see Periasamy & Muruganathan, 1985, on *Tabebuia*; Merril, 1986, on *Fraxinus*, etc.), as well as in the stipules that are linked with leaves and epiphyllous appendages in a morphogenetic continuum (Sattler & Maier, 1977; Rutishauser & Sattler, 1986). Such features support the "partial shoot" (Arber, 1950), or the caulome-phyllome leaf concept (Howard, 1974; Sattler, 1974b).

In the early seed plants, leaves are initially caulomic, that is, syntelomic. In the Carboniferous gymnosperms, the little modified syntelomic systems co-occurred with differentiated shoots bearing bifacial leaves. If descended from the latter, angiosperm leaves might be secondarily caulomic. However, even some Mesozoic plants seem to have retained the syntelomic leaf structures. In *Dinophyton*, a Triassic protognetalean

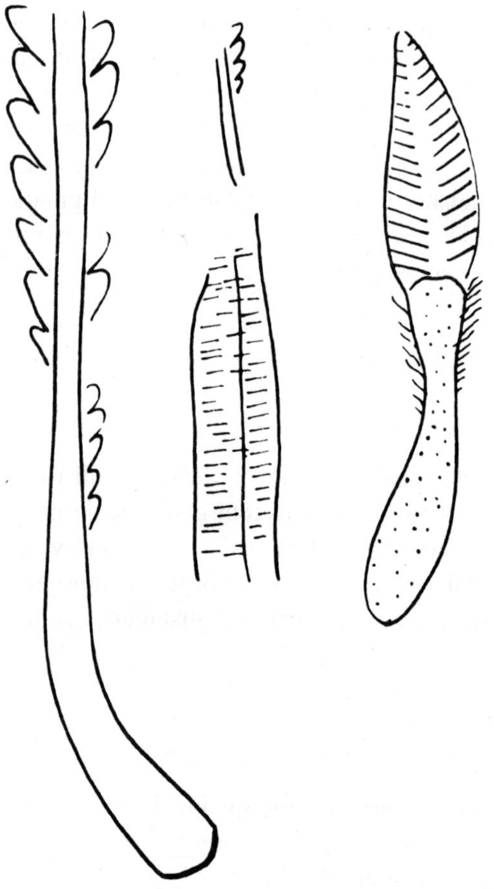

Fig. 44. Bennettitalean leaves and bracts bearing on phyllodic leaf concept: *Ptilophyllum* leaf with a swallen petiole (left) and *Cycadolepis* bracts with reduced blades (after Harris, 1969; Krassilov, 1982).

plant (Ash, 1968; Krassilov & Ash, 1988), the leafy shoots show archaic syntelome features, such as the cylindrical, occasionally forked or even pinnately branched lateral appendages ("leaves"), the uniform trichome and stomatal frequencies over the "leaves" and "axis", and the circinnate ptyxis of the whole shoot system.

In the geologically younger gnetalean proangiosperms, the photosynthetic function was transferred from the lateral appendages, almost lacking stomata, to the axis. The alternation of large and small vascular bundles in monocot-type leaves is characteristic also of the stem internodes of both extant *Ephedra* and of the Early Cretaceous *Leongathia elegans* (Krassilov & al., 1997b) and is, conceivably evidence of phyllomization. This model is further supported by the graminoid Early Cretaceous leaves, presently attributed to *Eoantha* or allied gnetalean proangiosperms (Plate23), Such parallelinerve blades could be formed by fusion of linear nodal leaves, with a few of them remaining distinct as auricles.

At the same time, the repeatedly suggested phyllodic origin of the monocot leaf type (De Candolle, 1827; Henslow, 1911; Arber, 1918; Kaplan, 1970) receives some support from the fossil record, but with a prototype among bennettites rather than dicots. In bennettites, the tendency to a phyllodic transformation involves, in the first place, their essentially parallelinerve or flabellinerve *Cycadolepis*-type cataphylls (Fig. 44) occasionally terminating in a residual pinnate or pinnatinerve blade (Harris, 1969: Krassilov, 1982b). One of the first appearing fossil angiosperm-like leaves, *"Dicotylophyllum" pusillum* (Fig.9) from the Early Cretaceous of Transbaikalia (Vakhrameev & Kotova, 1977; Krassilov & Bugdaeva, 1997) resembles a bennettitalean bract with a flabellate venation transformed into an oblique pinnatinerve one.

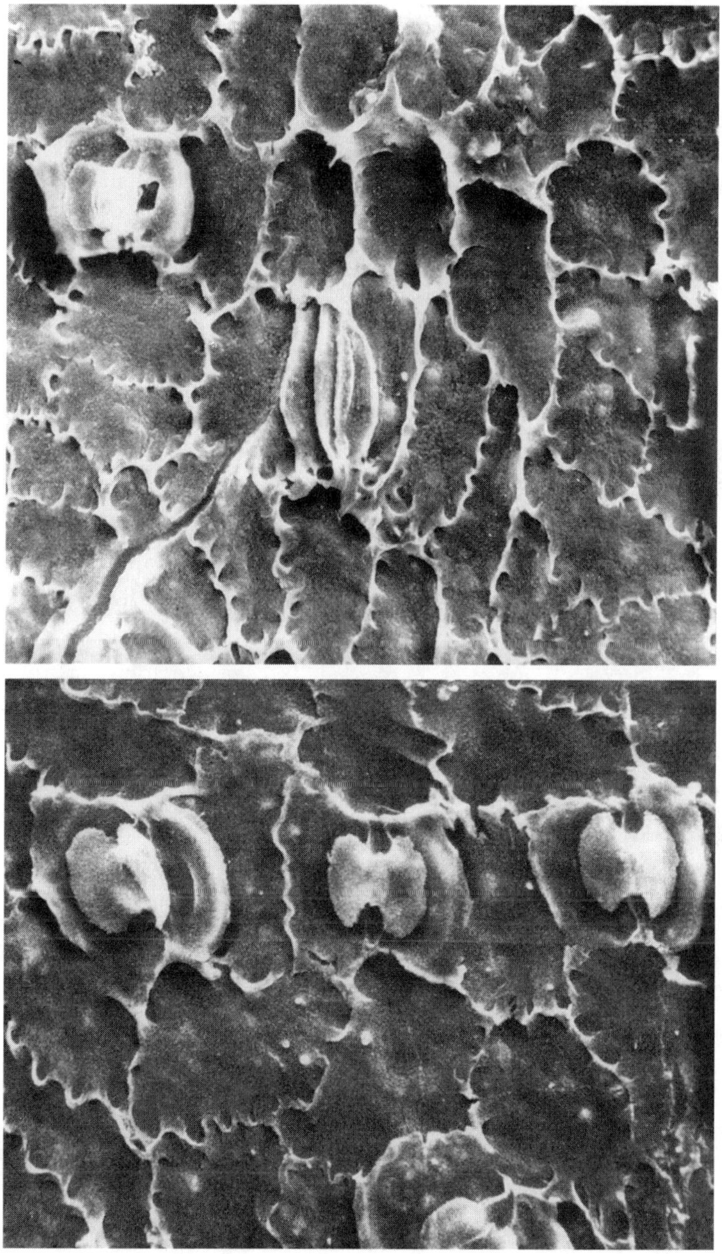

Fig. 45. Bennettitalean stomata: *Nilssoniolteris* sp., from the Late Jurassic of Bureya Basin, Far East, Russia, an aborted stoma (central in the top photograph) and normally developed syndetocheilic stomata (Krassilov, 1978), SEM, x500.

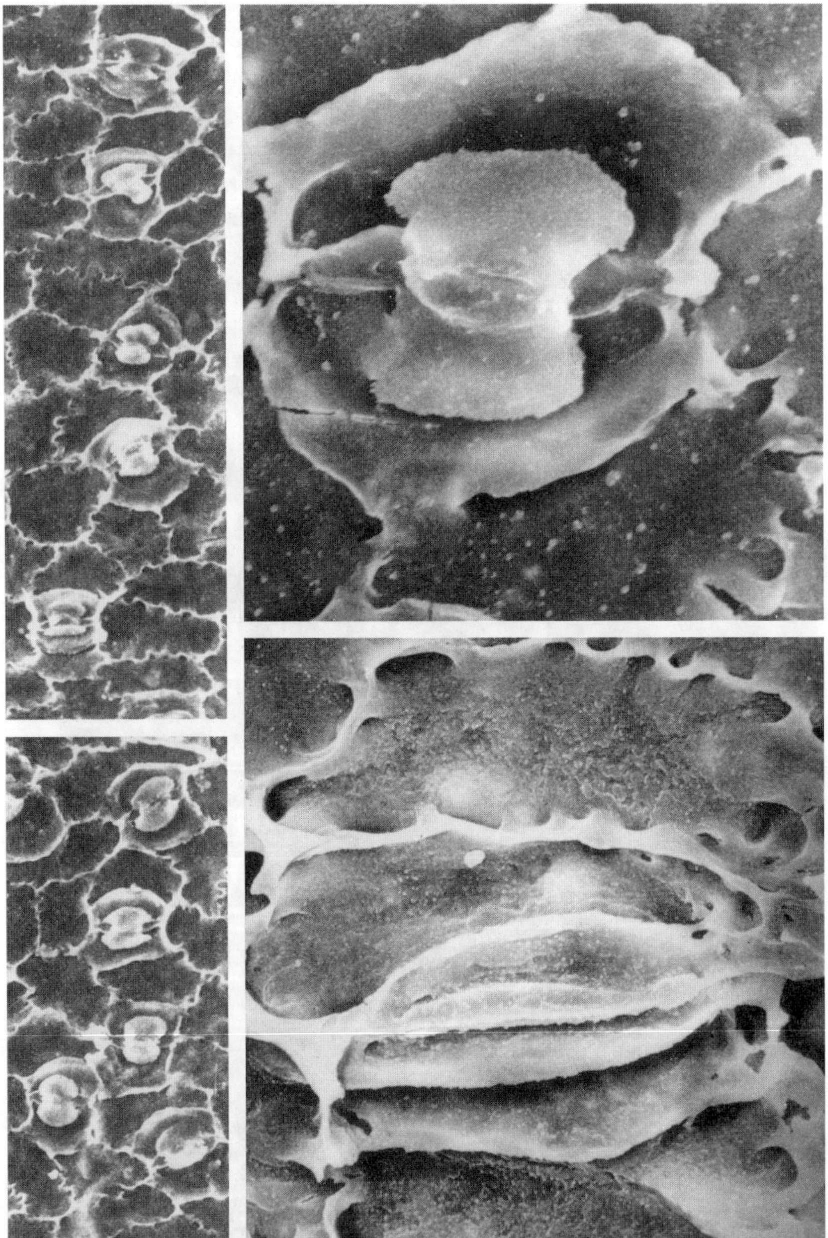

Fig. 46. Bennettitalean stomata: *Nilssoniolteris* sp., from the Late Jurassic of Bureya Basin, Far East, Russia, details of a normally developed stoma (top right) and an aborted stoma showing the derivation of both guard cells and lateral subsidiary cells from a single meristemoid; left photographs show stomatal patterns in the intecostal zones (Krassilov, 1978), SEM, x1 500 and 250.

Further bennettitalean foliar features in monocots are vessel elements with scalariform and porous perforations found in the leaf veins of the Early Cretaceous *Otozamites lacustris* (Krassilov, 1982b; Fig. 36). Such vessels are characteristic of the graminoid and palm leaves (Tomlinson & Wilder, 1984). The "syndetocheilic" laterocytic (tetracytic in cataphylls) bennettitalean stomato-type was developmentally mesogenous (as evidenced by aborted stomata in fossil leaves: Krassilov, 1978a; Figs. 45, 46), thus corresponding, both morphologically and developmentally, to one typical of monocots (Ras-mussen, 1983; Tomlinson, 1974; Wilder, 1985).

The dicot-type simple bifacial leaves with reticulate venation appeared in the Paleozoic glossopterids, at least some of them producing such leaves clustered on deciduous short shoots. Most Mesozoic seed plants of the temperate zone, in particular the proangiospermous czekanowskialeans, had their leaves similarly clustered (with the leaf numbers constant in the czekanowskialean species), whereas the caytonialean foliage (*Sagenopteris*) consisted of what appear as long-petiolate compound leaves of four leaflets that were occasionally arranged in two decussate pairs, betraying derivation from a short shoot with apically crowded decussate leaves. The leaflets were somewhat smaller, but nearly identical, in both their blade shape and venation, to the *Glossopteris*-type leaves.

Phyllomization of short shoots resulting in the transitional caulome-phyllome morphologies is evidenced by such Early Cretaceous forms as *Chankanella* (Krassilov, 1967) intermediate between the brachyblasts with planate clusters of *Glossopteris*-like simple leaves and the compound *Sagenopteris*-like leaves with palmate leaflets. Planation in the leaf clusters was accompanied by fusion of the leaf (leaflet) blades that evolved in the direction of the palmately lobed simple leaf blades (Fig. 47). Presently we have a

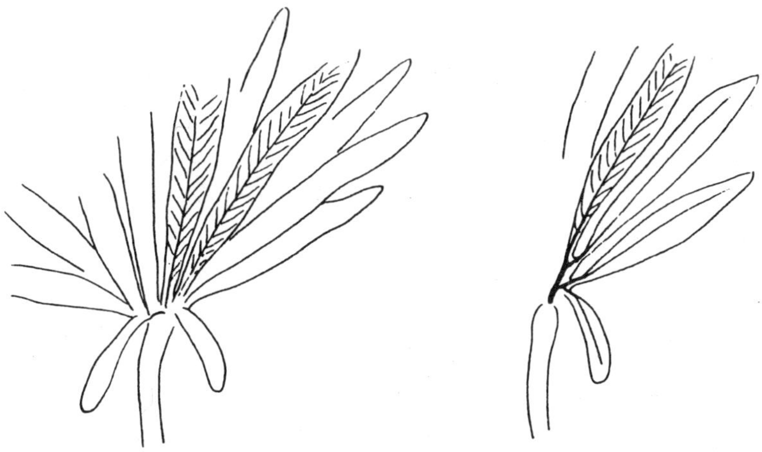

Fig.47 A suggested transition from the *Chankanella*-type short shoot (left, after Krassilov, 1967) to palmate leaf.

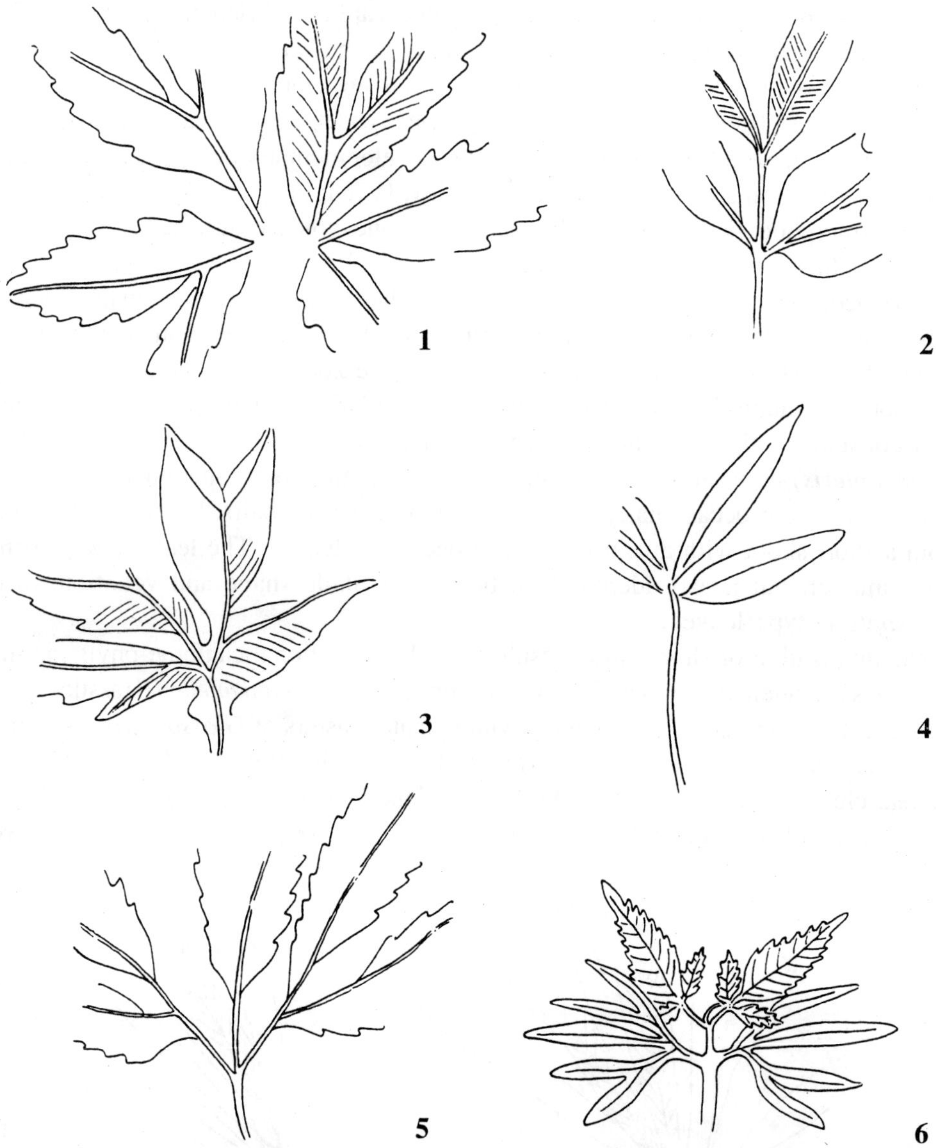

Fig. 48. Caulomic pinnate-palmate and palmate leaves with dichopodial branching of primary veins in proangiosperms and angiosperms: (1) *Scoresbya,* a Jurassic proangiosperm of caytonialean affinities, palmate leaf with forking segments, (2) *Imania,* a closely related Late Triassic leaf genus, pinnate morphotype with forking lateral segments, (3) *Sujfunophyllum,* a pinnate-palmate morphotype, supposedly proangiospermous, although found in association with Early Cretaceous angiosperms, (4) *Sagenopteris,* a caytonialean leaf, apparently palmate, but with a residual decussate pairing of the leaflets, (5) *Proteophyllum,* an angiosperm leaf of Late Cretaceous (Cenomanian) age, (6) *Garuya,* an extant angiosperm with dichopodially branching cotyledons and dissimilar leaves (after Harris, 1935, 1951a; Krassilov & Shorokhova, 1970; Knobloch, 1978; Rüffle, 1980a).

complete series of transitional forms, such as *Imania* with partly fused leaflets (Krassilov & Shorokhova, 1970; Plate 45), *Scoresbya* (Harris, 1932) with a dichopodial branching of primary veins that are asymmetrically laminaless (psilobasal) on one side, *Sujfunophyllum* and *Baikalophyllum*, the geologically younger forms of the same type (Krassilov, 1967; Bugdaeva, 1983) that co-occurred with the more conventional early angiosperm leaves (Fig. 48). Common among the latter were the morphologies intermediate between the compound and the lobed *Scoresbya*-like, dichopodially segmented leaves with psilobasal primary veins, as in *Proteophyllum, Fontainea, Manihotites, Araliophyllum, Anisophyllum,* etc. (Ettingshausen, 1868; Lesquereux, 1874; Velenovsky, 1889; Newberry, 1895; Fontaine, 1899; Berry, 1911; Velenovsky & Viniklar, 1926; Knobloch, 1978). They formed a distinctive group of archaic angiosperms (Krassilov, 1973a; Knobloch, 1978), their leaf architecture still found in some extant ranunculids and reiterated in the cotyledon morphology of some advanced angiosperms, such as *Garuya floribunda* (Rüffle, 1980a).

The morphological continuum formed by the above series of fossil links between the proangiosperm *Sagenopteris-Scoresbya* and the early angiosperm Proteophyllum leaf morphotypes may be of a certain phylogenetic significance. Anastomosing of the veins, variable in the *Sagenopteris-Scoresbya* group, is similarly developed in *Kingdonia*, an archaic extant ranunculid genus. In addition, the anomocytic stomata lacking stomatal ledges are characteristic of both groups.

The incipient areolate venation appears in gigantopterids and peltasperms in the course of marginal fusion of the pinnae pruducing an entire blade, but with a conservative venation pattern betraying its compound origin. A peltasperm species *Scytophyllum vulgare* has polymorphic leaves that show all transitions from pinnate with interstitial pinnules to lobed and, eventually, entire blades. In the latter, the lateral veins grade from pinnately branched, marking the formerly distinct pinnules, to fasciculate to forking. Remarkably, the end products are similar to the initial state: venation of the compound blade converges on that of its constituent pinnae, a process designated as retroconvergence and supposedly evidencing a Gestalt-holding capacity of genetic memory (Krassilov, 1995). Veins of the marginally fused pinnules approach each other, but are still separated by the narrow veinless stripes over the fusion meristem (derived from the marginal meristem that normally does not produce vascular tissues). In the course of further transformation, however, a successive looping of the veins occurs in the highly meristematic zone at the junction of adjacent pinnae with their interstitial pinnule, the latter overarched by the series of distal loops (Fig. 49). In the advanced stage, the interstitial pinnule is marked by a fasciculate vein filling the areole. The resulting pattern looks much as a scalariform tertiary venation of a dicotylid leaf.

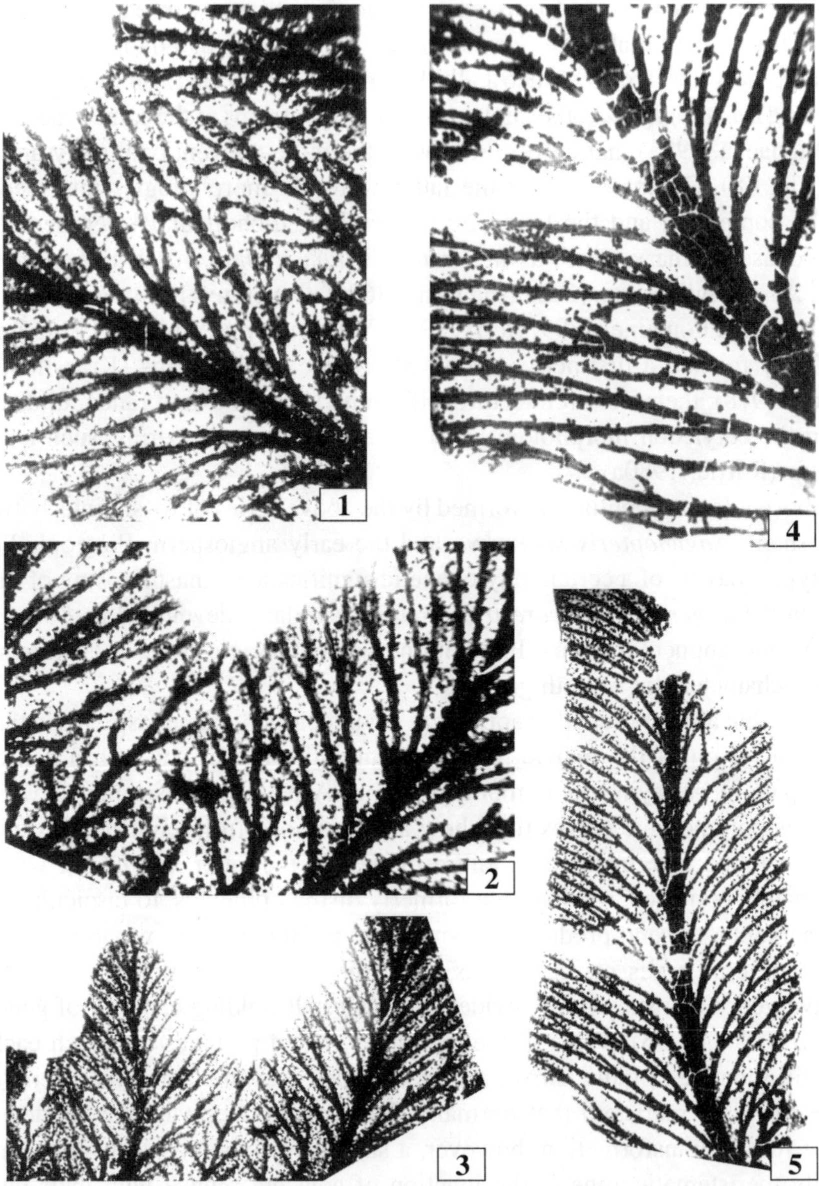

Fig. 49. *Scytophyllum vulgare* (Pryn.) Dobrusk., peltasperm leaves from the Late Triassic of Eastern Urals (Krassilov, 1995) showing marginal fusion of the pinnae and interstitial pinnules that are marked by the conservative venation pattern: (1-3) area of fusion incorporating an interstitial pinnule, its lateral veins approaching (2) or joining (3) those of the adjacent normal pinnules, thus forming a series of scalariform loops, with the proximal areole filled with short veins arising from the rachis, x3 (3) and 7 (1, 2); (4, 5) entire pinna formed of fused pinnules still marked by fasciculate lateral veins that are distally transformed into dichotomous venation; branches of the adjacent veins approach each other without actually joining, with occasional loops alone, x3 (5) and 7 (4).

Flower

Familiar by association with weddings and funerals, flowers have escaped a precise morphological definition. Historically, flower was not considered as a discriminating feature: Linné said that all plant species had flowers, even if concealed of our eyes (Linnaei, 1751). Jung, Ray, Tournefort and other XVII century botanists defined flower in relation to fruit as the precursory of the latter. In their time, any definition of a natural phenomenon was meant to reflect its essence, and the essence of flower revealed itself in fruit. The "fructists" based their classification systems solely on fruit characters (reviewed in Linnaei, 1751), with gymnosperms initially set apart (by Paul Hermann, Christoph Knaut and others) as plants having "naked fruits". But, insofar as fruits are defined in relation to the flower parts (or the whole flowers transformed into anthocarps or diclesia), the fructist definition is circular. Linné changed it by proclaiming anther and stigma the essentials of flower. In the thus established tradition, flower was conceived of as an assembly of stamens and/or carpels, with perianth as a less essential, even dispensable, attribute. However, since the definitions of the stamens and carpels are in their turn far from precise, it seems better to define flower independently of its however essential parts.

While we no longer believe that all plants have flowers and since we use the notion of flower discriminatingly as a special feature of angiosperms separating them from other plant groups, it seems logical to define this structure in relation to reproductive structures of non-angiospermous seed plants. In the majority of angiosperms, flowers consist of sporangiophores and/or ovuliphores assembled on the receptacles that may bear sterile perianth members proximally to the fertile zone. The latter develops from the floral meristem that is similar to the apical meristem of vegetative shoots but is more active mitotically and is of a less distinct zonal structure. Morphological transitions between carpels, stamens and petals may reflect a morphogenetic integrity of the floral meristem rather than their common phyllomic origins implied by the classical morphology. Thus, the typical flowers are specialized shoots comparable to those of *Ginkgo* where the sporangiophores and ovuliphores develope apically, surrounded by cataphylls. These latter shoots are diclinous, but the examples of similar monoclinous structures, such as *Irania*, are also provided by the fossil record (see under "Selected fossil links" above). Their apices are not fully fertile, however, bearing bracts or, in *Ginkgo*, fully developed leaves intermingled with ovuliphores.

Bennettites had similarly organized receptacles bearing ovuliphores and the so-called interseminal scales. The floral axis protrudes over the gynoecium and is crowned by sterile remnants of the second floral node forming a "corona". In *Gnetum*, the fertile organs and their subtending bracts are verticillate in the successive nodal zones, that can be reduced to two, as in *G. scandens* (Gifford & Foster, 1974). In *Eoantha*, the ovuliphores develope in the proximal fertile node, whereas the second node is reduced a tuft of linear bracts. In *Welwitschia*, the functionally polliniferous structures are flower-

1 2 3 4 5

Fig. 50. Preflowers, anthostrobili and primitive flowers: (1) an imaginary prototypal monoclinous floral shoot, based on *Irania* and *Dirhopalostachys*, (2) *Eoantha,* a bracteate gynoecial structure with protruding floral axis bearing a corona of bracts, (3) *Williamsoniella,* a monoclinous bennettitalean anthostrobilus, (4) *Liquidambar,* an extant preflower with interfloral phyllomes, (5) *Kingdonia,* an extant anthostrobilate flower (Krassilov, 1975; Schweitzer, 1977; Bogle, 1986; Hu & Tian, 1986).

like, weakly zygomorphous, with a cycle of introrse stamen-like sporangiophores. Their central secretory organ is formed of a solitary ovule crowned by a peltate protuberance.

Although differing from the typical flowers, these structures correspond to some anomalous flowers with bracts in the gynoecial zone and/or with a central sterile column - a protruding apical extension of the receptacle, or torus (Fig. 50). The latter occurs in the flowers of the Winteraceae, Piperaceae, Lauraceae, Illiciaceae, *Euptelea* and some ranunculids (*Aquilegia*), while in the strobilate magnoliacean flowers the vascular bundles ascend above the lateral gynoecial organs. Although strobilation may be either primary, as in *Austrobaileya* (Endress, 1983), or secondary, as in the Magnoliaceae (Tucker, 1960), the floral structures with sterile apices are anthostrobili rather than typical flowers.

Sterile phyllomic appendages in the fertile zone occur in *Eupomatia* where they form an "inner corolla" between the carpels and stamens (their interpretation as staminodes is inconvincing, since stamens tend to be sterilized on the outside of, rather than inside, the fertile zone). The less conspicuous inner phyllomes constantly occur in *Austrobaileya, Degeneria, Exbucklandia, Rhodolea, Altingia, Liquidambar* and some other genera. They are sometimes secretory, as in *Rhodolea*, but are mostly functionless. Their vascular anatomy differs from that of either pistils or stamens (Bogle, 1986).

The internally bracteate flowers might have a pseudanthial origin (Bogle, 1986) or they could result from an incomplete fertilization of the floral apex, as in bennettites. In the later case, they are preflowers (Krassilov, 1989, 1991) rather than true flowers. The closest approach to their prototypic form occurs in the extant *Cercidiphyllum* the pistillate structures of which consist of 2-7 decussate bracteate follicles. The more flower-like preflowers of primitive hamamelids and magnoliids are derivable from a similar structure, their inner phyllomes corresponding to subtending bracts of ovuliferous organs.

Thus, alongside with the typical flowers, here defined as apical aggregates of ovulate and/or staminate structures, there are at least two types of flower-like structures, the anthstrobili, or subapical aggregates of fertile organs crowned by a sterile extension of the floral axis, and the preflowers, or fertile organs intermixed with sterile phyllomes in the floral zone, both considered as transitional between the flowering shoots of gymnosperms and those of angiosperms.

Stamen

The antiquity of laminar stamens postulated by the phyllome theory has no support in the fossil record. They do not constitute a uniform morphotype but are instead diverse morphologically, with the anthers abaxial, adaxial or lateral, superficial or embedded, and with one (*Michelia*), two coalescent (*Austrobaileya*) or three vascular traces derived from both axial and cortical bundles (*Degeneria*) and, therefore, only superficially resembling leaf traces, or arising from petaline traces, as in *Nelumbo* (Moseley &

Uhl, 1985) or stelar, as in *Victoria* (Heinsbroek & Heel, 1969). They are highly specialized (Canright, 1972), in particular, for egg deposition by specific pollinators, such as Thysanoptera (Endress, 1986a). In addition, the laminar stamens occur in tropical species, whereas their relatives in temperate climates have filamentous stamens.

On developmental and vascular evidence, stamens are caulomic rather than phyllomic. Morpologically, their caulomic origin is betrayed by their occasional branching or forking, as in *Myrica, Ulmus, Carpinus*, etc. (Wilson, 1937). Fusion of stamens in the Chloranthaceae and Monimiaceae finds its analogy in gnetaleans (Pearson, 1929).

A caulomic origin is most evident in the case of fasciculate stamens that are often considered as derived from the more common solitary stamens. However, several lines of evidence suggest reverse phylogenetic relationships. In the first place, the fasciculate stamens are not rare. They are presently found not only in dilleniid families, but also in such a phylogenetically remote group as palms (Uhl, 1976). The typical fasciculate androecia develope from a few primary primordia, each giving rise to 3-10 secondary primordia. Paired stamens developing from split primordia, as in the Saururaceae (Tucker, 1985), apparently belong in the same category. However, in the ripe androecia, their fasciculate arrangement is less evident (Pauzé & Sattler, 1978). A fasciculate nature of the apparently individual stamens is sometimes betrayed by dendroid traces, typically developed in *Paeonia*. In a less distinct but still recognizable form, such traces occur in the "primitive" Degeneriaceae, Annonaceae and Papaveraceae, as well as in the Myricaceae (Stebbins, 1974), or the vascular bundles branch in the filaments, as in palms (Uhl, 1976).

Historically, the seed plant sporangiophores seem to have been fairly conservative, retaining a transitive caulome-phyllome morphology not only in the Paleozoic pteridosperms but also in their derivative peltasperms, corystosperms, nilssonialeans, etc., with the laminar fern-like sporophylls occasionally appearing in a few lines, such as the Permian gigantopterids, but short-lived in terms of geological time and scarcely prototypic for the angiosperm stamens.

Among the possible prototypes, *Meeusella* is a compound androclade of peltaspermid affinities, with spirally arranged lateral branches that are either simple and sterile or bearing a subapical pair of stalked peltate sporangiophores, each with a reflexed synangial head of 2-4 fused sporangia (Plate 46; Fig. 51). In occasional branches, the stalks are reduced rendering the sporangial heads sessile on the penultimate axis, the scale-like expanded apex of the latter protruding over them (Krassilov & Bugdaeva, 1988b). In the derivational model based on this structure, a shortening of the axis would result in a fasciculate arrangement of the lateral fertile branch systems evolving into fasciculate stamens, the sterile branches giving rise to staminodes. The paired ultimate branches bearing synangial heads are then precursory to the thecae of a typical anther, while the sterile scale-like apex of penultimate branch corresponds to a produced connective. Thus, a conventional filamentous stamen is conceived of as a condensed branching system, each component of the anther developing from a distinct morphological unit. A probable apotype of *Meeusella* would have a massive, distinctly lobed latrorse anther

Fig. 51. *Meeusella proteiclada* Krassil. & Bugd., a supposed prototype of fasciculate stamens from the Early Cretaceous of Transbaikalia: branching androclade (left) and its paraclades (one sterile) showing a transition from stalked to sessile staminate heads, each of the latter supposedly giving rise to the theca of an angiosperm anther (bottom), with the apophysis of the connective conceivably corresponding to the protruding paracladium apex (after Krassilov & Bugdaeva, 1988b).

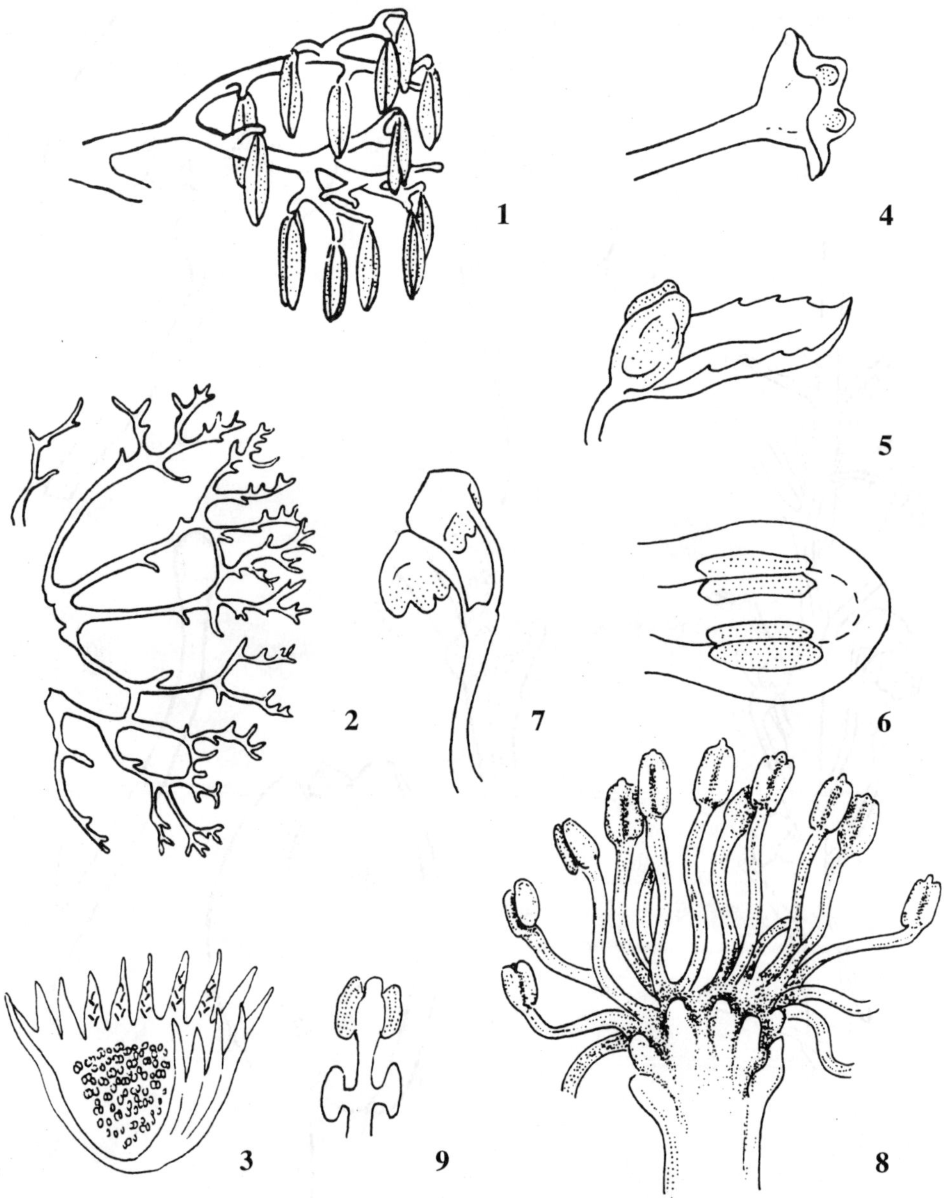

Fig. 52. Prototypical staminate structures: (1) *Caytonanthus,* a branched caytonialean androclade, (2) "dendroid" androecial traces of *Paeonia,* with the branching pattern resembling (1), (3) *Weltrichia,* a bennettitalean staminate preflower with proximal synangia transformed into nectaries, (4, 5) sporangiophores of *Ixostrobus* (czekanowskialeans) and *Dinophyton* (protognetaleans) with adaxial sporangia, (6) *Austrobeileya,* stamen with adaxial pollen sacs, (7) *Meeusella,* paracladium of a branched androclade, with stalked sporangial heads, presumably prototypic to bracteate stamens, as in the extant *Trochdendroides* (8) or with lateral appendages, as in *Monimia* (9) (after Harris, 1951a; Krassilov, 1972; Krassilov & Ash, 1988; Endress, 1986b; Krassilov & Bugdaeva, 1988b).

with a leafy, perhaps peltate connective and a distally branched vascular bundle. Such stamens can be found in a number of the early appearing angiosperm groups, notably the Chloranthaceae and Platanaceae.

A similar stamen morphology is derivable from the *Ixostrobus*-type sporangiophore (czekanowskialeans), but with the adaxial 4-lobed anther subtended by a reflexed vascularized connective. Little-modified stamens of this type occur in *Austrobaileya* and the Annonaceae.

Caytonanthus, a caytonialean androclade, is remarkable for its dichopodial "dendroid" branching reminding of a similar vascular pattern in the androecia of dilleniids and *Paeonia* (Fig. 52). This pattern was conceivably inherited from a dichopodial branch system that became incorporated in the receptacle (as another manifestation of vascular conservatism, see above), its free ultimate branches emerging as indistinctly fasciculate, perhaps forked, stamens with a slender filament and a relatively massive anther.

In *Dinophyton*, a Triassic protognetalean plant, the sporangiophores protrude from bracteate cupules and are laminar, with adaxial sporangia. The laminar part is interpreted as a bracteole fused to the simple or distally branched stalk with apical sporangia (Krassilov & Ash, 1988). This structure pertains to the origin of bracteate stamens (conventionally interpreted as 1-staminate flowers), as in *Hediosmum* (Leroy, 1983), and the bracteate clusters of such stamens, as in the staminate flowers of Myricaceae.

Pollen

Developmentally, angiosperm pollen grains seem different from those of gymnosperms in the patterns of primexinal matrix and sporopollenin deposition (Meyer, 1980; Pennell & Bell, 1986), although intermediate patterns occur in the Annonaceae (Waha, 1987). Notwithstanding the validity of such distinctions, their applicability is restricted to extant plants alone. Dispersed pollen grains of early angiosperms are recognized by their reticulate surface pattern and columellate infrastructure, the characters of the sporophytic self-incompatibility syndrome (Walker & Walker, 1984; Zavada, 1984b) that might have appeared independently in different lineages. Incidentally, both the perforate tectum and the columellate infrastructure appear in *Cornetipollis* (Vasanthy & Pocock, 1986), one of the costate morphotypes related to gnetaleans. At the same time, a number of angiosperm families, including the "primitive" Magnoliaceae, Degeneriaceae, Eupomatiaceae, Annonaceae, Nymphaeaceae and Myricaceae, have a granular infrastructure that is common also in various non-angiosperm groups. Moreover, a transitional granular-columellate, or pseudocolumellate, infratectum occurs both in angiosperms and gymnosperms. It has been described in *Ginkgo*, some conifers, and *Gnetum* (Surova & Kavadze, 1988) as well as in the fossil gnetalean pollen *Equisetosporites* and *Classopollis*. Such infrastructures are characteristic of some Late Cretaceous morphotypes (*Complexipollis*) of the then dominant *Normapolles* group (Kedves, 1982).

Although a non-lamellate endexine has been used as a major distinctive feature of angiosperm pollen morphology (Van Campo, 1971; Doyle & al., 1975), its practical use is complicated by pseudohomologies (Taylor, 1982), with a lamellate foot layer sometimes described as endexine, as well as by developmental flexibility of the nexinal structures in general. In angiosperms, primary lamellation often disappears in ripe grains, but sometimes persists in a less strongly condensed apertural endexine or is even discernible in the interapertural areas, as in *Austrobaileya* (Zavada, 1985), *Degeneria, Michelia fusca* and some Liliaceae. The lamellate nexinal layer of gymnosperms is developmentally different from the endexine of angiosperms (Zavada, 1984b), with a possible exception of *Ginkgo* (Rohr, 1977) and some fossil forms. On the other hand, a typical endexine is often lacking in extant monocots (Zavada, 1983) and "primitive" dicots (e.g. *Hedycarya*), whereas it was irregularly developed in some early angiosperm pollen grains (Walker & Walker, 1984) and the Late Cretaceous *Normapolles* group (Kedves & Párdutz, 1983b).

At least one of the diverse apertural types occurring in angiosperm pollen grains, the distally sulcate (anasulcate), characteristic of the Magnoliaceae and monocots, is also widespread among gymnosperms. Other apertural types pose a more difficult problem, however. In the early angiosperm pollen, such as *Clavatipollenites*, the anasulcate morphologies are closely linked to the trichotomosulcate ones, but the sometimes postulated transition from the latter to the likewise early appearing tricolpate type is not supported by any factual evidence. Inaperturate grains as an alternative starting point for all angiosperm apertural types (Muller, 1970) disagree with the obvious relatedness of anasulcate forms in gymnosperms and angiosperms.

The prevalence of the radially disposed apertures in angiosperms has been related either to a tetrad configuration (Wodehouse, 1935) or to germination on stigma (Eyde, 1971; Stebbins, 1974). Although multiple radial apertures certainly appeared before stigmas (e.g. in the Mesozoic gnetaleans), the latter, especially in conjunction with entomophily, might have increased the significance of the volume regulating (harmomegathy) structures. In addition, the pollen grain volume to surface ratio is affected by the ploidy level (Popova, 1971), typically being high in primitive angiosperms. Harmomegathy might involve germinal apertures and other structures, such as proximal triradiate scars, sacci and the residual folds of reduced sacci. There could be reversals from the harmomegathy to the apertural functions, with respective morphological transformations conceivably giving rise to the following apertural types:

(1) Bipolar apertures that appeared, in their incipient form, in *Classopollis*, a widespread Mesozoic pollen type produced by the hirmerellacean gnetophytes and showing both an apertural triradiate scar and a distal leptome.

(2) Zonosulcate apertures developing on the basis of monosaccate structures, as, incipiently again, in *Classopollis*, the equatorial girdle and its bordering subequatorial groove (rimula) of which might have been derived from a reduced annulate monosaccus (Krassilov & al., 1997a). Similar zonosulcate structures occur in monocots.

(3) Additional apertures developing from the lateral sacci. In gymnosperms, the saccate and asaccate pollen morphotypes diverged already at the level of Paleozoic pteridosperms, but saccate structures were lost in several Mesozoic groups, a tendency extended to angiosperms. Residual sacci are still occasionally discernible, as an atavistic feature, in the pollen grains of cycads and ginkgo (Shashi & Ueno, 1986). Among the Cretaceous pollen grains, a reduction of sacci can be traced from their vestigial forms, as in *Baculivesiculites* or *Microcahridites*, to harmomegathy folds, as in *Verrumono-colpites* (Zavada, 1984a), and to additional sulcate apertures, as in *Eucommiidites* (Fig. 53). A remarkable similarity between these gymnosperm pollen morphologies and the early *Tricolpites* may suggest a transformation of the harmomegathy structures into the transequatorial tricolpate apertural type.

(4) Polycolpate forms arising from polyplicate structures, as in *Welwitschia* and the fossil forms of the *Equisetosporites* type in which the intercostal pseudocolpi are rotated transequatorially in one of the hemispheres (Vasanthy & Pocock, 1986).

(5) Tricolporate structures derived from the tricolpate endocingulate. The possibility is prompted by *Sarbaya*, a mid-Cretaceous angiosperm (Krassilov & Shilin, 1995), whose tricolporate pollen grains show lalongate endoapertures forming a continuous weak zone,

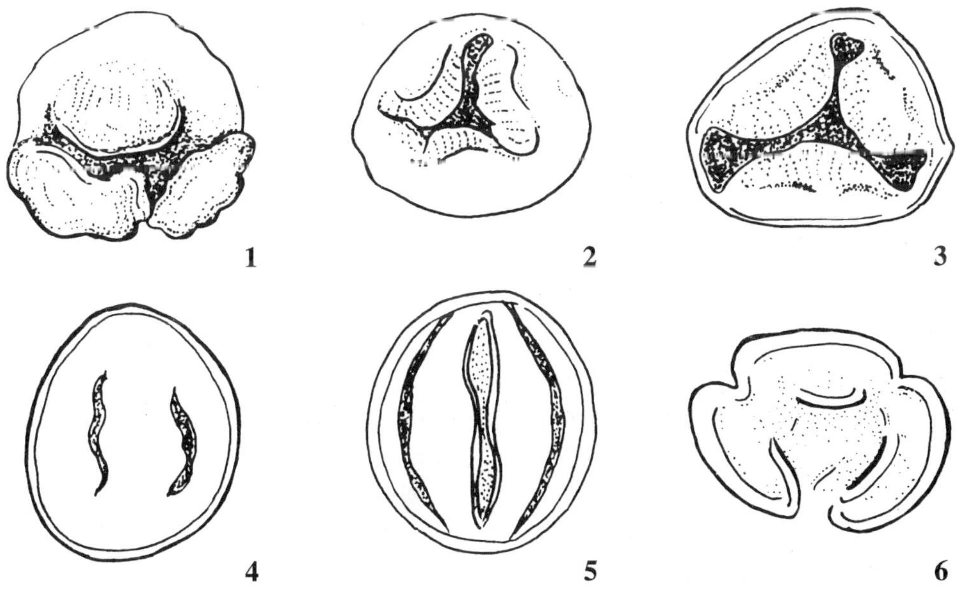

Fig. 53. Pollen grains illustrating a reduction of sacci as a pathway to additional apertures: (1) *Microcachridites antarcticus,* trisaccate grain, (2) *Trichotomosulcites subgranulatus,* trichotomosulcate grains, (4) *Baculivesiculites inchoatus,* showing folds of reduced sacci, (5) *Eucommiidites*, the subequatorial slits corresponding to reduced sacci, (6) *Tricolpites sagas*, tricolpate grain with polar folds (after Hughes, 1976; Zavada, 1984a; Dettmann, 1986).

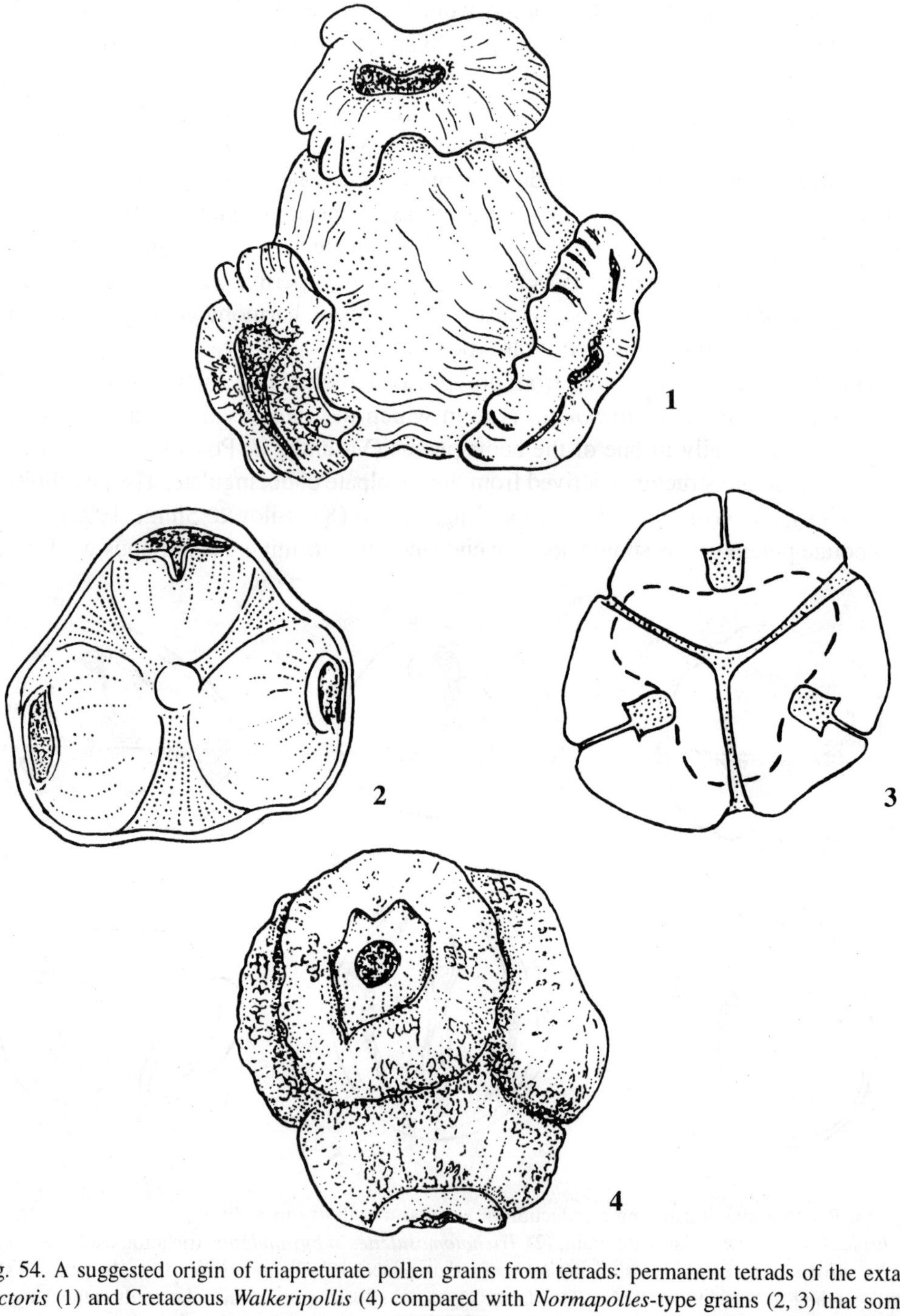

Fig. 54. A suggested origin of triapreturate pollen grains from tetrads: permanent tetrads of the extant *Lactoris* (1) and Cretaceous *Walkeripollis* (4) compared with *Normapolles*-type grains (2, 3) that sometimes show a tetrahedral outline of the nexine (after Médus, 1983; Walker & al., 1983; Kedves & Hegedus, 1983; Zavada & Taylor, 1986).

or endocingulum, conceivably inherited from the *Classopollis*-type equatorial girdle. The ores might have arisen at intersections of the colpi with such equatorial structures.

(6) Radially aperturate structures of tetrasporic origin (Krassilov, 1989, etc.). *Walkeripollis,* Early Cretaceous permanent tetrads with rimulate pores (Walker & al., 1983) resemble those of *Classopollis,* the distal pore (or leptome) of which is encircled by a supposedly monosaccus-derived girdle. In the former, however, the individual grains are calymmate, as in the Lactoridaceae, the only extant angiosperm family having saccate pollen grains with the reduced sacci encircling the apertures (Zavada & Taylor, 1986b). A further integration of such tetrads and a shift of meiosis one step back could have produced tetrasporic pollen grains that became trilobed triaperturate by the loss of a fourth member in symmetrical readjustment of a basically tetrahedral structure (Fig. 54), the latter still evident in the tetrahedral nexine configuration of some Cretaceous triporate pollen grains of the *Normapolles* type (Médus, 1983). The graminid pollen grains with rimulate pores and the cryptotetrads characteristic of the Cyperaceae might derive from the *Walkeripollis* prototype.

Gynoecium

Angiosperm gynoecia consist of ovuliferous organs, or carpels, that form, individually or jointly, the closed or semi-closed pistillate structures. As in the case of flowers, carpels have been defined morphologically in no other way as, circularly, the structural units of angiosperm gynoecia. That carpels are variously modified fertile leaves, or megasporophylls, is not a morphological observation but is an assumption of the now obsolete phyllome (Urpflanze) theory (see above). The foliar interpretation of carpels became a central dogma of typological morphology, the debates turning around the primordial type of leaf and the mode of its folding. Neither anatomical nor developmental studies have confirmed the foliar interpretation, however. Even the paradigmatic ranunculid follicle developes of an annular primordium (Rohweder, 1967) that is neither folded nor shoe-shaped as required by, respectively, the conduplicate and the involute foliar models, while the supposed "margins" are, in fact, ribs extending over the vascular bundles. In the typical "foliar" magnoliacean carpel with three traces, the dorsal trace arises from an axial bundle, while the two ventral ones depart from the cortical bundles, and the ovules are supplied with an interstitial vein ascending between the dorsal and ventral bundles (Skipworth, 1970; Tucker, 1975). In the supposedly even more "primitive" carpel of *Drimys*, the traces are in two pairs (Tucker, 1975) suggesting a bivalved pseudomono-merous structure.

Although carpels are conceived of as, primarily, the ovule-bearing structures, in a number of angiosperms the ovules are either borne laterally on the floral axis, as in *Illicium floridanum* or *Ochna atropurpurea* (Pauzé & Sattler, 1979), or are developed from the apex of the floral shoot, as in *Myrica gale* (Sattler, 1974a), Piperaceae (Kanta, 1962; Tucker, 1986), Juglandaceae and cyperid monocots (Cohen, 1970; Macdonald, 1977), and are separate from the carpels that are formed either of the lateral primordia as the ovule-subtending structures or of the annular primordia encircling the floral apex. In addition, the vascular supplies of the carpels and ovules in such gynoecia are disconnected, coming from different axial bundles. Some of such "acarpellate" gynoecia can be secondary (Philipson, 1985), but the assertion that they all are secondary lacks sufficient morphological evidence.

A foliar nature of carpels was once thought of to have been confirmed by the discovery of pteridosperms with their leaf-like ovuliphores. Presently we know that, initially, ovules were borne in non-foliar cupules formed of syntelomic three-dimensional branching structures. The bilateral cupules might have resulted from syncupuly, i.e. fusion of two or more radial cupules (Long, 1986). The leafy ovuliphores have appeared much later as a result of the gynoclade phyllomization. They were lost from the geological record at least 100 million years before the appearance of angiosperms, thus rendering the phylogenetic continuity unrealistic.

Among the morphological and chronological predecessors of angiosperms, *Leptostrobus* had bivalved cupules with the marginal stigmatic crests intruding the locule as inner stigmas and overhanging the anatropous ovules that were borne on submarginal placentae. A minimal morphological distance separates these structures from the pseudomonomerous pistils of the Winteraceae and perhaps other magnolialean "follicles" (Fig. 55).

The *Caytonia*-type cupule has been chosen by Thomas (1931) as prototypic for the ranunculid-type follicle. The ovules (reinterpreted in Krassilov, 1977b) are basal, anatropous, bitegmic, connected by presumably funicular tubes to a slit-like exostome. The latter is flanged by the free tip of subtending bract adnate to the cupule as a broad "lip". A minor elaboration is needed to transform this structure into an "ascidiform" carpel with the exostomal slit corresponding to the plicate region of the latter and the "lip" developing into a broad sessile stigma, as in *Paeonia* and its allied Cretaceous *Hyrcantha* (Krassilov & al., 1983).

Another Mesozoic proangiosperm, *Dirhopalostachys*, provides an example of a one-seeded beaked cupule comparable to the follicles of *Kingdonia*, an isolated extant genus remotely related to the Ranunculaceae (Hu & Tian, 1985). Superficially at least, the paired *Dirhopalostachys* cupules are similar to the likewise paired follicles of *Trochodendrocarpus*, a widespread Cretaceous to Paleogene fructification type (Krassilov, 1976, etc.).

At the same time, the secondary cupules of Mesozoic bennettites and gnetaleans were formed of interseminal scales or bracteoles connate around the ovule that was lateral or apical on the floral axis. Among them, the Triassic *Dinophyton* produced a 4-lobed bracteate cupule that, in the early developmental stages, appeared as an annular collar with incipient lobing, later expanding to enclose a solitary orthotropous ovule. The lobes bore tufts of long hairs probably assisting in pollination (Plate 7; Fig. 10).

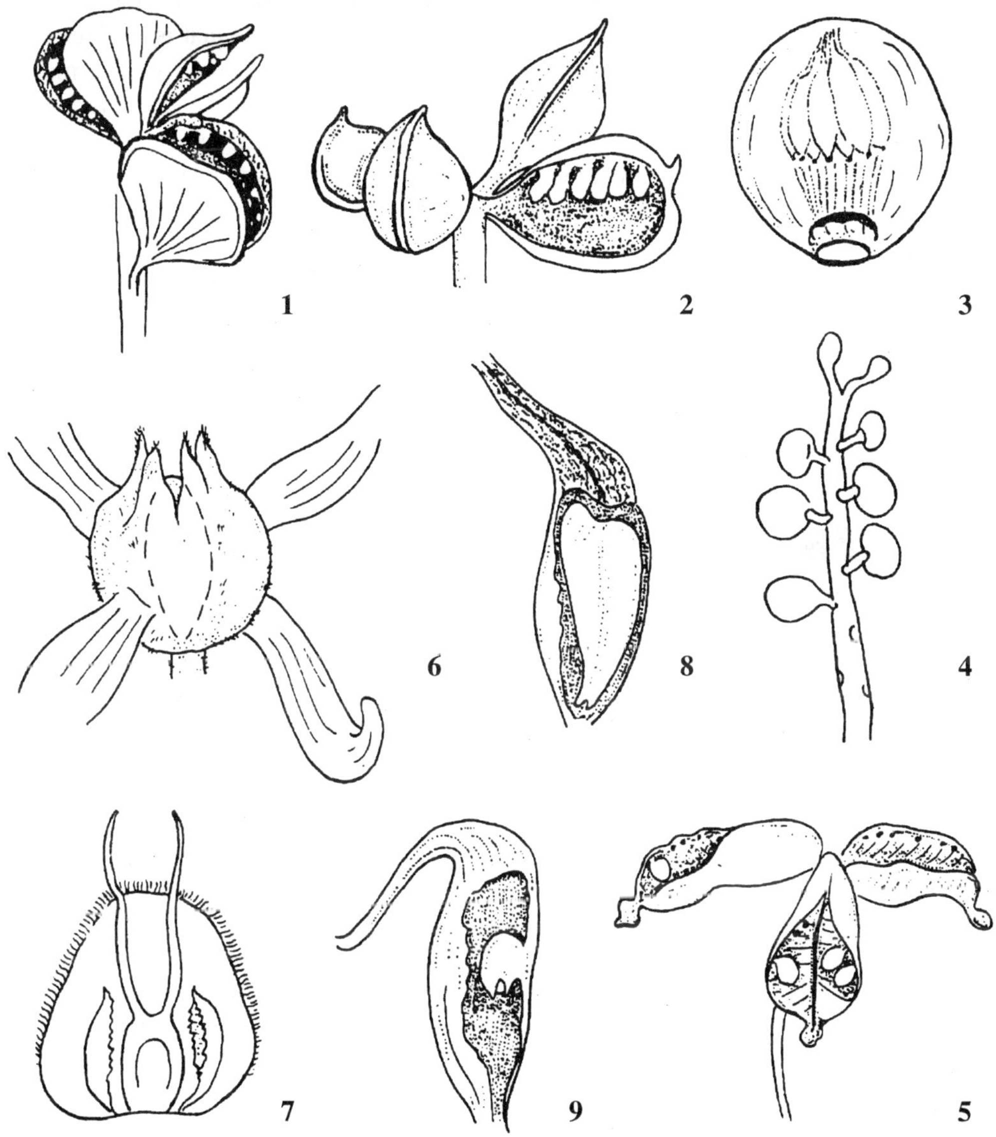

Fig. 55. Prototypic carpellate structures: (1) *Leptostrobus,* bivalved cupules with submarginal anatropous ovules and stigmatic marginal fringes, compared to (2) *Drimys,* extant, pseudomonomerous bicarpellate "follicles", (3, 4) *Caytonia,* gynoclade and cupule with the exostomal "lip" resembling leafy stigmas of the extant *Paeonia* (5), (6) *Dinophyton,* 4-lobed bracteate cupule with a solitary orthotropous ovule, comparable with (7) *Myrica,* extant, pistillate flower with bracteoles, (8) *Dirhopalostachys,* beaked cupule with a solitary anatropous ovule, similar to (9) follicle of *Kingdonia,* extant (after Harris, 1951a; Foster, 1961; Krassilov, 1972, 1975, 1984; Cronquist, 1981; Krassilov & Ash, 1988).

The Early Cretaceous *Baisia* illustrates a somewhat different cupule morphology, utriculate, enclosing a solitary ovule that is basal on the receptacle that is covered with bristles developing from the perianth bracts. The cupule appears to have been formed of connate interseminal scales, their tips joined into a terminal corona. The contemporaneous *Eoantha* had a 4-valved gynoecium, each valve with a basal ovule and with the floral apex protruding and terminating in a tuft of sterile bracts. These cupular structures relates to the 3-4-carpellate unilocular ovaries with basal ovules occurring in the Piperaceae, Myricaceae and glumiferous monocots. Transformations of the prototypic 4-lobed structures into the 3- or 2-lobed ones by the loss of a lobe or a pair of lobes occur in both the extant gnetaleans (Pearson, 1929) and the piperacean angiosperms, the latter having 3-4-carpellate gynoecia (Tucker, 1976, 1982).

Most of the early angiosperms had sessile stigmas characteristic of the forms with sporophytic self-incompatibility and morphologically traceable to the exostomal struc-

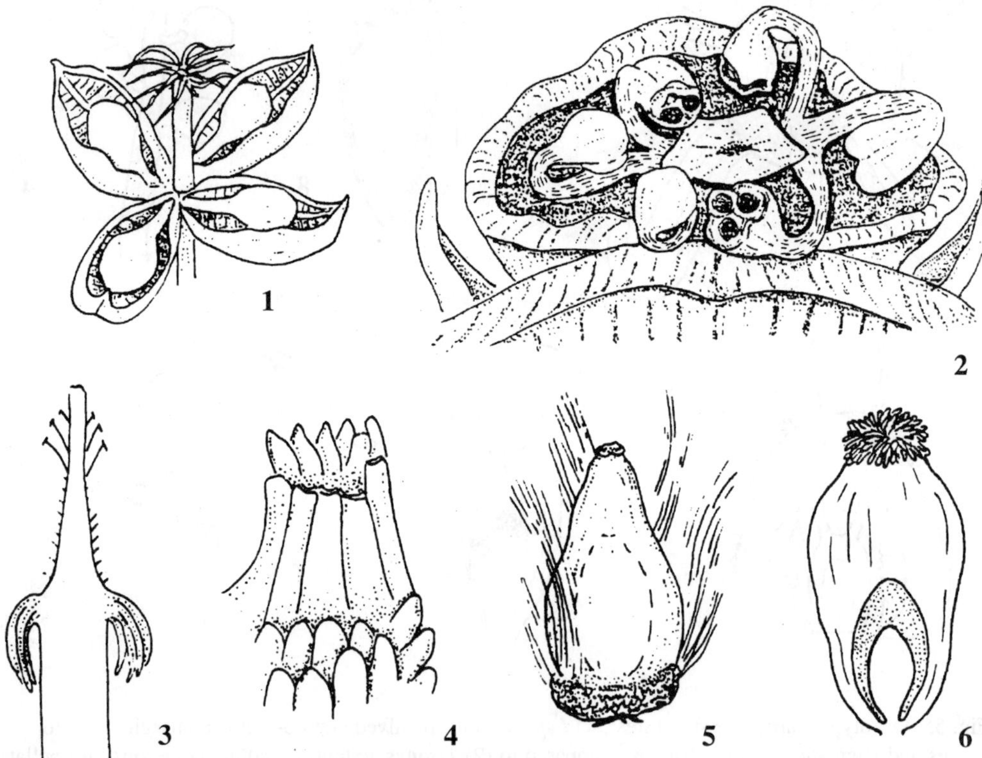

Fig. 56. Modifications of the second floral node in (1) *Eoantha,* cupulate preflower with a corona of bracts, (2) *Welwitschia,* extant, staminate preflower with a central peltate glandular structure, (3, 4) *Williamsonia,* gynoecium with a corona of vistigial interseminal scales, (5) *Baisia,* cupule with an apical corona, and presumably (6) *Piperomia,* extant, pistil with a bushy stigma (after Harris, 1969; Krassilov, 1972, 1986; Cronquist, 1981; Krassilov & Bugdaeva, 1982).

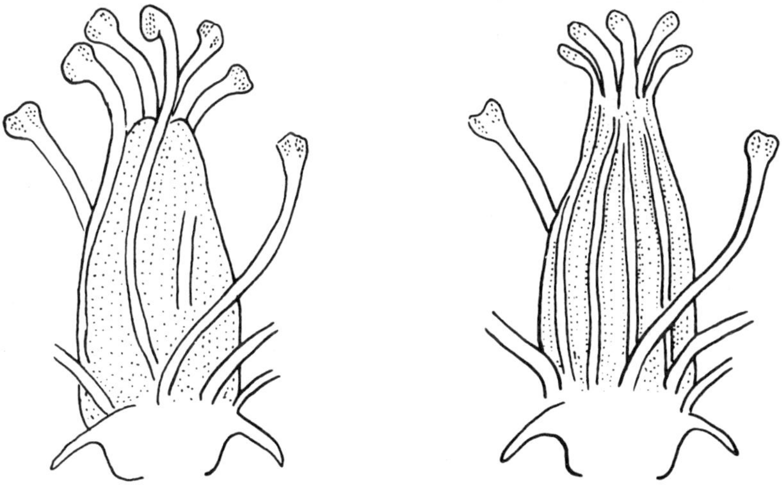

Fig. 57. A suggested derivation of free styles from stamens prompted by the filament to style similarity in *Sarysua,* a Cretaceous angiosperm (Krassilov & al., 1973).

tures found in the Mesozoic proangiosperms, such as the "lip" in *Caytonia*, the marginal papillate crests in *Leptostrodus*, the hairy cupule lobes in *Dinophyton*, the terminal corona of scale tips in *Baisia*, and the supposedly homologous tufts of sterile scales crowning the floral axis in *Eoantha* (Fig. 56). These structures are morphologically comparable with the foliar, decurrent, lobed, and bushy stigmas respectively.

The styles might have appeared in parallel in different angiosperm lineages in relation to gametophytic self-incompatibility and with no plausible prototypes in proangiosperms. However, in the Late Cretaceous *Sarysua*, the styles were remarkably similar to the filaments both of stamens and staminodia (Krassilov & al., 1983). Evidence of a conjoined developmental control is provided by examples of the similarly branched styles and stamens in the Euphorbiaceae, a correlation of the stylar and pollen polymorphisms in heterostylic species, and the occasional reversals of stigmas to anthers (Rao, 1935). These observations raise a possibility of styles originating from staiminodes adnate to the ovary (Fig. 57). Thus, in compound ovaries, the number of styles may not correspond to the number of carpels.

Ovule and dispersal

The ovule is perhaps the only reproductive organ indisputably shared by gymnosperms and angiosperms, thereby substantiating a morphological unity of all seed plants. Presently there is little doubt that ovules evolved from sporangia borne not on leaves, but rather in syntelomic branching systems. *Lenlogia* with paired sporangia surrounded by the whorled sterile appendages may illustrate one of the possible prototypes (Krassilov

& Zakharova, 1995), with the nucellus arising of a pair of fused sporangia and with the integument, or two successive and nearly homomorphous integuments, formed simultaneously of the whorls of indusial appendages.

In primitive gymnosperms, the integuments are the apically open, lobed structures serving for protection and perhaps for trapping air-borne pollen grains that settled on top of the nucellus. Subsequently, the nucellus has developed receptive structures that emerged over the integumental lobes. The cupule might protect a young ovule and was overgrown by the integument at maturity or, alternatively, it could overgrow the integument and overtake its functions. In the latter case, the integument was set upon the track of reduction and of the proximal or even wholesale fusion with the nucellus, perhaps contributing to the apical structures of the latter (see above under pteridosperms).

Both ovules and seeds may serve as dispersal units. The former mode is considered primitive and is sometimes described as "prephanerogamous". However, its chronological occurrence, as well as its correlation with other phylogenetically important characters are inconsistent in both gymnosperms and angiosperms, some of the latter also "prephanerogamous". Notably, in some Mesozoic ginkgoaleans, attached seeds contained the more advanced embryos than in the living ginkgo (Krassilov, 1972b). Early gymnosperms have dispersed both cupulate and excupulate ovules.

A functional change occurred with the appearance, in more than one lineage, of micropyles as integumental, or also cupular, pollen receptive structures. This change marked a transition from the archaic to the advanced gymnosperms whose nucelli lost their apical receptive structures (or retained their rudiments functioning as glands). In the advanced gymnosperms, cone scales and even the entire seed cones serve as dispersal units or, less frequently, the seeds are furnished with arillate structures. Transition from the radial to the bilateral symmetry, also parallel in different lineages, seems to have been related to the cone scale dispersal, while the secondary seed dispersal might restore the radial symmetry.

Angiospermy has led to the even more radical functional changes. The integuments have lost their pollen receptive and partly also protective functions, at the same time acquiring a pollen tube-conducting function. The nucelli have lost their nutritional "endosperm" function that has been passed over to the embryonic endosperm. Furthermore, in angiosperms, the periovular structures are consistently involved in dispersal, forming various fruits.

Fruit dispersal may seem a decisive factor of seed evolution in angiosperms. However, morphologically as well as functionally, the fruit is a dispersal unit formed of periovular structures of some kind(s), such as ovaries, perianth bracts, whole inflorescences, etc., enclosing seeds (critical comments on alternative definitions see above under "flower"). In this sense, the samaras of *Hirmerella* or *Welwitschia* are likewise fruits, the production of which being scarcely restricted to angiosperms. In particular, such dispersal units are characteristic of proangiosperms that produced samaroid cupules with persistent bracts, as in *Dinophyton* or *Hirmerella*; fleshy cupules with many seeds, as in *Caytonia*, perhaps endozoochorous; dry achene-like cupules, as in *Dirhopalostachys*; utriculate bristled cupules, as in *Baisia*; closed preflowers with persistent fleshy receptacles, as in some bennettites, etc. The dispersal mechanisms presently restricted to angiosperms had first appeared in proangiosperms, e. g., the parachute of coronal bristles in *Problematospermum*, or the cotton-grass-like pappus of *Baisia* (Plates 5, 6; Fig. 43) conceivably assisting in anemochorous or epizoochorous dissemination, or both.

Yet it is difficult to decide how far the functional changes were responsible for the morphological distinctions. In angiosperms, seeds are more diverse morphologically, especially in respect to dispersal mechanisms, their symmetry (radial versus bilateral) is more variable, sometimes even within a species, and the lower seed size limit is considerably lower than in gymnosperms. The latter distinction, as well as the neotenic gametophytes, seem to be related to developmental acceleration in the early angiosperms in turn associated with a pioneer ecology and, alternatively or additionally, with epizoochorous dispersal mode. Actually, the hooked appendages in the ceratophyllacean achenes from Koonwarra (Douglas, 1969; Dilcher & al., 1996) or the spiny capsules of *Tyrmocarpus* (Krassilov, 1977a, 1989) point to epizoochory as an early appearing dispersal mode in angiosperms.

However, the prevalence in angiosperms of anatropous bitegmic ovules is less amenable to a functional explanation. On the ground of their occurrence in some allegedly primitive angiosperms, these features are often considered as primary for the group. Then the orthotropous type should be secondary, whereas an unitegmic condition might arise by fusion as well as reduction either of the outer or, more often, of the inner integument that could be developmentally retarded or shifted to the apex by endochalazal growth (Bouman, 1984; Bouman & Calis, 1977). These processes are inferred on the basis of comparative morphology. Thus, in the Liliaceae having both bitegmic and unitegmic ovules, the latter apparently lost their inner integument in fusion with a comparatively massive nucellus (Sterling, 1973). Alternatively, priority is given to orthotropous unitegmic ovules as more common in gymnosperms (e.g. Meeuse & Bouman, 1974), with the anatropous condition considered as derived in relation to siphonogamous pollination mode (Takaso & Bouman, 1984). The bitegmic condition might then arise by acquiring either the outer integument of an arilloid origin or the inner integument of a nucellar origin, the latter functioning as a pollen tube-conducting structure (De Boer & Bouman, 1972; Heslop-Harrison & al., 1985), or, in orchids, as a germination-mediating structure (Lucke, 1984). Any generalizations are risky, however, in the light of the great variability of the ovular characters in angiosperms.

Incidentally, the anatropous condition can arise either by a reversion of the funicle or by a developmental shift of the raphe toward the micropyle, or sometimes by inversion after fertilization, or even by reversion of the carpel (Vijayaraghavan, 1964). Both integuments can be of a dermal origin or they arise from non-homologous - dermal and subdermal - initials, as in most of the ranalean angiosperms. In 75 families, both integuments contribute to the micropyle (sometimes zigzagged, with the outer, or exostome, not in line with the inner). The micropyle is formed either of the inner or of the outer integument alone in 38 and 4 families respectively, while in 79 families this character varies (Tilton, 1980). Vascular bundles either terminate in the chalaza or extend to one or, rarely, both integuments, as in the Euphorbiaceae. The cutinization of the integuments and the development of stomata (in 25 families: Rugenstein & Lersten, 1981) are likewise variable. In the Capparaceae, the outer integument occasionally encloses two ovules. At least some of these variations are paralleled by gymnosperm ovules.

As is convincingly shown in the early seed plants, the orthotropous ovules are primary in spermophytes, while two or more homomorphous or heteromorphous integumental structures can arise simultaneously, the outer forming a cupule that contains one or more ovules. The cupules enveloping solitary ovules can have their exostomes not in line with the integumental micropyles, as in the zigzagged micropyles of angiosperms. In pteridosperms, the inner integument, or integument proper, tends to be lost as a distinct morphological structure in fusion with the nucellus, resulting in the massive vascularized nucelli, as in trigonocarps and cycads. However, the bitegmic (or occasionally tritegmic) condition was retained in the gnetaleans and some bennettitaleans that advanced to the proangiospermous level with orthotropous ovules, as in *Baisia, Eoantha* and *Gnetum*. In the latter genus, the cupule, or outer integument, occasionally contains more than one ovule (Keng, 1965) and the integuments develop in acropetal succession, the inner and the middle ones from the dermal, and the outer one from both the dermal and subdermal initials. They are vascularized, and the outer ones are stomatiferous, while the inner one is archaically lobed. Furthermore, the middle integument is developmentally retarded, whereas the inner integument is shifted to the apex by endochalazal growth (Takaso & Bouman, 1986), the tendencies that can be traced to many angiosperms.

In *Baisia*, a derived representative of the Early Cretaceous bennettites (Krassilov & Bugdaeva, 1982), the cupule was utriculate and comparable with the utricles of the Late Cretaceous lemnoid monocots (Krassilov & Makulbekov, 1995). Both the baisian cupules and the lemnoid utricles enclosed a solitary orthotropous ovule arising from the bristled receptacle in the former, and with a funicle in line with the fruit pedicel in the latter (Plates 47, 48). The appearance of anatropous ovules in several gymnosperm lineages, such as corystosperms, nilssonialeans and conifers, correlates with strobilation and anemophily, and is perhaps explicable by aerodynamics of the airborne pollen grains whirled around the cone axis (destrobilation is accompanied by reappearance of a secondarily orthotropous - pseudoorthotropous - ovule polarity, as in *Ginkgo, Cycas* and some podocarps). At the same time, the anatropous condition correlates with dispersal of the seed-bearing cone scales (in conifers) or cupules (in czekanowskialeans and caytonias) that is precursory to fruit dispersal in angiosperms.

Furthermore, while the bitegmic condition tends to be lost in the exposed ovules developing thick coats, both integuments tend to be retained as distinct, although delicate, coats in the enclosed ovules.

At least *Caytonia* has advanced to the proangiospermous level with anatropous bitegmic ovules (Krassilov, 1984a) that are strikingly similar to those of a mid-Cretaceous ranunculoid angiosperm, *Caspiocarpus* (Vakhrameev & Krassilov, 1979; Krassilov, 1984a) in general shape and dimensions, as well as in showing a median raphe trace, and in having the integuments a few cells thick, of which the inner one wedges out slightly above the midlength.

The historically deep-rooted divergence of the orthotropous and anatropous ovule morphotypes in proangiosperms may thus be phylogenetically significant, for these types are traceable to different groups of early angiosperms, such as the piperoid - juglandioid and the ranunculoid lines respectively.

Double fertilization and the embryo

Double fertilization is often considered as a major, if not the only, diagnostic feature of angiosperms. It has to be kept in mind, however, that angiosperms had been well-established as a higher taxon long before the phenomenon of double fertilization was discovered (by Navashin in 1898) and asserted by a representative sample of angiosperms from different families that revealed a considerable variability of the process (see above). In spite of that and of some very atypical forms of double fertilization, as in the Onagraceae, where the term is scarcely justified at all, none of the known angiosperms has yet been excluded from the group on the pretext of having no double fertilization.

For extinct plants, there is no direct evidence of fertilization, but, insofar as the double fertilization associates with accelerated gametophyte development, an indirect evidence may come from the size of the ovules. Although this character is primarily under ecological control (see above), with the early successional plants tending to have more numerous and smaller seeds, in gymnosperms there is a lower seed size limit of about 1 mm, apparently set in by the gametophyte development. In angiosperms, an accelerated gametophyte development, with archegonia reduced to the clusters of nuclei, have opened a perspective of seed miniaturization down to dust size, as in orchids. Actually, the advent of angiosperms in the mid-Cretaceous is marked by the appearance of fossil seeds well under 1 mm (Hall, 1963; Colin, 1973; Binda & Nambudiri, 1983; Knobloch, 1983, etc.), perhaps suggesting the angiospermous level of female gametophyte reduction. Remarkably, even such proangiosperms as the caytonialeans and czekanowskialeans have produced smaller seeds than did the bulk of gymnosperms.

Double fertilization is a complex morphological as well as physiological phenomenon, the components of which occur in non-angiosperm seed-plants, too. Polyembryony of various kinds is widespread among gymnosperms. Likewise, the sperm-bearing pollen tubes and the involvement of both sperms in fertilization are not restricted to angiosperms, but occur, systematically, in conifers and, sporadically, in other non-angiospermous groups. Such steps on the way to double fertilization as (1) the appearance of archegonial complexes and the ensuing diminution of individual archegonia, as in *Ephedra* and Taxaceae, and their complete reduction in the end members of the not closely related gnetalean lineages, (2) the functioning of both sperms, as in the Taxaceae, Cephalotaxaceae, *Ephedra*, etc., and (3) the sexualization of the embryo sac nuclei, appearing, again independently, in *Gnetum, Welwitschia* and angiosperms, are parallel developments taking place in different seed plant divisions.

Actually common for all properly studied angiosperms is not so the highly variable double fertilization process as the lack of morphologically distinct archegonia, but this feature is shared with at least two extant non-angiospermous genera, *Gnetum* and *Welwitschia*. The embryo sac nuclear homology between these genera and the typical angiosperms is difficult to decipher, however, for, typically, the two former have much more numerous nuclei, an indefinite number of which are sexualized and can potentially function as eggs, while syngamy frequently occurs both before and after fertilization (Maheshwari & Vasil, 1961; Martens, 1971). Yet in the third extant gnetophyte genus *Ephedra*, the archegonia are morphologically distinct, joined in the archegonial complexes and are conceivably on their way to fusion into a multinuclear structure. Characteristically, in angiosperms the sperms are discharged in a synergid corresponding to an archegonial neck cell in *Ephedra*. In the latter, one of the sperms reaches to the ventral canal cell, the other to the egg that sinks deeper in the venter (Herzfeld, 1922; Khan, 1940; Moussel, 1978). A mass of cells is produced of the fertilized ventral canal cell (Pearson, 1929). On the other hand, the nuclear configuration in a typical 8-nucleate embryo sac is suggestive of two oppositely oriented archegonia whose potential neck cells differentiate as synergids and antipodal cells respectively. A potential ventral canal cell of the micropylar complex forms the egg, while the potential eggs form the polar nuclei that, like archegonial eggs, sink in the common venter and meet there, still presenting a target for the second sperm.

In other words, what we conventionally call the egg cell in angiosperms can be homologous to the archegonial ventral canal cell, while the archegonial eggs correspond to the polar nuclei (Cocucci, 1973; Krassilov, 1989, 1991). Double fertilizations in *Ephedra* and in typical angiosperms can thus prove to be developmentally homologous (Fig. 58).

At the same time, the most common free-nuclear endosperm development in angiosperms is like a typical embryo development in gymnosperms, and even the less common type of an initially cellular and then free-nuclear endosperm development may have its parallel in *Ginkgo* (see above). In contrast, the embryonic development in angiosperms, with a sole exception of *Paeonia*, is cellular, while in gymnosperms it is free-nuclear, the polyploid *Sequoia* being a single anomaly. Thus, to all appearances,

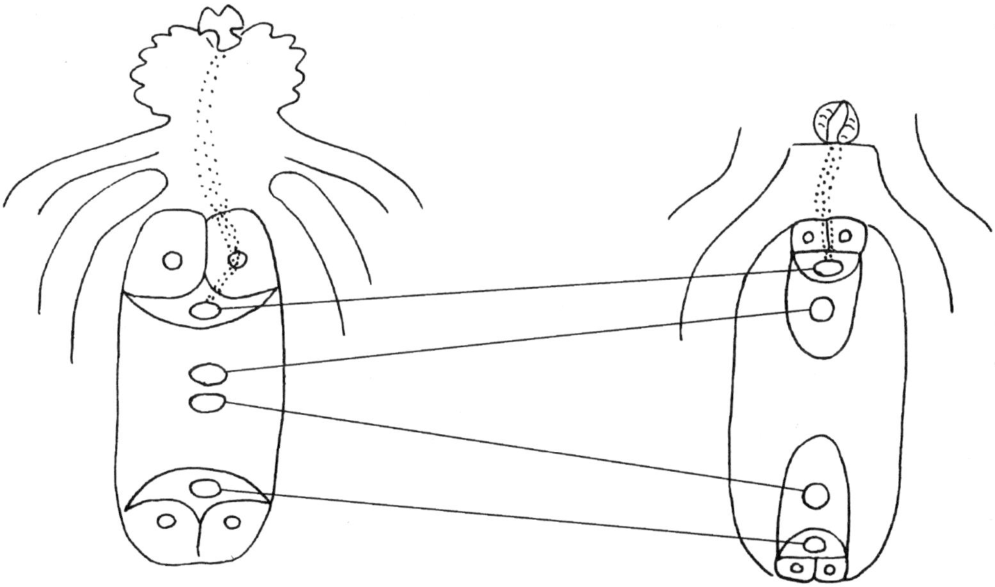

Fig. 58. Possible homologies of angiosperm egg and polar nuclei (left) to the ventral canal cell and egg, respectively, of a gymnosperm archegonium.

the embryo of gymnosperms is homologous to the endosperm, rather than to the embryo, of angiosperms.

As for the intragroup and outgroup homologies of embryological structures, it has to be reminded that the idea of cotyledons being not, or not always, homologous in monocots and dicots, first put forth by Arber (1925), is presently confirmed in a number of case studies. Dicotyledonous embryos are common in several groups of gymnosperms, some of them having heteromorphous cotyledons and showing a tendency of cotyledon fusion, the tendencies extending both to dicotylids and monocotylids (e.g. *Piperomia*, with the cotyledons heteromorphic, and one of them haustorial). In some monocots, the cotyledon becomes pseudoterminal by replacing the primary apex (e.g. in palms: Guignard, 1984), as is occasionally the case in dicots as well (*Claytonia*). However, in a number of monocot genera, cotyledons proper seem totally lacking (Gerasimova-Navashina, 1972).

The grass embryo alone has been exempt from homologization by a noncommittal descriptive nomenclature of its structures. The attempts at homologization of, e.g., the grass scutellum with hypocotyl or fused cotyledons or their parts or prophyll and other structures of a conventional embryo (see Negbi & Koller, 1962; Cocucci & Asregiano, 1978; Tsvelev, 1987) have not been entirely successful yet. Actual homologies may be found outside the group (Mestre & Guédès, 1983), perhaps involving such haustorial structures as the feeder of the gnetalean embryos.

Trend types

The principles of phylogenetic analysis applied above to gymnosperms are valid for angiosperms as well, but gymnosperms differ from their precursory progymnosperms in the presence of ovule, a single charismatic character. In contrast, angiosperms are defined by a number of loosely correlated characters that often form intersecting morphoclines.

In the following analysis, phylogenetic relations are deduced on the basis of multiple homologies, that is, homologies of different, independently evolving, organ systems. In plants, lacking a distinct herm line, evolution has been, to a considerable extent, directed by functional adaptations. Although a growth form may affect pollination and seed dispersal, the functional correlation of the vegetative and reproductive systems is too weak for impelling their coordinated evolution. Both the vegetative and reproductive shoot systems are integrated developmentally by morphogenic fields that occasionally inflict a morphological convergence of their constituent structures, as in the petaloid or carpelloid stamens, but developmental correlation of the vegetative and reproductive organs does not preclude their independent evolution under different selection pressures. We have to distinguish between the general selection pressures, such as environmental instability, inducing whole plant trends, such as developmental acceleration, and the specific selection pressures, such as a pollen vector efficiency, inducing morphological trends in the respective structures alone. Both trend types can be "fashionable" for a particular geological time interval or they can be long-lasting, traceable through geological times and across taxonomic boundaries of successively appearing plant groups. The gametophyte reduction trend belongs to the latter category (see above under "gametophytic characters"), while phyllomization of leafy shoots seems to have been "fashionable" among Mesozoic proangiosperms, resulting in the highly flexible caulome-phyllome systems characteristic of angiosperms as an adaptive type. Such single character (semophyletic) trends are, by themselves, insufficient for substantiating phylogenetic relationships that can be more convincingly demonstrated by the concurrent trends involving different organ systems.

Since trend recognition is crucial for our analysis, the following terminology is proposed for different trend types:

- Morphocline, or semophyletic trend: pertaining to a particular organ or organ system;
- Corporal, or whole plant, trend: manifested in the evolution of different organ systems, such as developmental acceleration;
- Concurrent trend: involving functionally independent organ systems;
- Chronocline: chronological succession of character states;
- Time-fashion trend: a trend involving concomitant plant lineages;
- Transgrade trend: a trend extending from one morphological grade to another across a conventional grade boundary.

The sequential stages of phylogenetic analysis then will be:
- Recognition of the corporal and semophyletic trends;
- Selecting the chronologically relevant time-fashion trends;
- Singling out the continual transgrade chronoclines;
- Ensembling the concurrent transgrade chronoclanes as a material for phylogenetic reconstructions.

A practical realization of this ideal scheme runs into certain difficulties related both to morphocline and chronocline ambiguities. Trends deduced from the data of comparative morphology alone may have little phylogenetic significance, whereas chronoclines are, for most part, fragmentary. In effect, phylogenetic reconstructions are hypothetical to be verified or falsified by the more convincing homologies, as well as by the more complete transgrade chronoclines.

Proangiosperms as a morphological pool

The proangiosperm grade as defined above became apparent due to a series of palaeobotanical discoveries (Krassilov, 1969, 1973b, 1975, 1977b, 1982a, b; 1984, 1986; Krassilov & Bugdaeva, 1982, 1997; Schweitzer, 1977) that brought to light the seed plants chronologically preceding or contemporaneous with the earliest angiosperms and showing character states typical of the latter, but lacking in some critical characters upon which the conventional recognition of angiosperms is based. Among the typical angiosperm features, the following appeared in one or more proangiosperm groups as described in the preceding chapters:
- Vessels of the axial and/or foliar vascular systems in gnetophytes and bennettites;
- Paracytic stomata in bennettites;
- Graminoid leaf morphology in gnetophytes;
- Compound-palmate to lobate-palmate leaves with reticulate venation in the *Sagenopteris-Scoresbya* group (the Caytoniales and allies);
- "Dendroid" androclades, anticipating fasciculate androecia, in *Caytonanthus*;
- Pollen grains with a protocolumellar infrastructure and with zonal, as well as porous, protoapertures in *Classopollis* (the hirmerellean gnetophytes);
- Cupular gynoecia in *Dirhopalostachys* of a cycadophyte descent (with a solitary anatropous ovule) and *Baisia* of a bennettitalean descent (with a solitary orthotropous ovule);
- Ascidiform cupules with many ovules in *Caytonia*;
- Four-membered cupulate gynoecia with a bracteate perianth in *Eoantha* (the gnetophytes);
- Syncupulate capsules with stigmatic crests in *Leptostrobus* (the Czekanowskiales), etc.

Different proangiosperm groups are thus complementary in forming a morphological pool of incipient angiosperm characters. Moreover, in the Cretaceous representatives of the principle proangiosperm groups, the reproductive structures show the tendencies of condensation and fusion (as, e. g. in the pollen organs of *Basianthus*, see above), with the chimeric organs derived from the fusion meristems having great potentials for morphological innovation (e. g. in the case of areolate venation, see above under "Leaf"). The early angiosperms seem to have been likewise condensed and highly meristematic plants.

But no single proangiosperm lineage might have conceivably given rise to the basic angiosperm diversity evidenced by the mid-Cretaceous fossil record. Rather the diversity of angiosperm morphologies could have arisen by a recombination of the diversity of proangiosperm morphologies, involving both vertical and horizontal transmissions of genetic material. The latter pathway is assumed for certain palynological characters (see above) but is probably of a more general significance for parallel evolution (Syvanen, 1994). The angiosperms as a whole might thus have evolved from the proangiosperms as a whole, although some particular lines of descent may still be discernible across the proangiosperm-angiosperm grade boundary.

Concurrent homologies

A semophyletic trend is significant for the whole plant phylogeny if it is supported by other semophyletic trends pointing in the same direction. So, as the next step of our analysis, we shall select the concurrent homologies of the functionally independent structures. On the basis of the receptacular orthotropous versus the appendicular anatropous ovule morphologies, both proangiosperms and early angiosperms can be arranged in two series:

Receptacular orthotropous	**Appendicular anatropous**
Bennettites	Dirhopalostachians
Gnetaleans	Czekanowskialeans
	Caytonialeans
Ceratophyllids	Ranunculids
Platanoids	Hamamelids
Piperoids	Magnoliides
Myricoids	Rosoids
Graminoids	Alismatoids

If these series have a phylogenetic meaning, then a number of the features shared by their juxtaposed members are homoplasies. This has to be verified by the concurrent homologies of the phyllodic versus caulomic leaves, the sulcus-derived versus sac-derived pollen apertures, the bract (bracteole)-derived versus ovuliphore (seed-scale)-derived cupules, etc.

The gneto-bennettite line

Both gnetophytes and bennettites have been used as outgroups in angiosperm phylogeny (Wettstein, 1907; Arber & Parkin, 1907). However, until recently, the gnetaleans had no fossil record, whereas the angiospermoid features of their extant representatives, particularly those related to gametophyte reduction, could be ascribed to parallel developments. The bennettite - angiosperm comparisons have involved, in the first place, the strobiloid flowers that, in both groups, are related to cantharophily and, therefore, homoplastic. New evidence has come from the recently discovered fossil gnetophytes, starting with *Eoantha* that showed transitional angiospermoid gynoecial (Krassilov, 1986) and foliar (Krassilov & Bugdaeva, 1997) structures. It was soon followed by other forms of diverse vegetative and reproductive morphologies. At the same time, angiospermoid characters have been found in a few derived bennettitalean forms, such as *Problematospermum* and *Baisia* (Krassilov, 1973b; Krassilov & Bugdaeva, 1982). Summarily, these findings have provided a factual basis for analyzing the gneto-bennettite to angiosperm concurrent homologies.

Gnetophyte to monocot trends

The bifacial leaves with a several-rank areolate venation, the vessel elements with scalariform perforation plates (Muhammad & Sattler, 1982), the sieve elements companion-like cells (Pearson, 1929), and the fertilization pattern anticipating double fertilization (Herfeld, 1922; Keng, 1940; Moussel, 1978) are the features shared by the extant gnetalean genera with angiosperms rather than with each other. However, these genera are relicts of a previously much more diverse group.

The Early Cretaceous gnetophytes, appearing together with the earliest angiosperm records, were small, sedge- or grass-like plants evidencing the miniaturization and condensation corporal trends in their more robust early Mesozoic precursors. The gametophyte reductions leading to the archegonia-less embryo sacs, as in *Gnetum* and *Welwitschia*, were conceivably associated with an accelerated sporophyte development. Both trends are continued in angiosperms.

The dicotylid-type leaves, as in *Gnetum*, are lacking among the pre-angiosperm records, therefore considered as parallel developments, perhaps appearing earlier in angiosperms than in gnetophytes. Fossil gnetophytes show diverse foliar morphologies, from the poorly differentiated caulome-phyllome systems, as in *Dinophyton*, to the reduced sheathing leaves in the *Leongathia-Ephedra* line, their stomatiferous stems taking up the photosynthetic function. A similar pattern of alternate large and small vascular bundles in the shoot internodes of these gnetophytes and in graminoid leaves suggests a semophyletic relatedness, with the morphologically graminoid caulomic leaves of *Eoantha* plant as a fossil link. The Late Cretaceous graminoid leaves assigned to *Dammarophyllum* (Shilin, 1986) are associated with *Taldysaja*, a graminoid floral structure (Krassilov & al., 1983).

The floral structures of gnetophytes are typically decussate or whorled at two or more bracteate nodes of spicate inflorescences. In *Eoantha*, they are solitary pedicellate bracteate preflowers corresponding to a single floral node of a typical gnetalean spike, with the sterile second node reduced to a tuft of bracts on a short rachylla. Protruding rachyllae occur also in *Taldysaja* and extant graminoid florets. The granular to irregular rugulate-columellate pollen grain infrastructures are nearly identical in the gnetophytes and extant graminoids (Zavada, 1983), while the porous apertures of the latter have their prototype in the Mesozoic *Classopollis* (the hirmerellacean gnetophytes), the rimulate pollen grains with a distal cryptopore. Some other apertural types might have been derived from the saccate or plicate exinal structures of extinct gnetophytes.

Developmentally, the gynoecia of graminoid monocots are similar to the gnetalean cupules arising by tubular growth from the annular primordia (Takaso & Bouman, 1986). In *Eoantha*, a fossil intermediate form, the gynoecium consisted of a four-lobed cupule bearing basal orthotropous ovules. Tetramerous gynoecia occur as a plesiomorphic feature in the extant graminoids, while the more common di- to trimerous forms might have been derived by the loss of one or two members, a widespread mode of morphological transformation. In addition, there are embryological features, not available in fossils, compliant with a phylogenetic proximity of gnetophytes and graminoids, such as the tetrasporic embryo sac, coleorhiza, sheathing cotyledons, and haustorial appendages of the embryonic axis - the feeder in gnetophytes and scutellum in grasses.

Bennettite to monocot trends

Common in bennettites were phyllodic cataphylls with parallelodromous or flabellate venation, occasionally terminating in a vestigial pinnate or entire pinnatinerve blade (Harris, 1969; Krassilov, 1982b). They suggest a leaf blade reduction trend that continued in the phyllodic leaves of monocotylids (Arber, 1918), with intermediate morphologies represented by the Baisian leaves or cataphylls (Fig. 9) with oblique, irregularly cladodromous lateral veins (Krassilov & Bugdaeva, 1997). A mesogenous developmental pattern is confirmed for the "syndetocheilic" bennettitalean stomata on the evidence of their aborted initials (Krassilov, 1978a; Figs 45, 46). In the bennettitalean cataphylls, the stomata were typically tetracytic, as in the graminoid leaves.

Vessels in the leaf veins occur in a Cretaceous bennettite *Otozamites lacustris* (Krassilov, 1982b; Fig. 36) and are characteristic of graminoids and palms.

Typical bennettitalean flowers resemble those of magnoliacean angiosperms in the relatively massive receptacles with an apically protruding corona (a vestige of the second floral node), the spiral arrangement of polymerous floral parts, and the dimorphic - stomatiferous sepaloid and stomata-less petaloid - perianth bracts (Harris, 1969). However, these similarities are related to cantharophilous adaptations. In the advanced bennettitalean forms, there was a tendency to aggregation of preflowers alongside with oligomerization of the floral structures. *Baisia*, a numerically dominant genus of the Baisian proangiosperm assemblage, is conceived of as an end-product of these functionally correlated trends. Its gynoecial preflowers produced a solitary erect ovule apical on the receptacle and enclosed in the utriculate cupule (Plates 5,6), the latter conceivably formed of connate interseminal scales. These structures were dispersed as bristled disseminules with persistent receptacles (Krassilov & Bugdaeva, 1982). Their similarity to cyperacean achenes may be more than superficial, for in both the bristles arise from perianth bracts, whereas the gynoecial structures are essentially cupular, developing from the floral apices.

Gnetophyte-bennettite features in other angiosperm groups

Among the groups formally assigned to dicotylids, but of obvious monocotylid alliances, the Piperales shares with the gnetophytes a decussate arrangement of lateral organs, the tracheary elements with both scalariform of many bars and porous perforations, orthotropous ovules arising form the floral apex, gynoecial structures developing from annular primordia encircling the apex (in *Piperomia*. Tucker, 1986), and the heteromorphous cotyledons (in *Welwitschia*).

Antiquity of the piperoid line is evidenced by the earliest dispersed angiosperm pollen morphotype *Clavatipollenites* at least partly assignable to the Chloranthales, an order of the widely divergent archaic genera recently separated from the Piperales (Dahlgren, 1983). A peculiar androecial feature of both extant and Cretaceous chloranthaleans is fusion of the stamens forming synandria with different numbers of sporangia on the median and lateral lobes. Fusion of ovuliphores occurs in *Gnetum* as well as in the Cretaceous gnetalean plant *Baisianthus* (Krassilov & Bugdaeva, 1997; Plates 16-19). Moreover, there is a similar variation of sporangial numbers in the central and lateral synangia.

Worth considering in this respect are two other groups of archaic angiosperms, the Platanaceae, a monotypic family usually placed in hamamelids, and the Nymphaeales, a heterogeneous aquatic group of ambiguous dicot/monocot affinities. Platanoids differ from hamamelids primarily in the orthotropous ovule morphology shared with the gnetoid line and, among nymphaeids, with *Barclaya* and *Ceratophyllum*, the latter genus ranking among the earliest angiosperms, with a precursory in the Early Cretaceous (Aptian) Koonwarra Assemblage (Dilcher & al., 1996). Notable for further morphological comparisons are such nymphaealean features as the stigmatic disk developing from the floral apex and apparently homologous to the "corona" of bennettitalean preflowers or the apogynoecial bracts of *Eoantha*, or other modifications of the second floral node (see above, Fig. 56). In the pollen grains of *Nymphaea*, the equatorial harmomegathy zone and the rim of the operculum (Kupriyanova, 1976) topologically correspond, respectively, to the girdle and the rimula of *Classopollis*, a Mesozoic gnetalean pollen morphotype.

In the Myricales, the staminate "flowers" are small strobili of bracteolate sporangiophores that are occasionally fused and are adnate to their supporting bracts, resembling both the Mesozoic *Dinophyton* and extant *Ephedra*. The pollen wall infrastructure is granular to shortly columellate (Zavada & Dilcher, 1986), as in *Gnetum*, the endexine is lacking (in the phylogenetically close Juglandaceae it is developed as a uniform layer scarcely thickened under the apertures, with similar endexinal structures found in monocots: Zavada, 1983). The carpellate "flowers" are provided with two lateral bracteoles, as in *Welwitschia*. The gynoecium is essentially cupular, 2 (3-4)-carpellate, surrounding a solitary ovule developed of the floral apex.

The fossil fruits *Gurvanella* (Krassilov, 1982b) and *Ramonicarya* (Krassilov & Dobruskina, 1995) provide evidence of an early (Aptian) appearance of myricoids. They resemble also the samaras of *Welwitschia* and of the extinct hirmerellean gnetophytes, but are as yet insufficiently studied for being considered as semophyletic links.

Caytonialean to ranunculid trends

Phylogenetic links with angiosperms and, in particular, with ranunculids have been postulated for caytonialeans by their discoverer already (Thomas, 1931), but they were rejected by the leading plant morphologists of the time in favour of, in the words of Agness Arber (1950), the elegant simplicity of the classical carpel concept. It must be recalled that ranunculid follicles were then conceived of as the paradigmatic involute sporophylls. They are not considered as such anymore (see Rohweder, 1967).

Yet, with a few exceptions (Stebbins, 1974), the caytonialeans are not discussed as plausible angiosperm ancestors. Their morphological revision undertaken by Harris (1951 and elsewhere) tended to discredit the earlier angiosperm comparisons, while, on the other hand, the leading system-makers, with a notable exception of Hutchinson (1959), tended to place the ranunculid orders in the Magnoliidae and to consider them as descending from the magnolialean plexus rather than as a stem group (e.g. Takhtajan, 1966; Cronquist, 1981). In the present author's opinion, the taxonomic distances between the ranunculids, magnoliids and hamamelids are artificially diminished by unwarranted assignments of some ranunculid taxa to magnoliids (in angiosperm systematics, such misassignments often create a false impression of morphological continuity, see above). The magnoliids have thus been turned into a collection of archaic forms, with any non-magnoliid group derivable from them either directly, or with a few intermediaries. Most fossil forms have fitted in the magnoliids in the above sense, thus strenghtening their interpretation as a stem group.

My studies of the Yorkshire material and the new caytonialean records in the Far East (Krassilov, 1977b, 1984a), as well as the discovery of the early Cretaceous ranunculids (Vakhrameev & Krassilov, 1979; Krassilov & al., 1983; Krassilov, 1984a) seem to support a phylogenetic relatedness of caytonialeans to ranunculids, with the following concurrent trends:

(1) The typical caytonialean leaves (*Sagenopteris*) are compound-palmate of four lanceolate reticulinerve leaflets that resemble simple leaves of the Paleozoic *Glossopteris* and are occasionally arranged in decussate pairs betraying their origin from a short shoot with a terminal leaf cluster, as in glossopterids. Even more caulomic are some satellite genera, such as *Chankanella* (Krassilov, 1967), with similar, but more numerous, apically clustered leaves or leaflets (Figs. 47, 48). Concurrent to phyllomization of the short shoots, there was a leaf blade fusion trend that led, through a series of intermediate forms, to the *Scoresbya*-type palmately lobed leaves with the primary veins asymmetrically laminaless on the outside (Harris, 1935). These peculiar leaf morphologies are traced to the mid-Cretaceous transitional compound to palmately lobed *Vitiphyllum-Cissites* group, with similar leaf morphotypes occurring in *Trollius, Aconitum, Anemone* and other extant ranunculacean genera. Such leaves were found attached to the shoot axis of *Caspiocarpus*, an early ranunculid genus (Vakhrameev & Krassilov, 1979), while the pinnate-ternate leaves *"Leguminosites" karatcheensis* with *Sagenopteris*-like lanceolate leaflets and an irregular oblique cladodromous venation came in association with *Hyrcantha*, a paeonid fructification (Krassilov & al., 1983).

(2) A caytonialean (*Sagenopteris*) leaf venation is typically reticulate with frequent anastomoses. In the satellite genera *Scoresbya, Imania, Chankanella* and *Sujfunophyllum*, the anastomoses are sparse or even lacking. This venation type can be traced through the mid-Cretaceous *Diplophyllum* (Knobloch, 1978) to the archaic extant ranunculids *Kingdonia* and *Circaeaster*.

(3) The caytonialean stomata are peculiar among the Mesozoic gymnosperm stomatotypes in the leveling of the guard cells with the unspecialized subsidiary cells, and the lack of stomatal ledges. Prof. T. Harris, generally critical of any links between caytonialeans and angiosperms, has admitted (Harris, 1951 and personal communication, 1975) that the caytonialean stomatal structures are definitely angiosperm-like. Such stomata are characteristic, among angiosperms, of the ranunculid complex including the Cretaceous *"Leguminosites" karatcheensis* (see above) and *Menispermites* (Upchurch, 1984; Krassilov & al., 1988).

(4) The dichopodial ("dendroid") branching of a caytonialean androclade *Caytonanthus* is analogous to the branching pattern of the "dendroid" androecial traces in *Paeonia*, an archaic genus linked to ranunculids by the mid-Cretaceous *Hyrcantha* (Krassilov & al., 1983). Such traces are characteristic of fasciculate androecia that might

have been derived from the *Caytonanthus*-type branching system, with individual stamens corresponding to the terminal branchlets of the latter.

(5) The caytonialean pollen grains, proto-bisaccate to monosaccate with laterally connected bladders (Krassilov, 1977b; Zavada & Crepet, 1986) may seem quite different from any apertural types occurring in ranunculids. However, a reduction trend involving several Mesozoic seed plant lineages might have led to a transformation of protosaccate structures into harmomegathy folds and, further, into transequatorial colpi (see above, under "Pollen").

(6) The caytonialean cupules are basically ascidiform, with a plicate exostome and a transcurrent fertile zone, thus corresponding to the typical ranunculid carpel (Rohweder, 1967). Broad sessile stigmas, as in *Paeonia* and its precursory mid-Cretaceous *Hyrcantha*, could have been derived from a morphologically similar exostome appendage ("lip") of *Caytonia,* in turn derivable from a reduced subtending bract (Fig. 16). Further developments in the direction of modern ranunculids are evidenced by the mid-Cretaceous *Caspiocarpus* having shortly beaked dorsiventricidal follicles, as in extant *Glaucidium*. An elongation of the stylar region might have been conceivably induced by a transition from the pollination drop mechanism to extraovular pollen germination that concurrently led to reduction of sacci as a pollination drop-related feature.

(7) Both in the caytonialeans and ranunculids, the ovules are anatropous, ranging from the basal to basiventral, bitegmic with thin integuments. There was a size reduction from about 1 mm in the Jurassic *Caytonia* to about 0.5 mm in the Cretaceous *Caspiocarpus*, but otherwise their ovules were quite similar, platyspermic, with a short micropyle formed by the outer integument that was two to several cells thick and thinly cutinized, while the inner integument was 2-3 cells thick at the chalaza, distally wedging out above the midlength of the ovule (Plates 28, 36). The nucelli were likewise similar, relatively massive, shortly beaked, with a median line reflecting the raphe (Krassilov, 1984a). On the other hand, the ovule morphologies found in *Caspiocarpus* resemble, among extant ranunculids, *Circaeaster* in the extremely thin integuments, and *Aquilegia, Hydrastis*, etc. in the outer integument forming the micropyle.

Other proangiosperm contributions

As suggested above, the strobiloid flowers with interfloral phyllomes, as in the primitive hamamelids (*Liquidambar*) and magnoliids (Degeneriaceae, Himantandraceae, Eupomatiaceae, and Austrobaileyaceae), are derivable from the short shoots bearing terminal clusters of androclades and/or gynoclades, subtended by, or intermingled with, scaly cataphylls. Such shoots occur in *Irania*, as well as in the czekanowskialean and dirhopalostachyacean proangiosperms (Plate 4; Fig. 7). Furthermore, a prototype of the winteracean pseudomonomerous follicles with sutural stigmatic crests (Vink, 1978) can be found in the bivalved cupules of *Leptostrobus stigmatoideus*, an advanced species of czekanowskialean gynoclades, in which the sutural papillate crests intrude the locule as the "inner stigmas" that onlap the sutural ovules (Krassilov, 1969, 1972a). At the same time, the lateral ovuliphores of *Ixostrobus*, a czekanowskialean androclade, are comparable with certain stamen morphotypes of the extant magnoliids, such as *Austrobaileya*, on account of the adaxial anthers and the scaly sterile apophyses (Krassilov, 1969, 1972a). In addition, an androclade origin of the stamens in magnoliids (in the restricted sense, discussed above) was suggested on the basis of their vascular anatomy.

Pairing of cupules is characteristic of the Mesozoic Dirhopalostachyaceae (Krassilov, 1975a). Their gynoclades are strikingly similar to the Late Cretaceous and Paleocene hamamelid fructifications *Trochodendrocarpus* (Krassilov, 1976 and elsewhere) in having a loosely racemose structure, as well as in the external morphology of the paired cupules and follicles, both showing the ventral sutural ridges, reflexed beaks and an oblique lateral striation (Plates 4, 44). However, a wide chronological discontinuity makes direct phylogenetic links between these two forms unlikely.

Such similarities may indicate a certain degree of phylogenetic relatedness or, at least a genetic contribution of the abovementioned proangiosperm groups to the respective angiosperm stalk lines. Convergence is a poor explanation in the case of characters of an obscure adaptive meaning. Alternatively, such similarities might result from a spurious donation of genetic material by horizontal transfers from one proangiosperm group to another within a proangiosperm assemblage that formed a communal source of angiosperm origins.

Concluding remarks

Angiosperm origins are here traced from the Devonian, passing through a series of gymnosperm grades that show evolutionary trends and patterns more fully expressed at the angiosperm level. Several gymnosperm lineages had acquired certain angiosperm features, thus collectively giving rise to the proangiosperm grade.

Due to the discovery of proangiosperms, the advent of angiosperms may seem less enigmatic than it appeared in the Darwinian time. This does not mean that we know a direct ancestor(s) or the precise place and time, but these may not be strictly definable at all. What we do know is a common source of angiosperm characters, a morphological

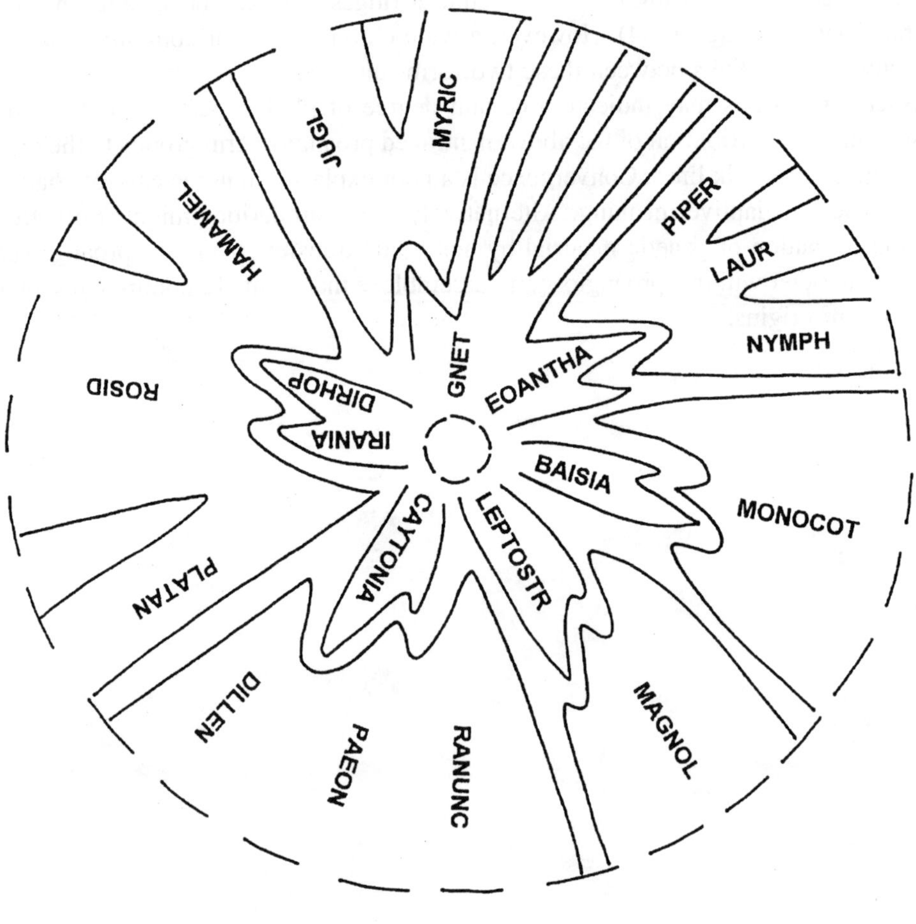

Fig. 59. Suggested phylogenetic relations of the major proangiosperm and angiosperm groups.

pool containing these characters in their incipient form. A few morphological trends are recognized as linking the proangiosperm and angiosperm levels. Gaps remain, but at least we know where to look for them.

In deciphering the angiosperm to proangiosperm homologies, it has been necessary to clarify our concepts of the flower, stamen, carpel, double fertilization and other supposedly typical, but as yet vaguely defined, angiosperm traits. Their origins are shown to have been due primarily to developmental acceleration and functional integration of the respective prototypic structures found in proangiosperms. These processes have determined the concurrent trends in the evolution of separate morphological domains. The above-suggested phylogenetic relationships (Fig.59) are based on such concurrent trends, with some recent finds substantiating the gneto-bennettite to graminid and the caytonialean to ranunculid affinities, the other proangiosperm groups considered as possible contributors of genetic material.

Notably, angiosperms appeared in the assemblages containing various proangiosperms. Hence, their origins seem to have been the product of a communal breakthrough rather than a single event. Evolution of the cradle communities may shed some light on the angiospermization process occurring over the whole their ranges, with multiple diversification centres. Angiospermization involved a number of lineages trending in the same direction. Whether the angiospermy was achieved by a single or several of the parallel lineages, is a matter of secondary importance, although both the present-day angiosperm morphology and the early angiosperm diversity seem to suggest more than one successful contender. Angiosperm validity as a class may depend not so on a solution of this dilemma as on their structural integrity that is based on the shared evolutionary trends and the ensuing structures of a major adaptive significance. Characters appearing in a single lineage are often lost in the course of subsequent evolution, but parallelism ensures their consistent development in a large group of organisms. The angiosperm story is thus of some general interest for understanding the nature and the origins of higher taxa.

Acknowledgements

I am grateful to my former collaborates in the Paleobotanical Laboratory of the Institute of Biology and Pedology, Vladivostok and to my present collaborates in the Paleontological Institute, Moscow. My special thanks are due to Eugenia Bugdaeva and Natalia Maslova who contributed some valuable material and information, to my wife Sophia for the encouragement and advice, to Ludmila Volkova for technical help and to photographers Claudia Novikova and Alexander Masin. I also acknowledge the linguistic corrections by the editor Dr. Sergei Golovatch. This work was financially supported by the Russian Foundation of Basic Research, grant No. 95-04-11863 and the International Science Foundation, grant No. N6V000. The publication costs have been met by Pensoft *Publishers* and the H. Rausing Foundation.

Plate 1

Cycandra profusa Krassil. & Delle, a giant cycadalean pollen cone from the Jurassic of Georgia, Caucasus, showing medullosalean synangial structures: (1) pollen cone (the numbers refer to detached sporophylls on the slab), x0.7, (2) sporangium, the distal part of the endothecial membrane with a beak filled with pollen grains, x240, (3) pollen grains, x500, (4) sectioned pollen grain showing sulcus (arrow), TEM, x5 000.

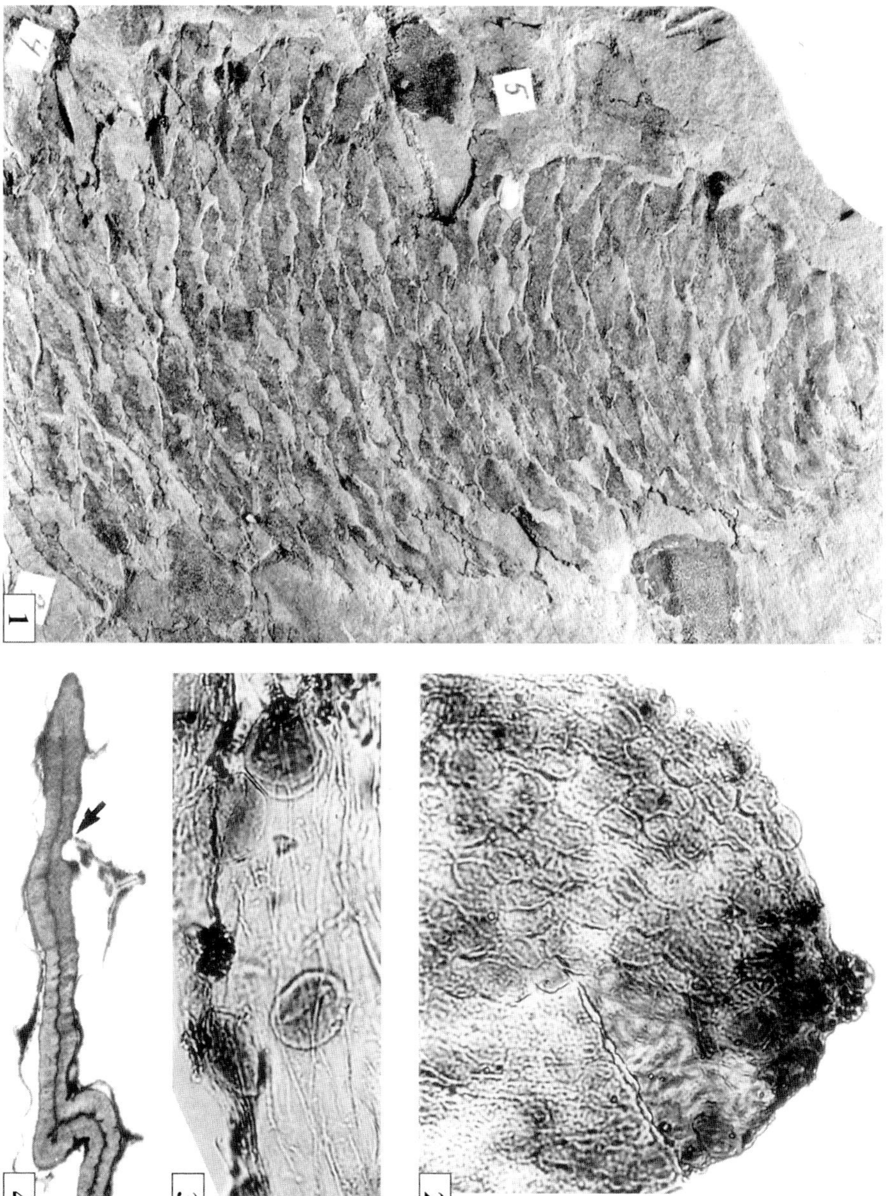

Plate 2

Cycandra profusa Krassil.& Delle, a giant cycadalean pollen cone from the Jurassic of Georgia, Caucasus, showing medullosalean synangial structures: (1) heads of sporangia as seen on the rock surface, SEM, x70, (2) microsporophyll with a sterile fringe and a central synangium of numerous sporangia in concentric rings, x7, (3) files of vertical, marginally adnate sporangia, the synangial cuticle with meshes of sporangial heads seen at the left margin, SEM, x90.

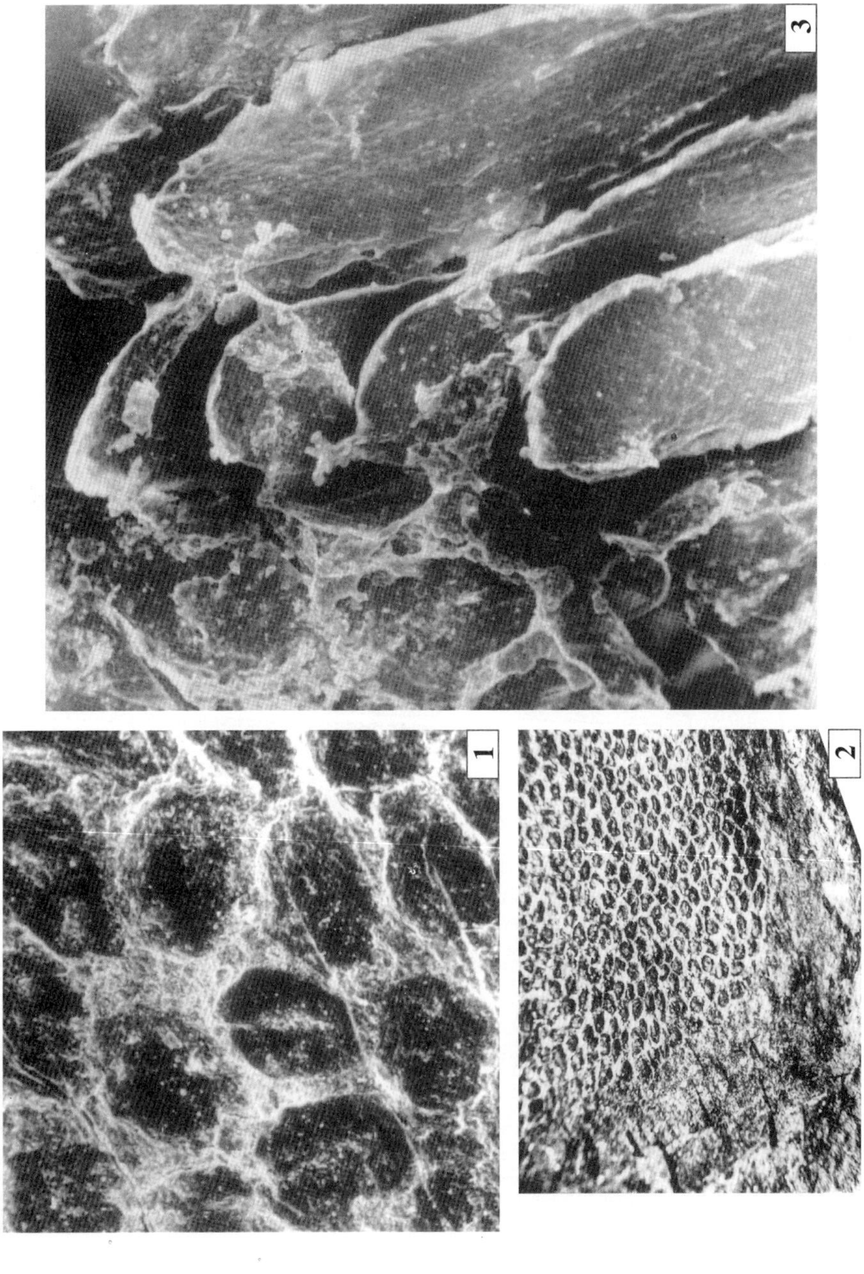

Plate 3

Semionogyna bracteata Krassil. & Bugd., a proto-cycadean ovulate spike from the Early Cretaceous of Transbaikalia (Krassilov & Bugdaeva, 1988c): (1) spike with immature ovules, distal bracts empty, x1.5, (2) spike with ripe ovules, x1.5, (3, 4) ovules showing the outlines of a nucellus, x10, (5) bract supporting a juvenile ovule, x10, (6) pollen organ of presumably the same plant, x10.

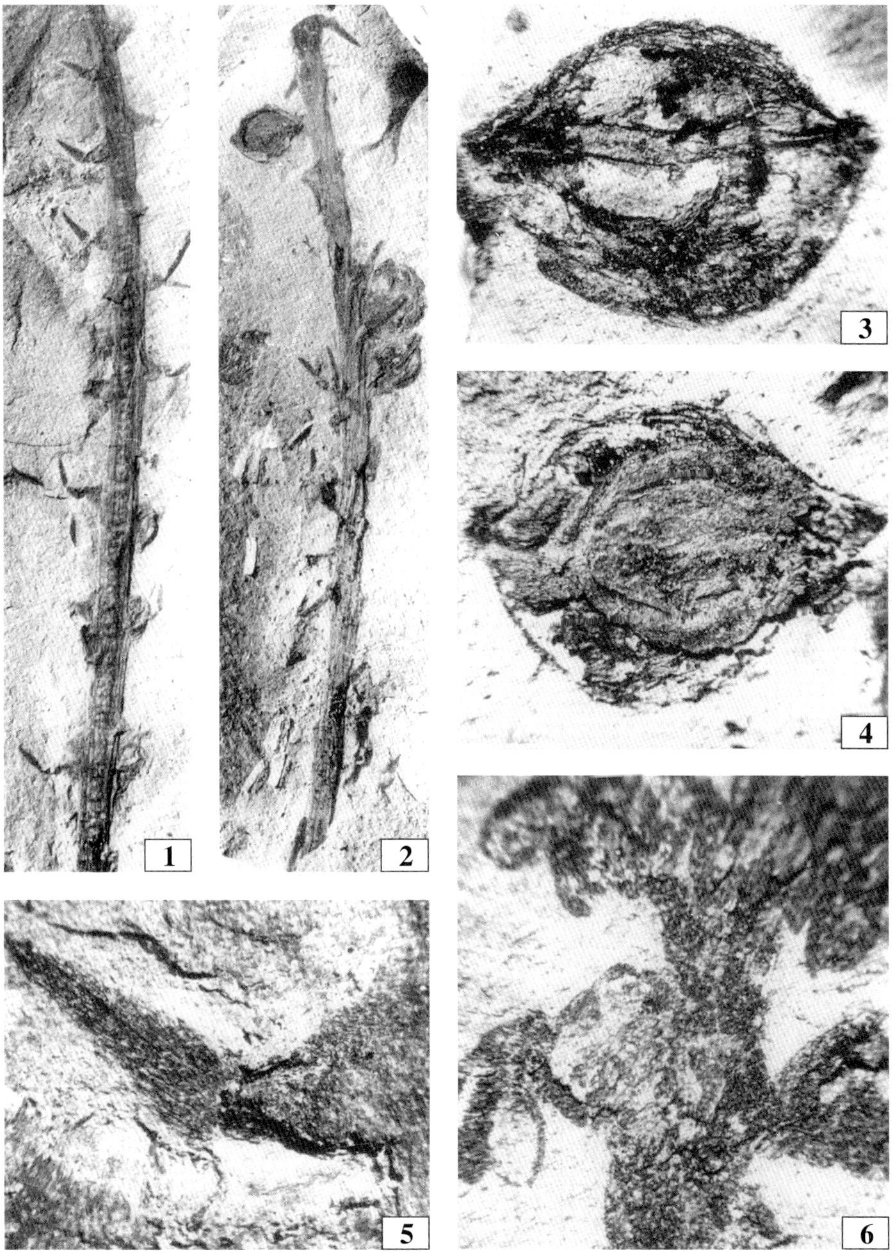

Plate 4

Dirhopalostachys rostrata Krassil., a proangiosperm from the Late Jurassic of Bureya Basin, Far East, Russia (Krassilov, 1975): (1) gynoclade with lateral branches bearing paired follicle-like cupules, x1, (2) cast of a cupule showing suture, x10, (3. 4) cupule, with the ventral wall removed in (3) to show a solitary anatropous ovule, x10, (5, 6) distal parts of dehiscent cupules, with beaks slightly gaping, x3 and 10.

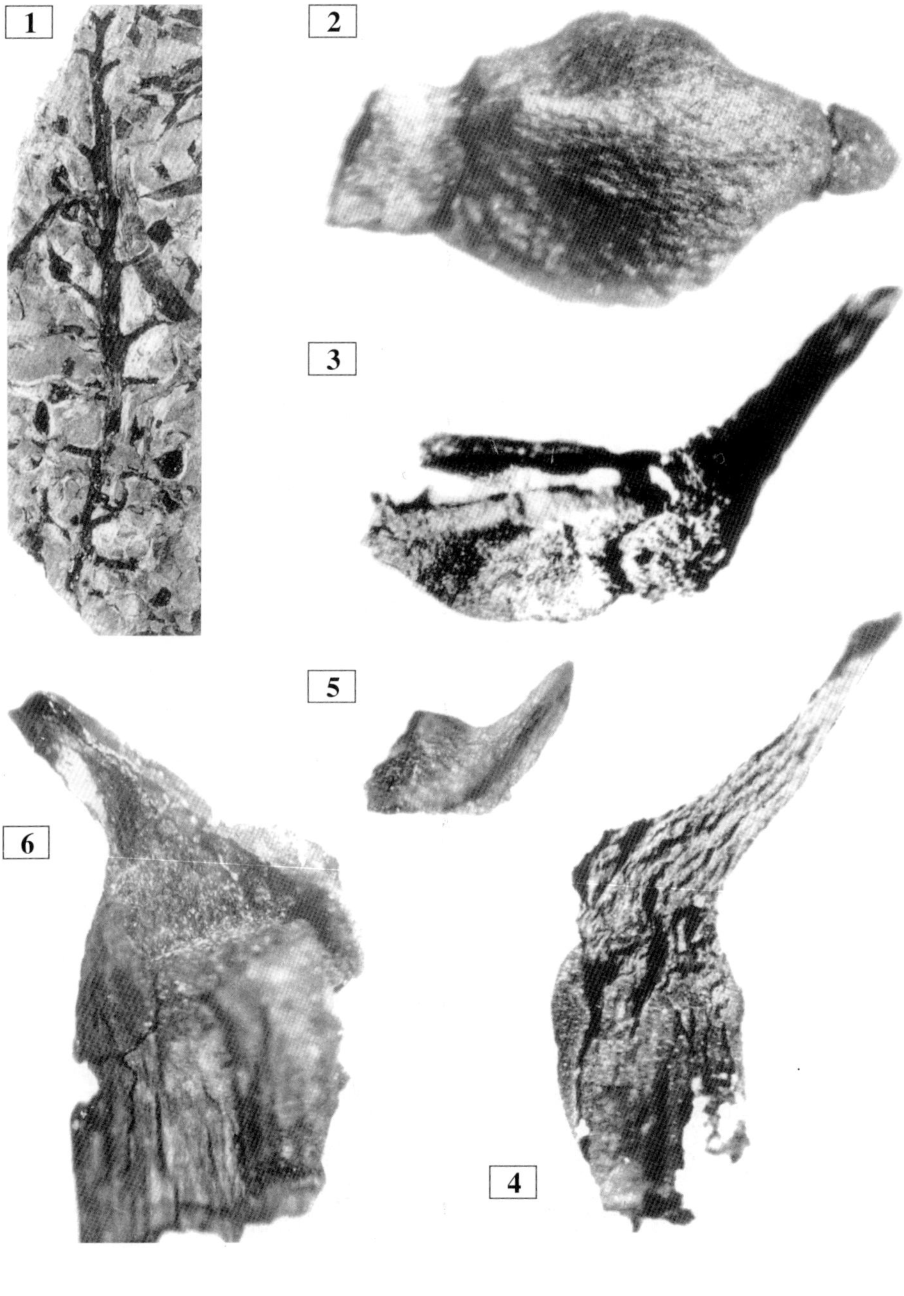

Plate 5

Baisia hirsuta Krassil., a proangiospermous plant of bennettitalean affinities from the Baisian Assemblages, Early Cretaceous of Transbaikalia (Krassilov & Bugdaeva, 1982): (1) dispersed cupule with a persistent receptacle bearing a tuft of long bristles; apical corona is distinctly marked off, x12, (2, 3) opened cupules showing seeds, x12, (4) receptacle, proximal aspect with short bracts, x140.

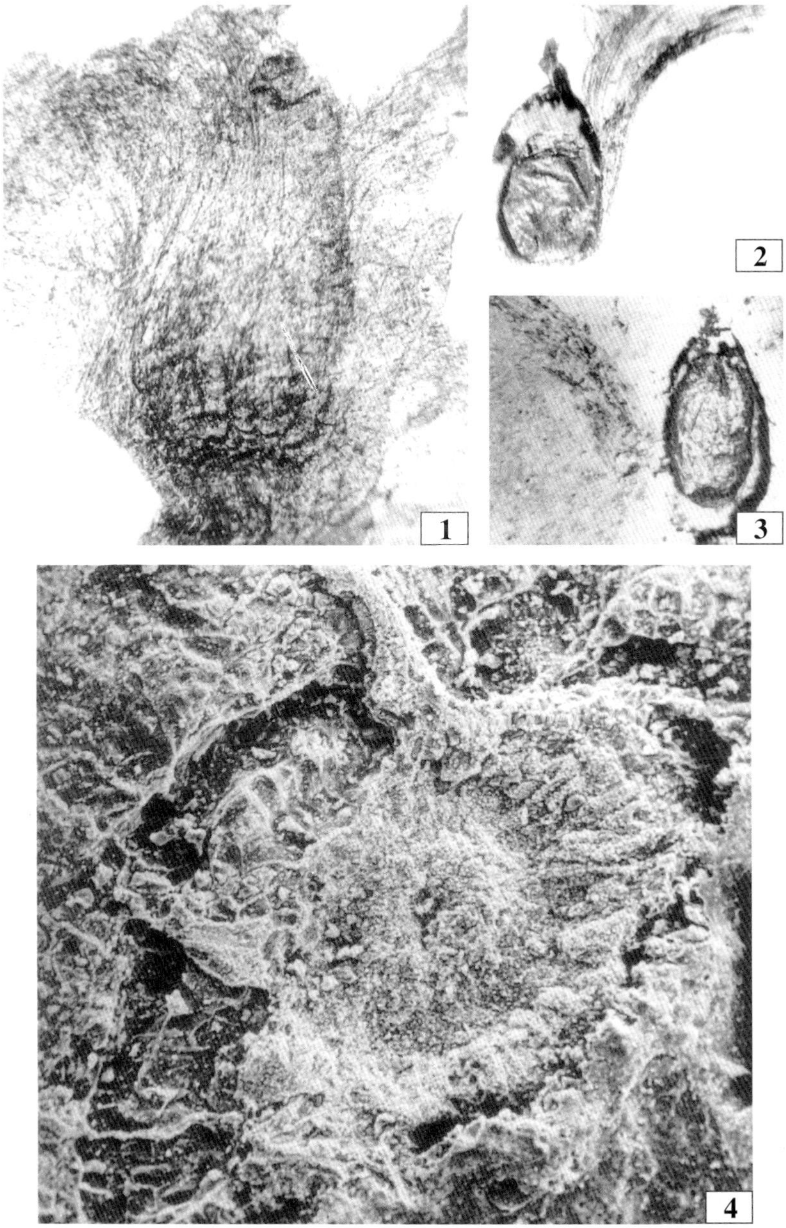

Plate 6

Baisia hirsuta Krassil., a proangiospermous plant of bennettitalean affinities from the Baisian Assemblages, Early Cretaceous of Transbaikalia (Krassilov & Bugdaeva, 1982): (1) dispersed cupule, x12, (2) detached receptacle, x12, (3) ovule removed from the cupule, x15, (4,5) pollen chamber with pollen, SEM, x400, and light micrograph, x300.

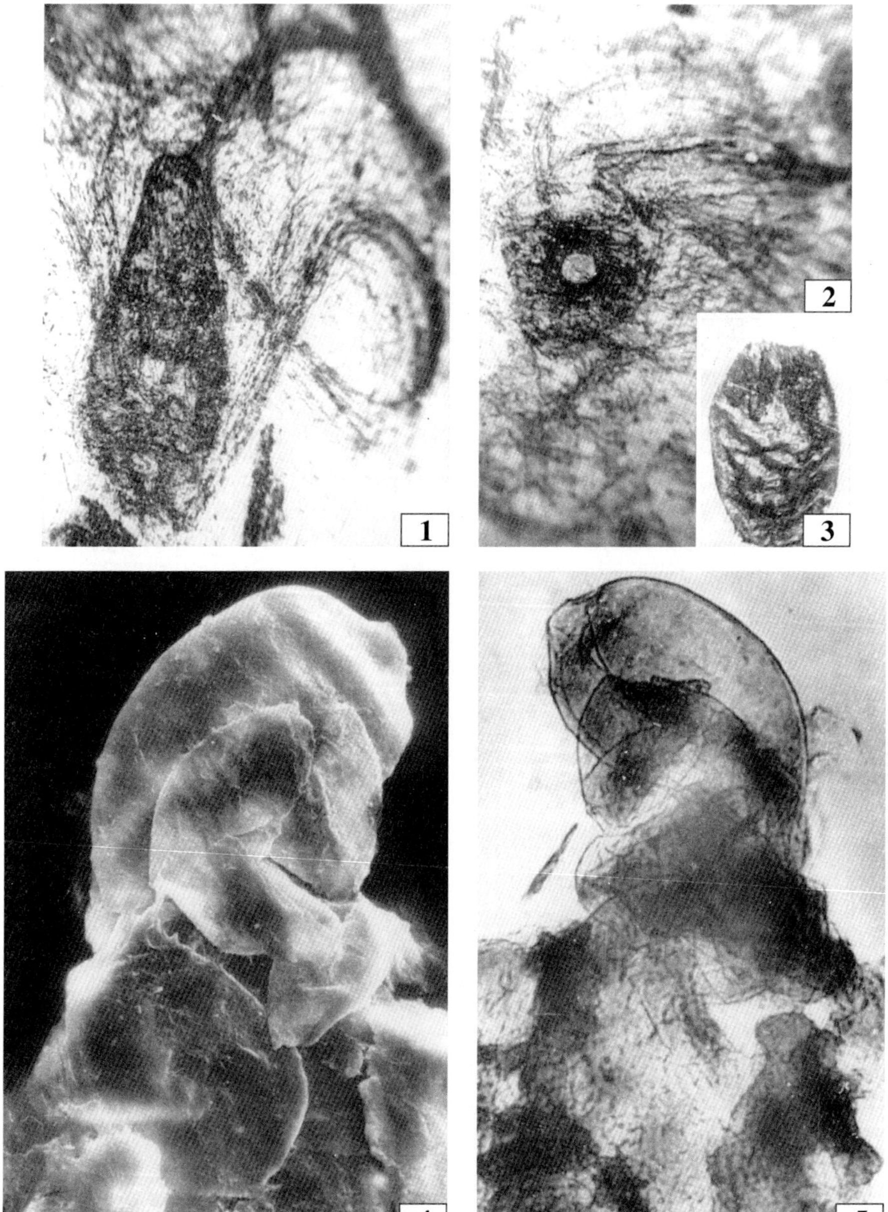

Plate 7

Baisia hirsuta Krassil., a proangiospermous plant of bennettitalean affinities from the Baisian Assemblages, Early Cretaceous of Transbaikalia (Krassilov & Bugdaeva, 1982): (1) pollen grain from the pollen chamber, x600, (2) infrastructure of the pollen grain seen in the areas where the tectum was erased by the pollen chamber exudates, x15 000.

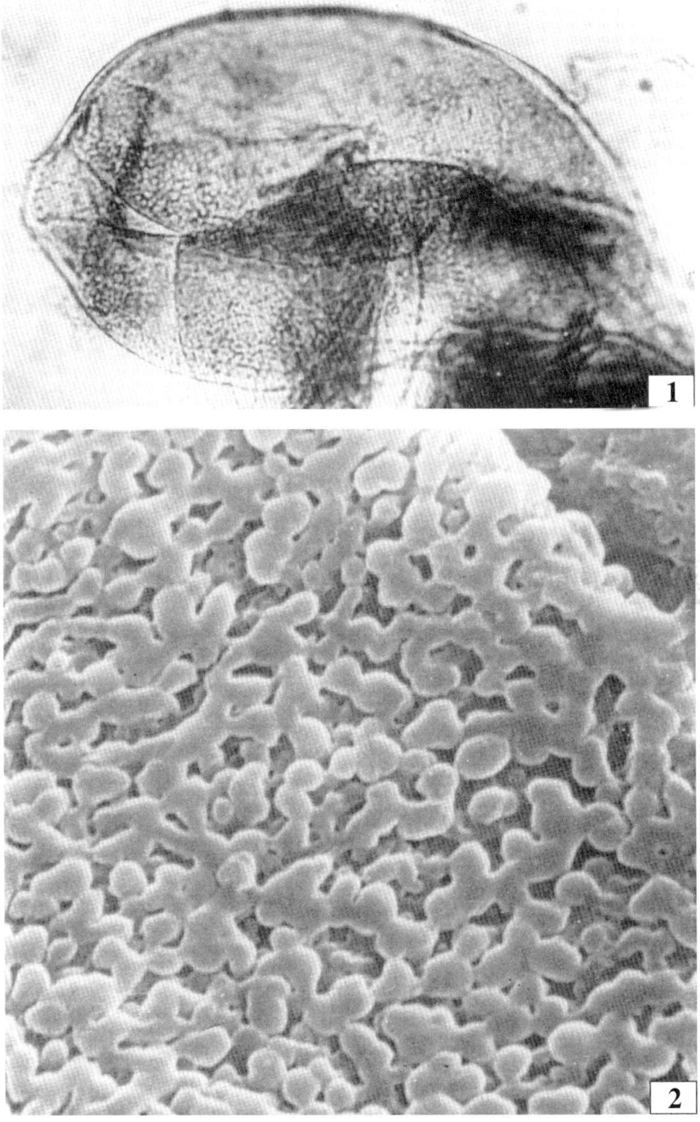

Plate 8

Dinophyton spinosum Ash, a protognetalean plant from the Late Triassic of Arizona, U.S.A. (Krassilov & Ash, 1988): (1) partly cleared ovulate cupule with a four-lobed propeller-like wing formed of decussate bracts, x15, (2) cupule opened to show the ovule, x12, (3) cupule showing pubescent lobes, x15, (4) nucellus, trigonal at base, x17.

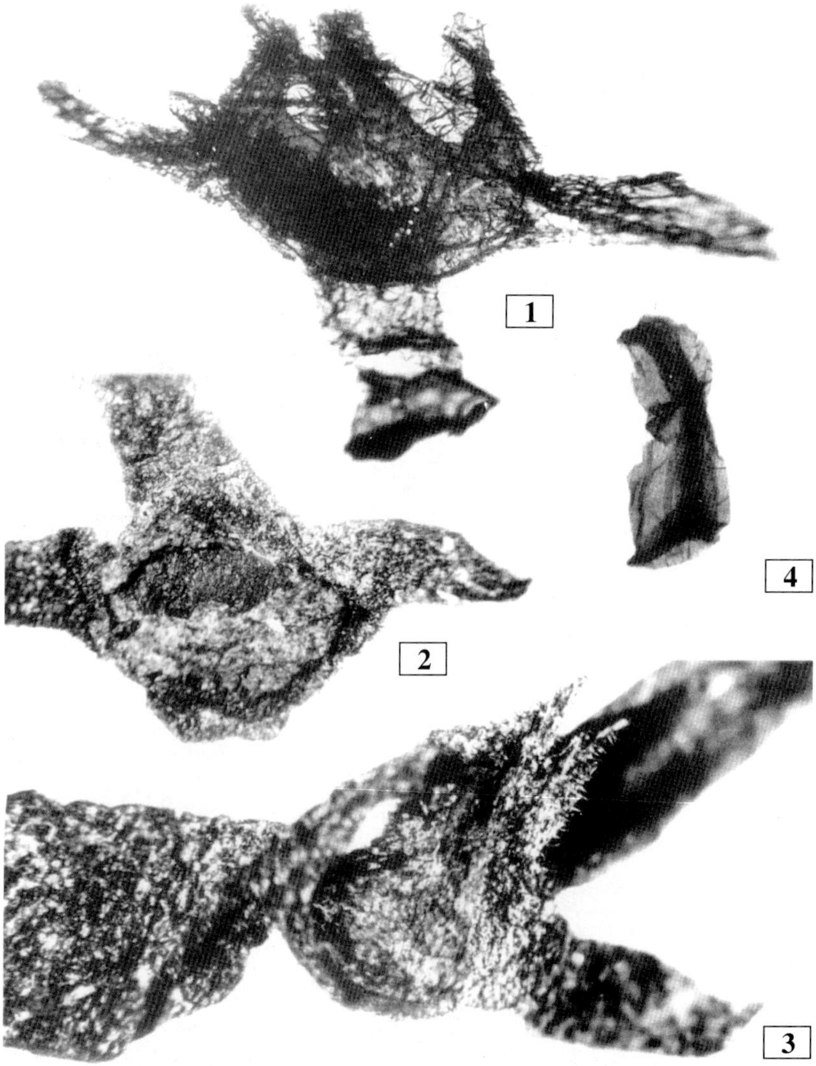

Plate 9

Dinophyton spinosum Ash, a protognetalean plant from the Late Triassic of Arizona, U.S.A. (Krassilov & Ash, 1988): (1) part of pollen cupule with bract, SEM, x70, (2, 3) laminar sporangiophores with a basally attached sporangium, SEM, x110, (4) pollen grain, SEM, x1 000.

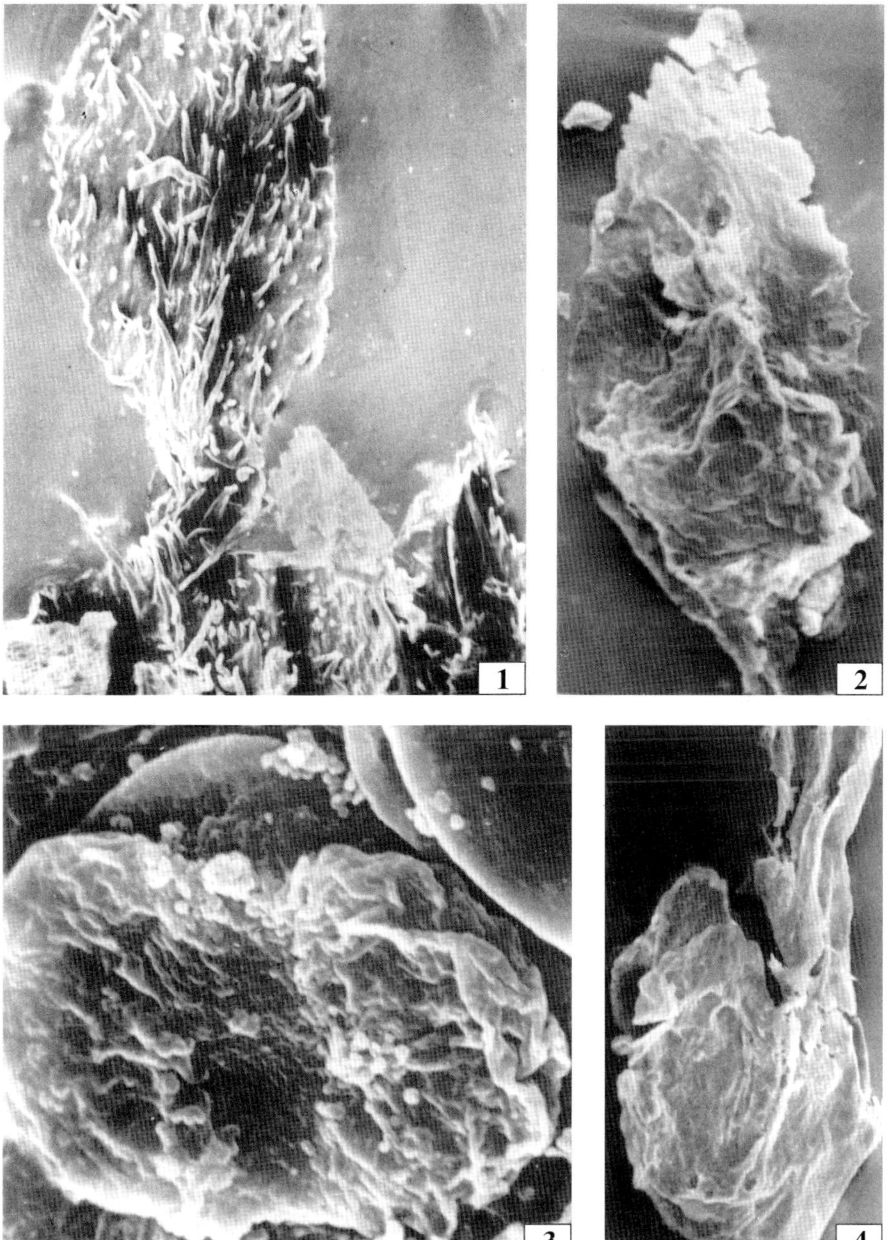

Plate 10

Hirmerella sp., the ovuliferous organ of a gnetalean proangiosperm from the the Early Jurassic of Poland (Krassilov, 1973c): (1, 2) samaras with leaf-like wing lobes, x5, (3) wing cuticle with stomata, SEM, x600, (4) double cuticle showing intersecting cell files of the locule and the seed coat, x170, (5) two nucelli inside the samara shown in (1), x15, (6) *Classopollis* pollen grains on the papillate cuticle of the same samara, SEM, x500.

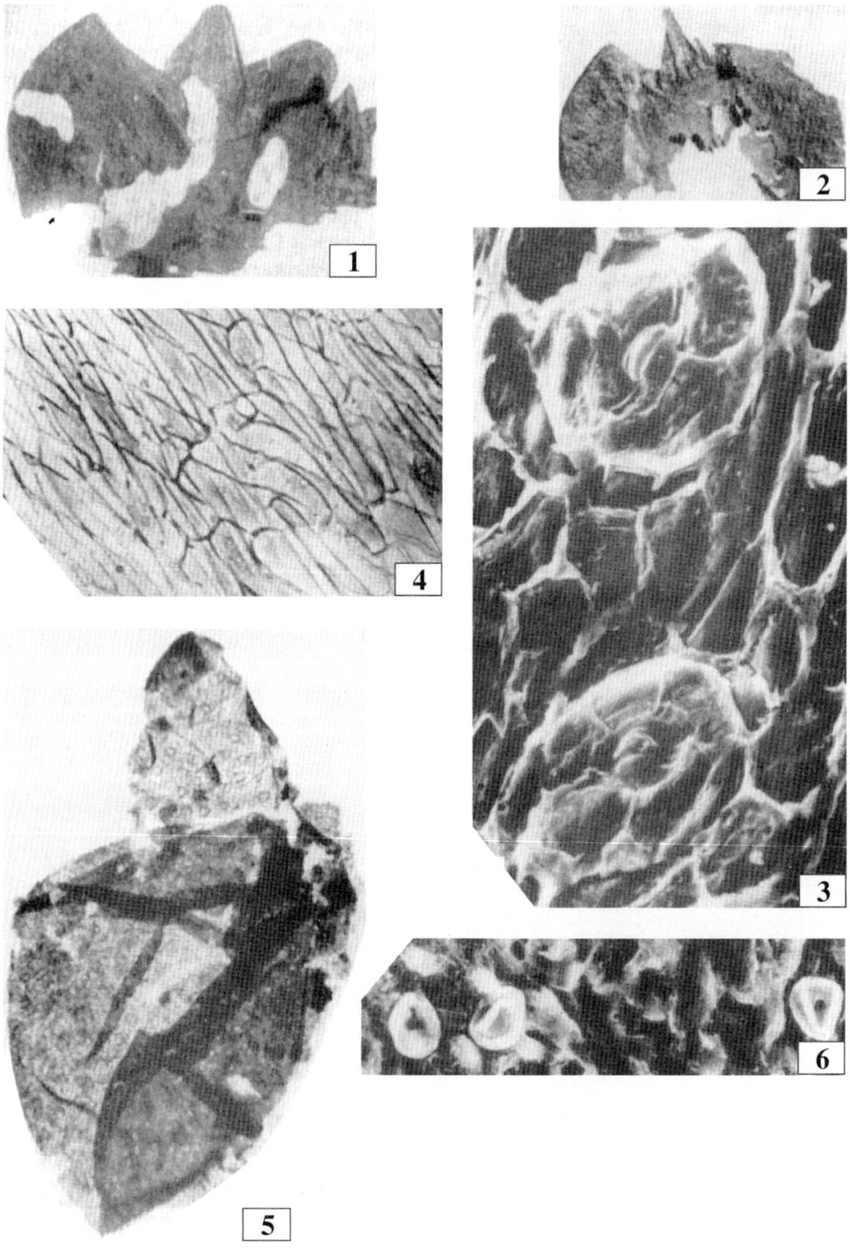

Plate 11

A gnetalean plant from the Middle Jurassic of Ust-Baley, East Siberia (Krassilov & Bugdaeva, 1987): (1, 2) *Angarolepis odorata* Krassil. & Bugd., cone bracts showing an attachment scar and resin bodies, x4, (3-6) *Heerala antiqua* (Heer) Krassil., samaras showing a median ridge and resin bodies, x4 and 10.

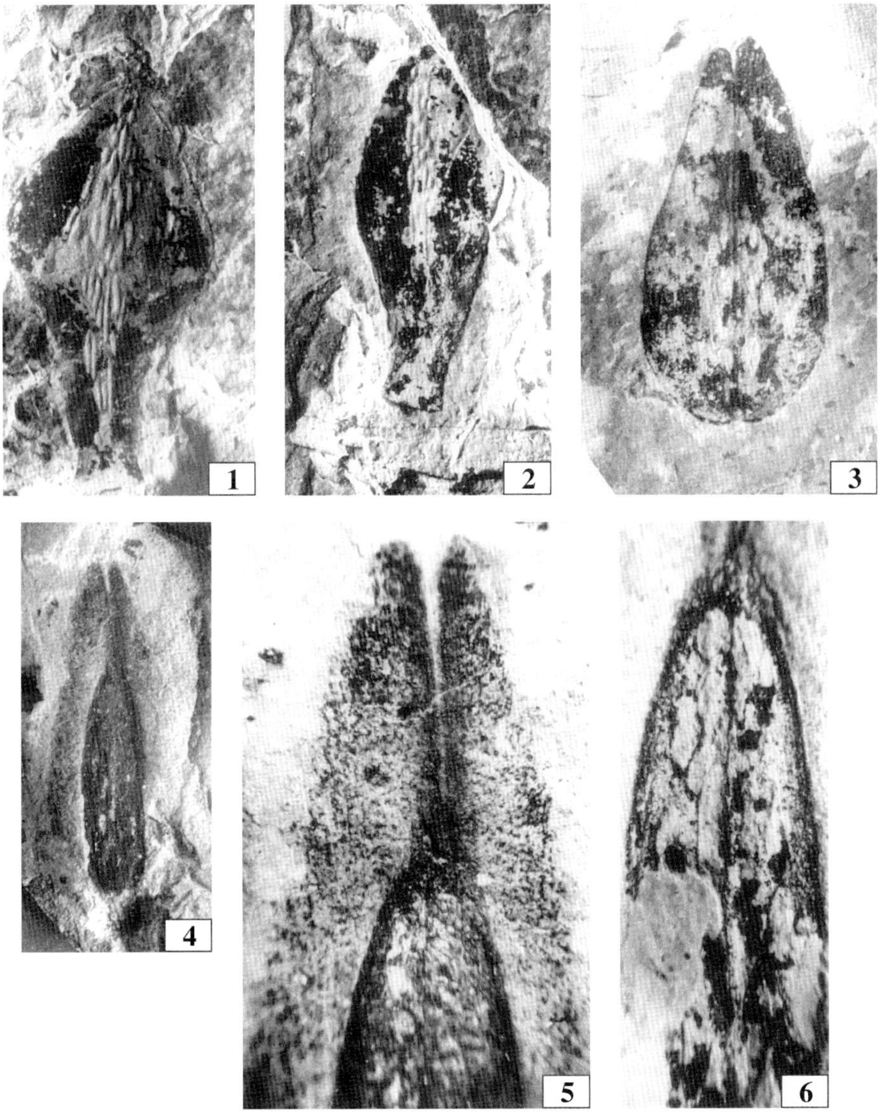

Plate 12

Pollen organs of presumably the same gnetalean plant that produced *Angarolepis-Heerala* bracts and samaras from the Middle Jurassic of Ust-Baley, East Siberia (Krassilov & Bugdaeva, 1987): (1) *Aegianthus sibiricus* (Heer) Krassil., pollen cone, x1.5, (2) sporophyll heads with lateral facets, x3, (3) sporangia, x10, (4) pollen grain, SEM, x1 800.

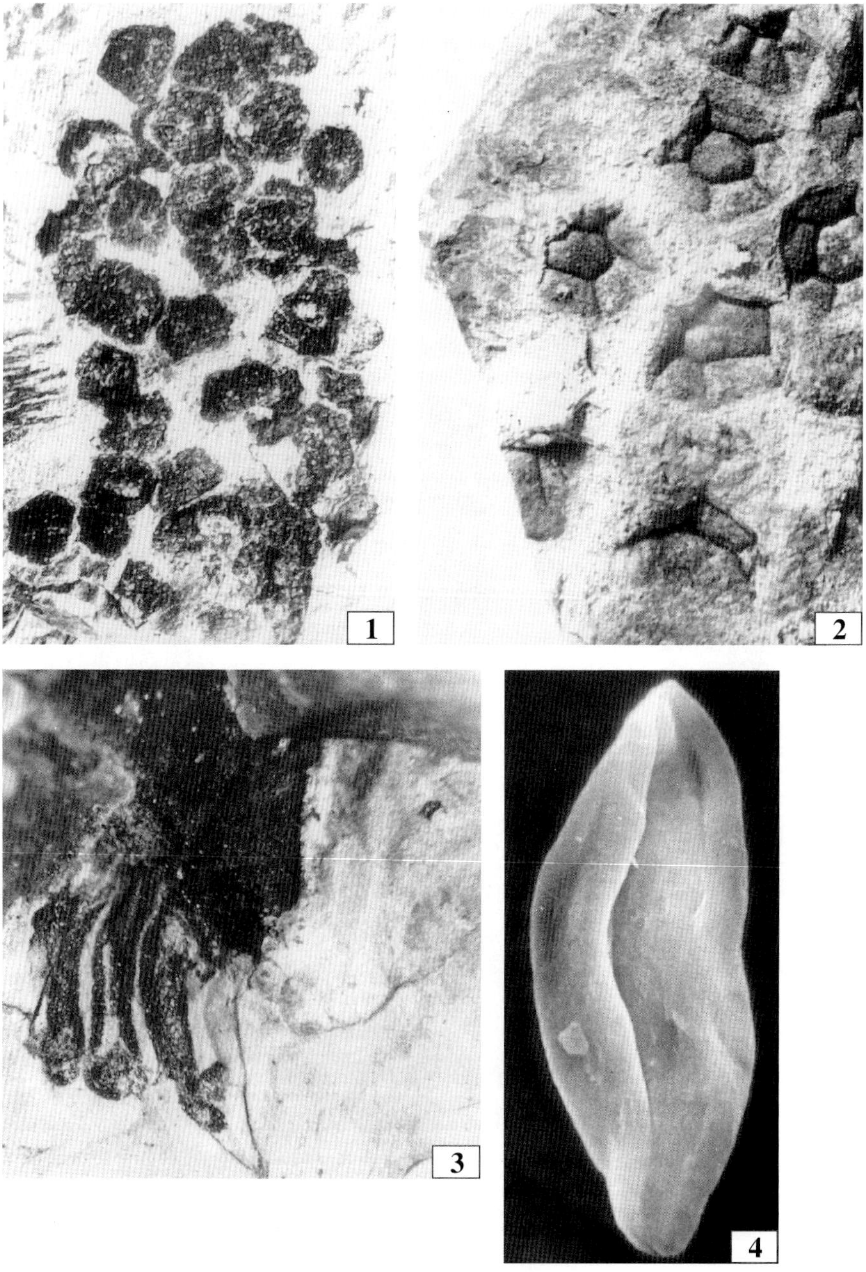

Plate 13

Leongathia, ephedroid plant from the Koonwarra Fossil Bed Assemblage, Early Cretaceous of Victoria, Australia (Krassilov & al., 1997): shoots with decussate leaf whorls; note that terminal internodes are much shortened and sometimes actually condensed as in (1), x12 and 14.

Plate 14

Leongathia, ephedroid plant of the Koonwarra Fossil Bed Assemblage, Early Cretaceous of Victoria, Australia (Krassilov & al., 1997), SEM: shoots differently preserved as subcrustation (mineral film infiltrated under the cuticle) or incrustation (surface mineral deposits) and, respectively, showing either the pattern of vascular bundles (1, 3), x30 and 100, or a ribbed surface with transverse striation reflecting a sclerenchymous network of the cortex (2), x38, and stomata (4), x1 800, arrow in (1, 3) marking a small vascular bundle ascending between the larger bundles at a deeper level in the vascular system.

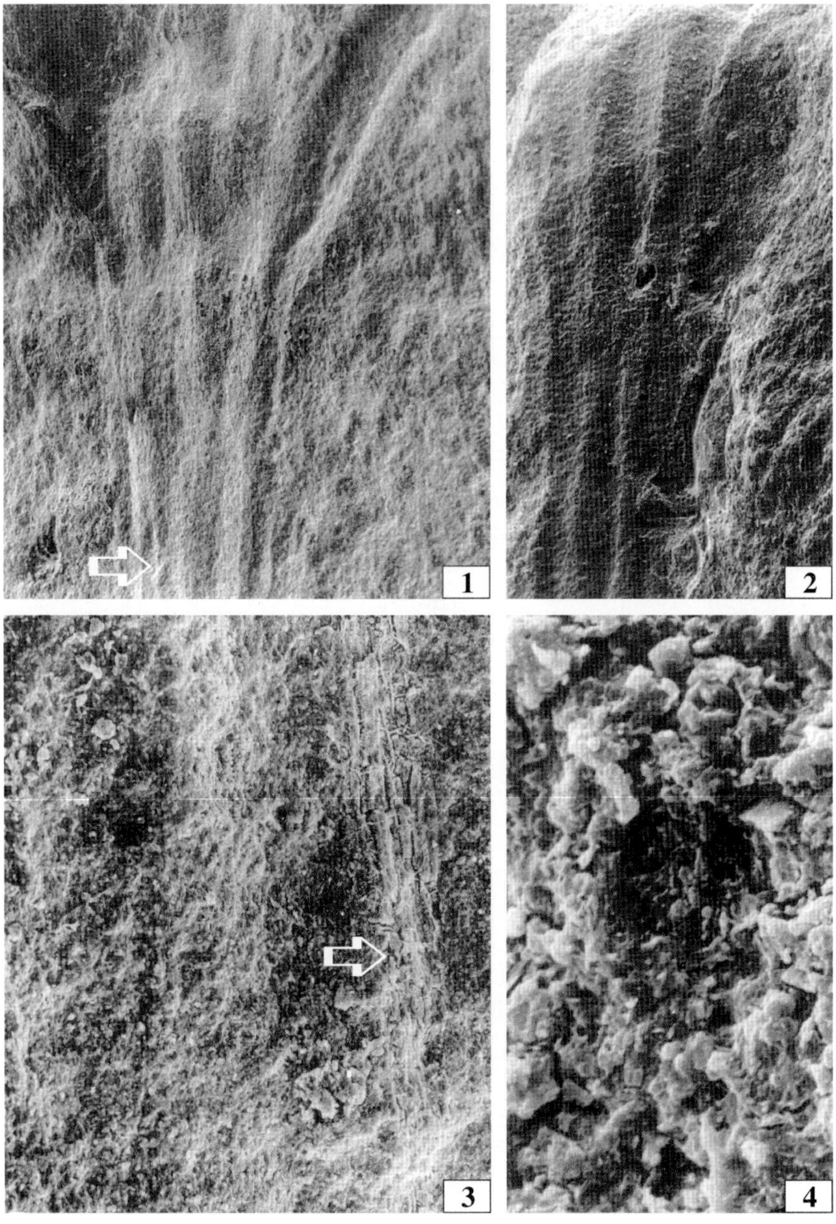

Plate 15

Leongathia, ephedroid plant of the Koonwarra Fossil Bed Assemblage, Early Cretaceous of Victoria, Australia (Krassilov & al., 1997), SEM: (1) interrib stomatal zone, x500, (2-4) tracheary elements with helical and annular thickenings and with pits, in (4) showing perforation plate at the lower end (arrow); also discernible are solid elements of a cortical sclerenchymous network overlapping the vascular tissue, x2 000 (2), 2 500 (4) and 5 000 (3).

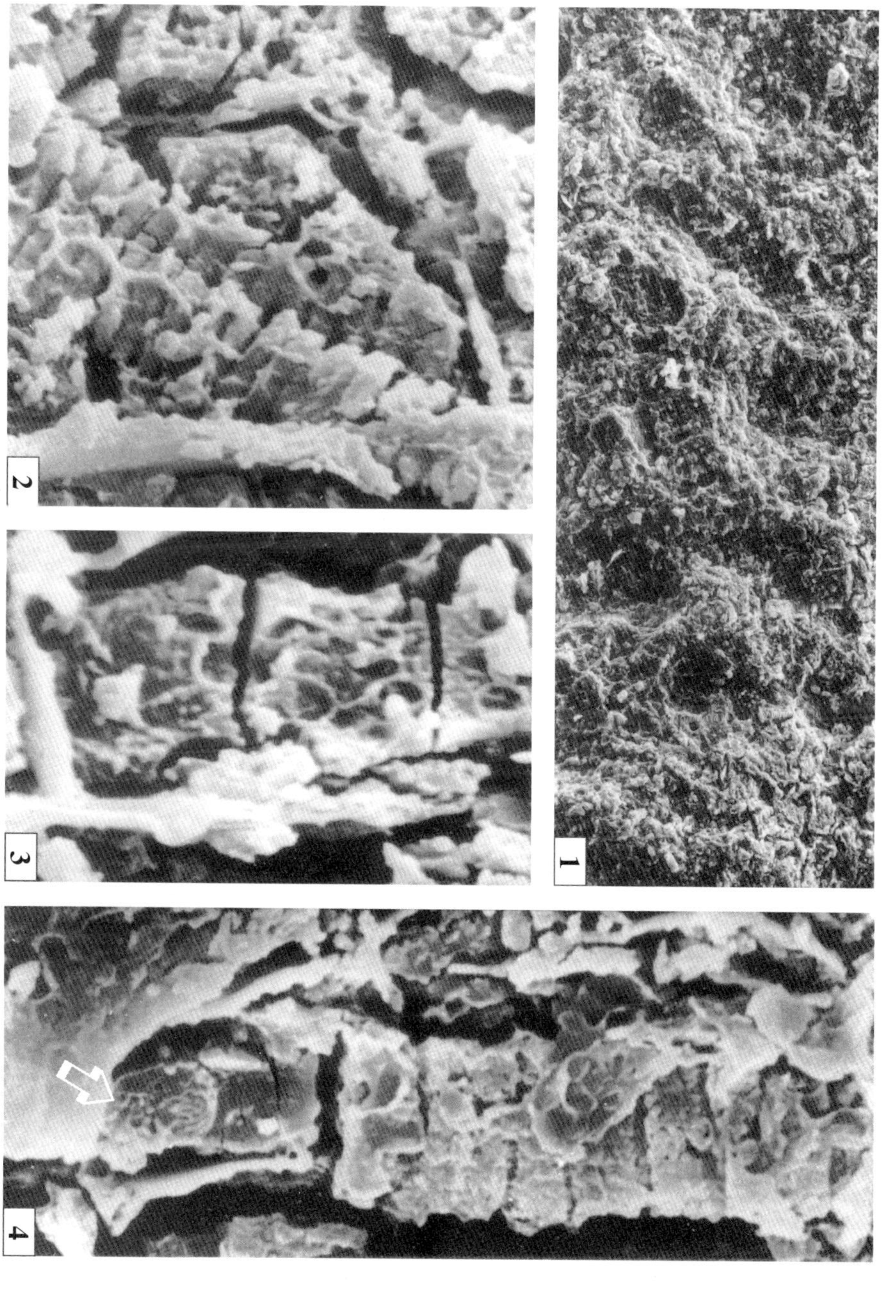

Plate 16

Baisianthus, a gnetalean androclade of the Baisian Assemblage, Early Cretaceous of Transbaikalia (Krassilov & Bugdaeva, 1997): branching shoots bearing whorls of sporangiophores subtended by linear bracts; lateral branches are decussate, arising at bract axils (1), and the fertile whorls arise at some distance above the branch node, but are more crowded distally (2), x10.

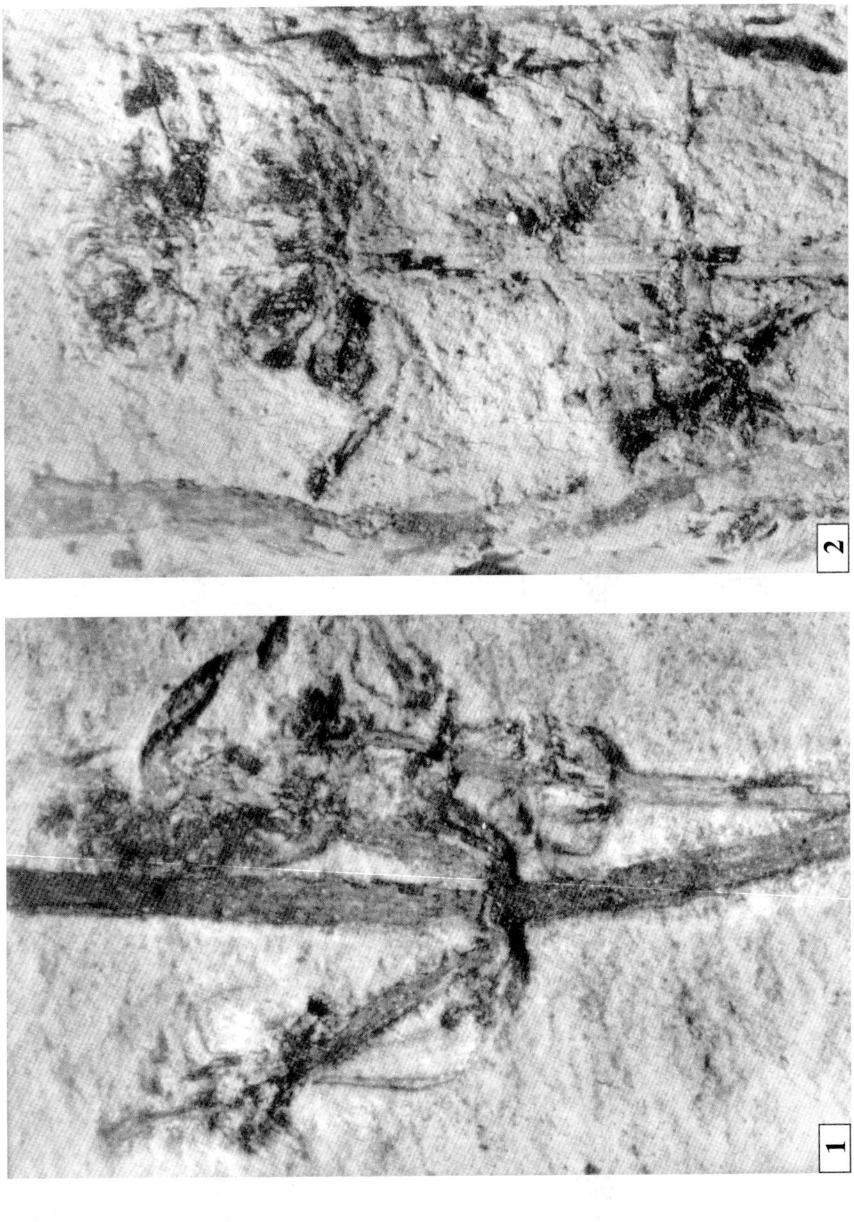

Plate 17

Baisianthus, a gnetalean androclade of the Baisian Assemblage, Early Cretaceous of Transbaikalia (Krassilov & Bugdaeva, 1997): tracheary elements of vascular bundles of the axes, note septate elements with porous perforations and helical thickenings transformed into irregular scalariform perforations of what appears as a long perforation plate (arrow), SEM; x900 and 2 100.

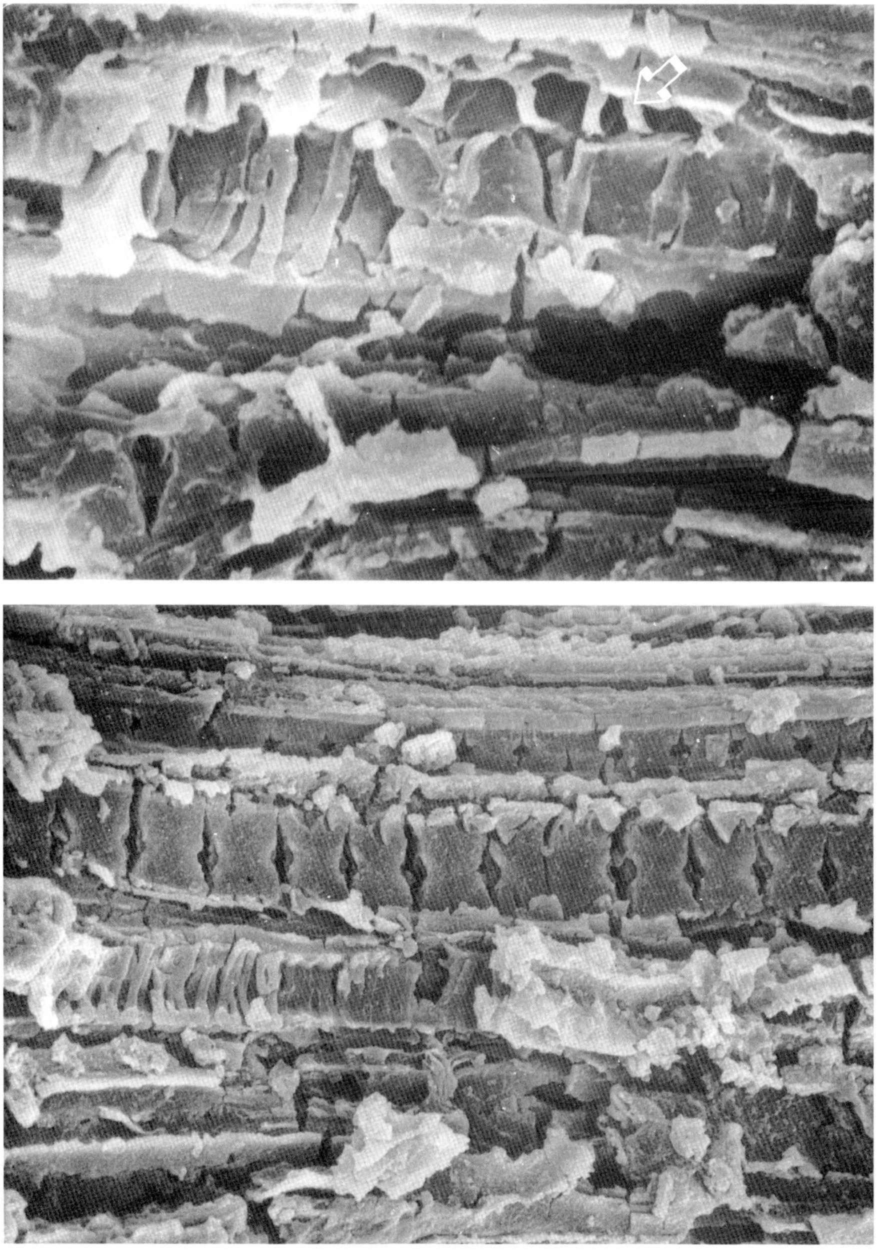

Plate 18

Baisianthus, a gnetalean androclade of the Baisian Assemblage, Early Cretaceous of Transbaikalia (Krassilov & Bugdaeva, 1997): (1) detached sporangiophore bearing synangia on lateral branches and terminal on the main axis, x40, (2) a pair of cupules containing sporangiophores, one of them discernible in the upper cupule, x25, (3) sporangiophore branch with sporangia, SEM, x130.

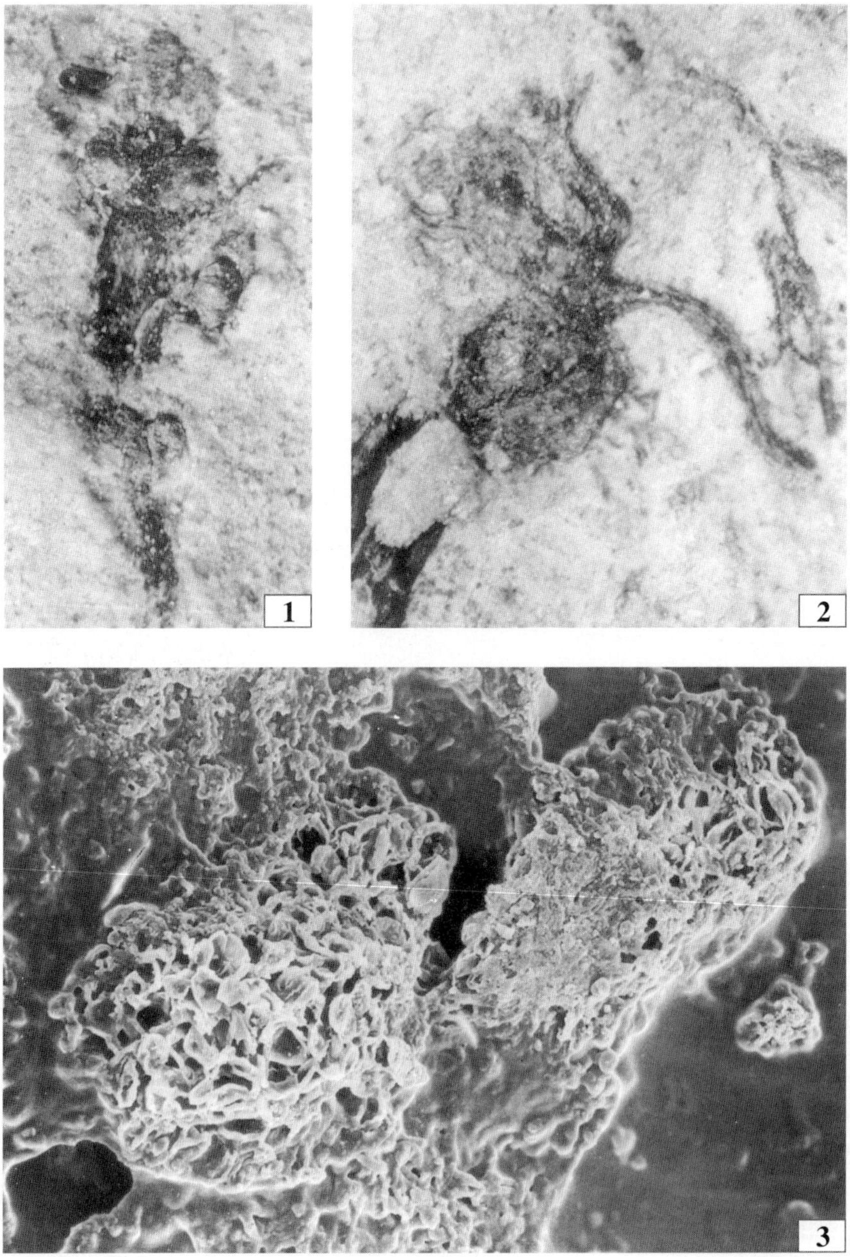

Plate 19

Baisianthus, a gnetalean androclade of the Baisian Assemblage, Early Cretaceous of Transbaikalia (Krassilov & Bugdaeva, 1997), SEM: (1) synangium of three sporangia, x210, (2) pollen mass of a sporangium, note intact tetrads at the left marin (arrow), x450.

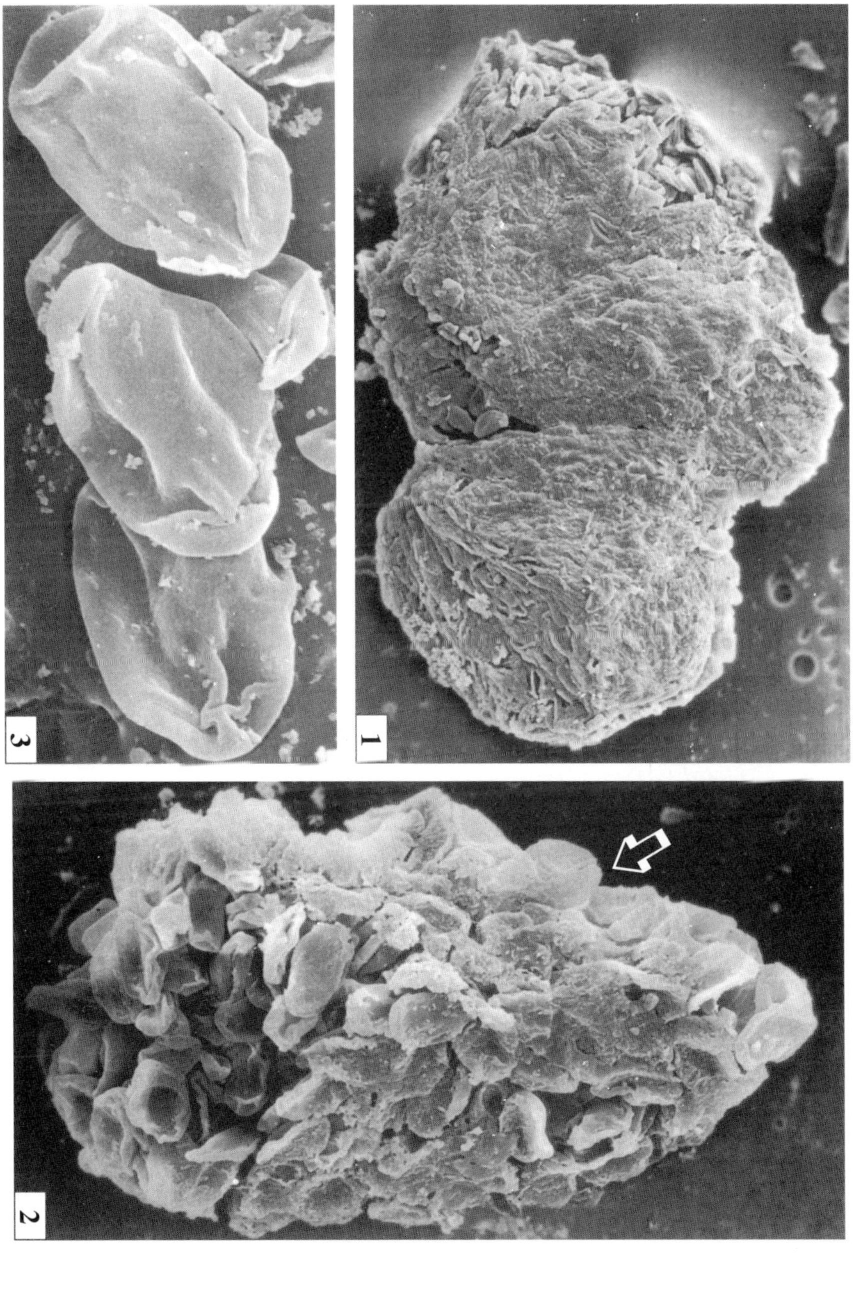

Plate 20

Eoantha zherikhinii Krassil., a proangiospermous gnetalean plant of the Baisian Assemblage, Early Cretaceous of Transbaikalia (Krassilov, 1986): (1, 2) opened preflowers with radially spread perianth bracts and gynoecial lobes, each of the latter bearing a solitary orthotropuos ovule, x3 and 8, (3, 4) closed preflower showing obliquely spreading perianth bracts, one of the gynoecial lobes with a characteristic transverse striation, and a toral extension of the floral apex with a tuft of bracts raised above the gynoecium, x3 and 8.

Plate 21

Eoantha zherikhinii Krassil., a proangiospermous gnetalean plant of the Baisian Assemblage, Early Cretaceous of Transbaikalia (Krassilov, 1986): (1) macerated ovules showing obovate, shortly beaked nucelli with broad pollen chambers (arrow on a clump of pollen grains in the nucellar beak), x70, (3-5) clump of polyplicate pollen grains marked in (2) and individual grains, x300, 600 and 1 000.

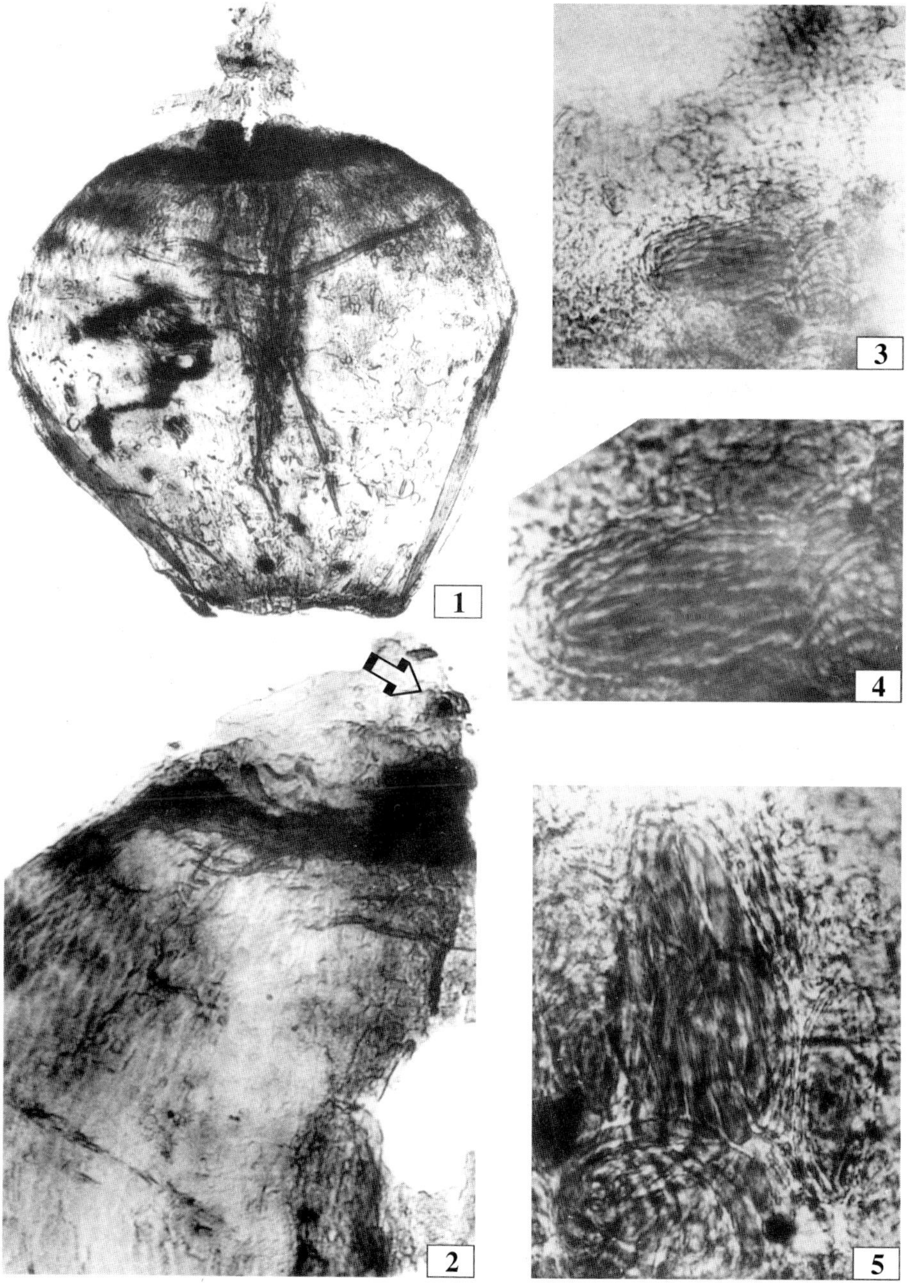

Plate 22

Eoantha zherikhinii Krassil., a proangiospermous gnetalean plant of the Baisian Assemblage, Early Cretaceous of Transbaikalia (Krassilov, 1986; Krassilov & Bugdaeva, 1987), SEM: (1) cleared ovule of the specimen illustrated in Text - Fig. 14, showing a low platform with canals of vascular traces at base, x50, (2, 3) well-preserved granular megaspore membrane of the latter ovule, x650 and 300, (4) vascular strand of the pedicel, tracheary elements with helical thickenings and pits, x1 000.

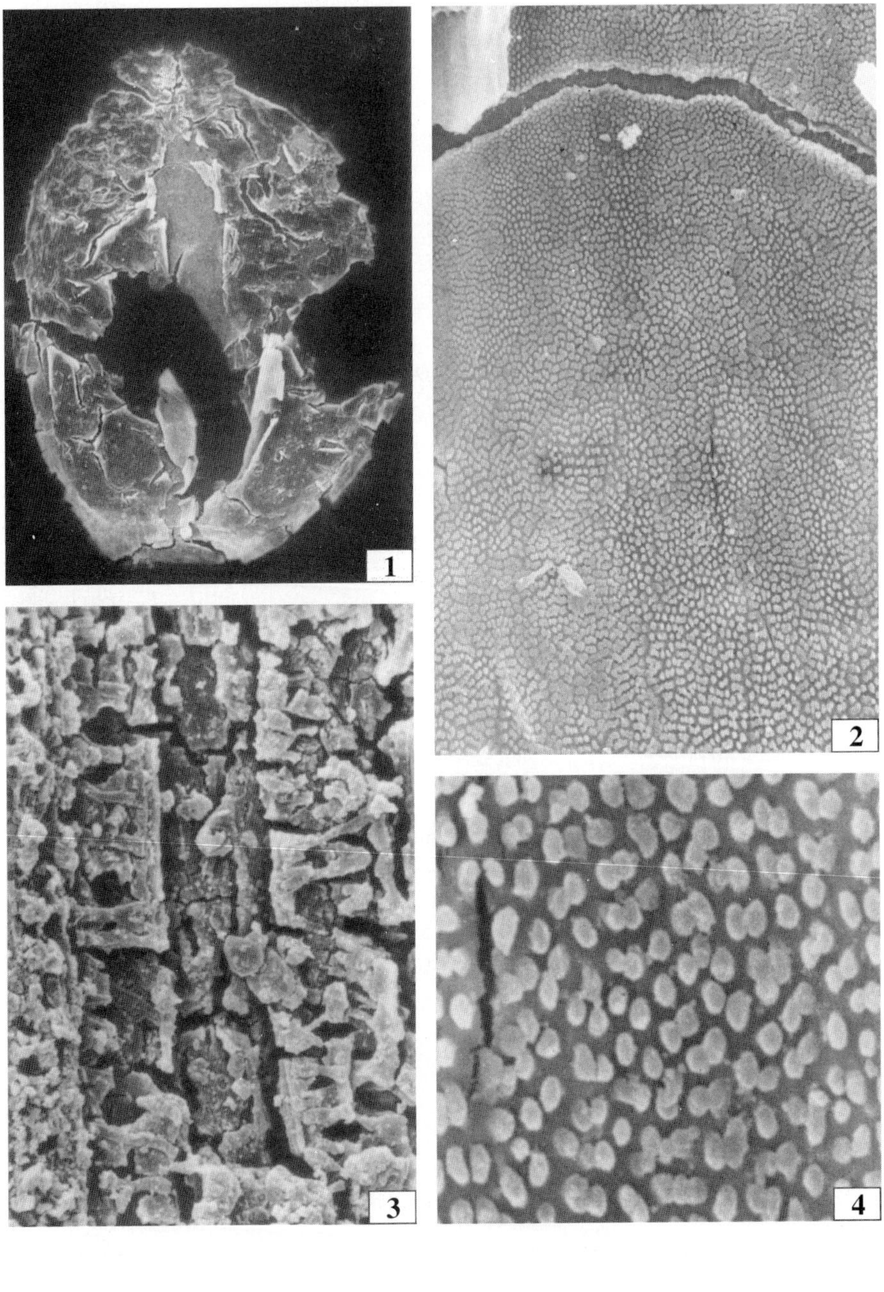

Plate 23

Graminoid leaf presumably of the *Eoantha* plant, the Baisian Assemblage, Early Cretaceous of Transbaikalia (Krassilov & Bugdaeva, 1997): (1) leaf with a lacerate blade and minute appendages at the node, x8, (2) cross vein in a cleared leaf compression, SEM, x320, (4, 5) stomatal files of intercostal grooves, those in (3) apparently stretched by leaf elongation, SEM, x690 and 750.

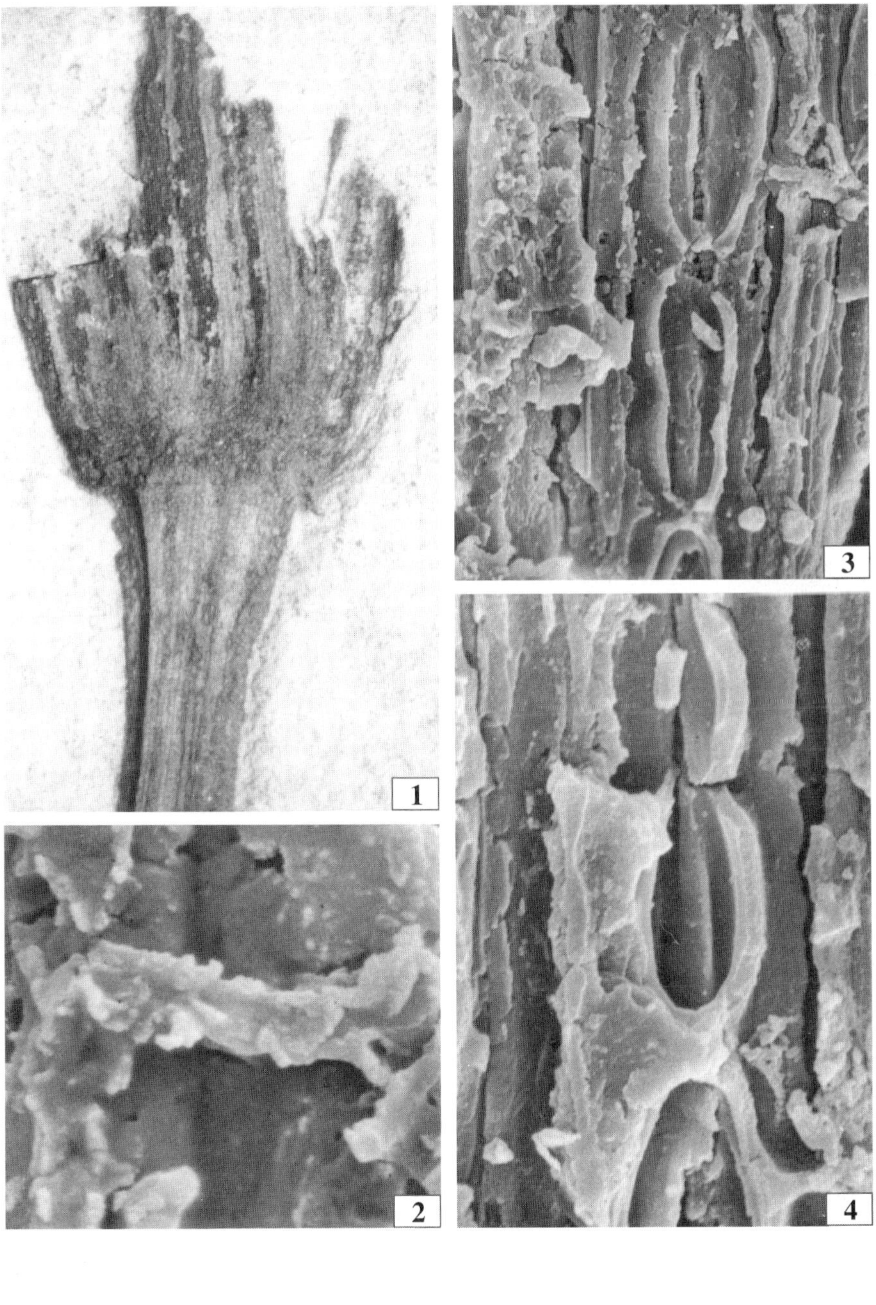

Plate 24

Graminoid leaf presumably of the *Eoantha* plant, the Baisian Assemblage, Early Cretaceous of Transbaikalia (Krassilov & Bugdaeva, 1997): vascular tissue showing fibres with small pits and tracheary elements with helical thickenings and the larger pits, x1 400 (1) and 3 000 (2).

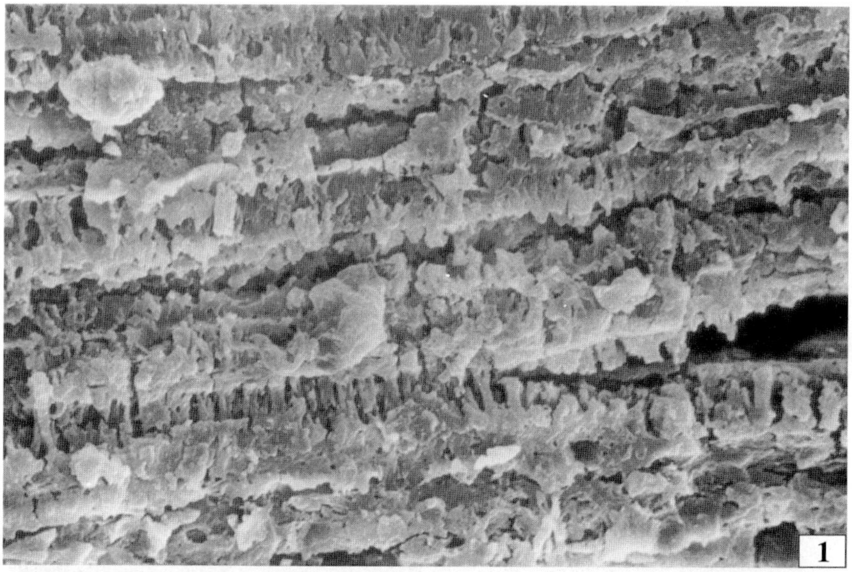

1

2

Plate 25

Vitimantha, a preflower of the Baisian Assemblage, Early Cretaceous of Transbaikalia (Krassilov & Bugdaeva, 1997), showing four bracteate tepals with anastomosing veins and the follicular structures (arrow), pistils or cupules, of which one to the right is fully preserved, with an attenuating apical region, x6 and 10.

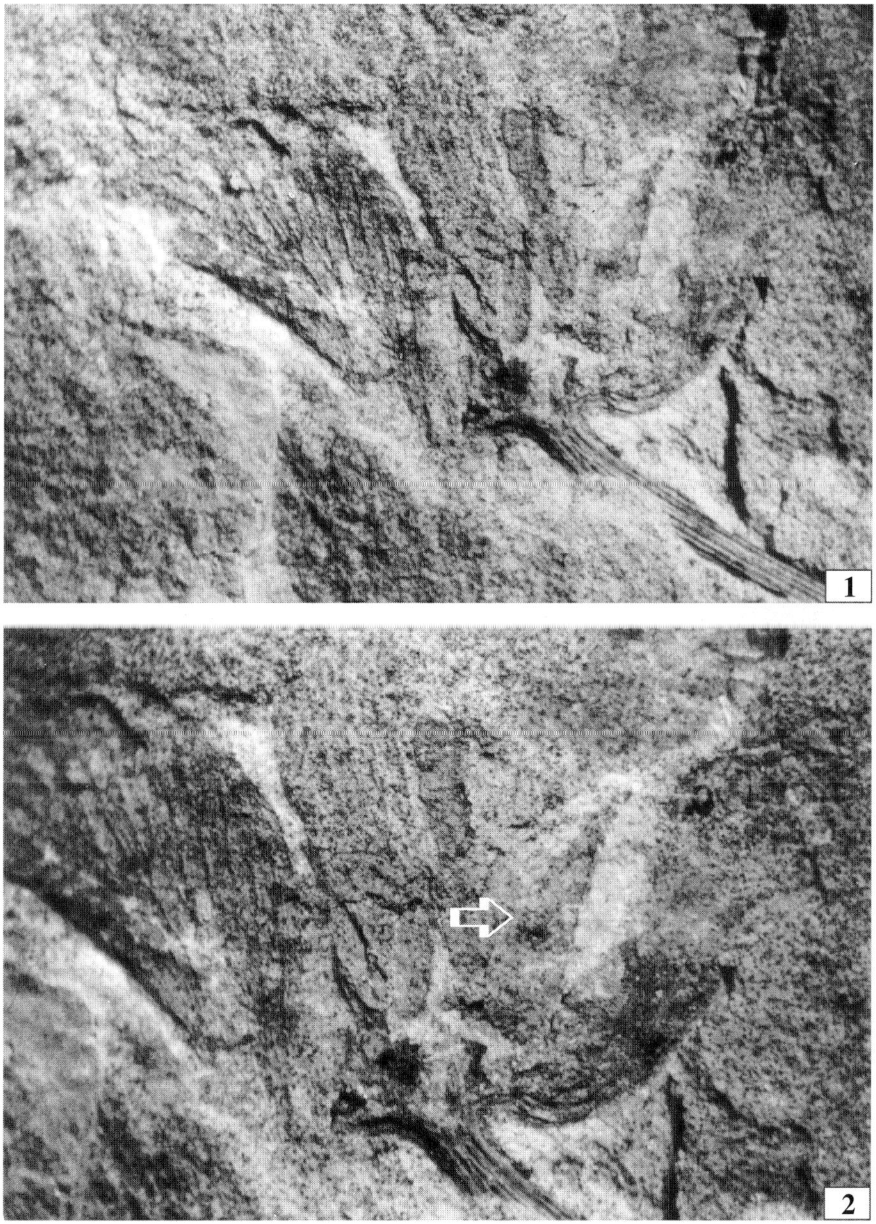

Plate 26

Vitimantha, preflower of the Baisian Assemblage, Early Cretaceous of Transbaikalia (Krassilov & Bugdaeva, 1997), SEM: (1) sutural papillate fringe of follicular organ (marked by arrow in Plate 25), x900, (2) polyplicate ephedroid pollen grains stuck to the marginal fringe, x1 500, (3) helical tracheids from a trace of follicular structure, x220.

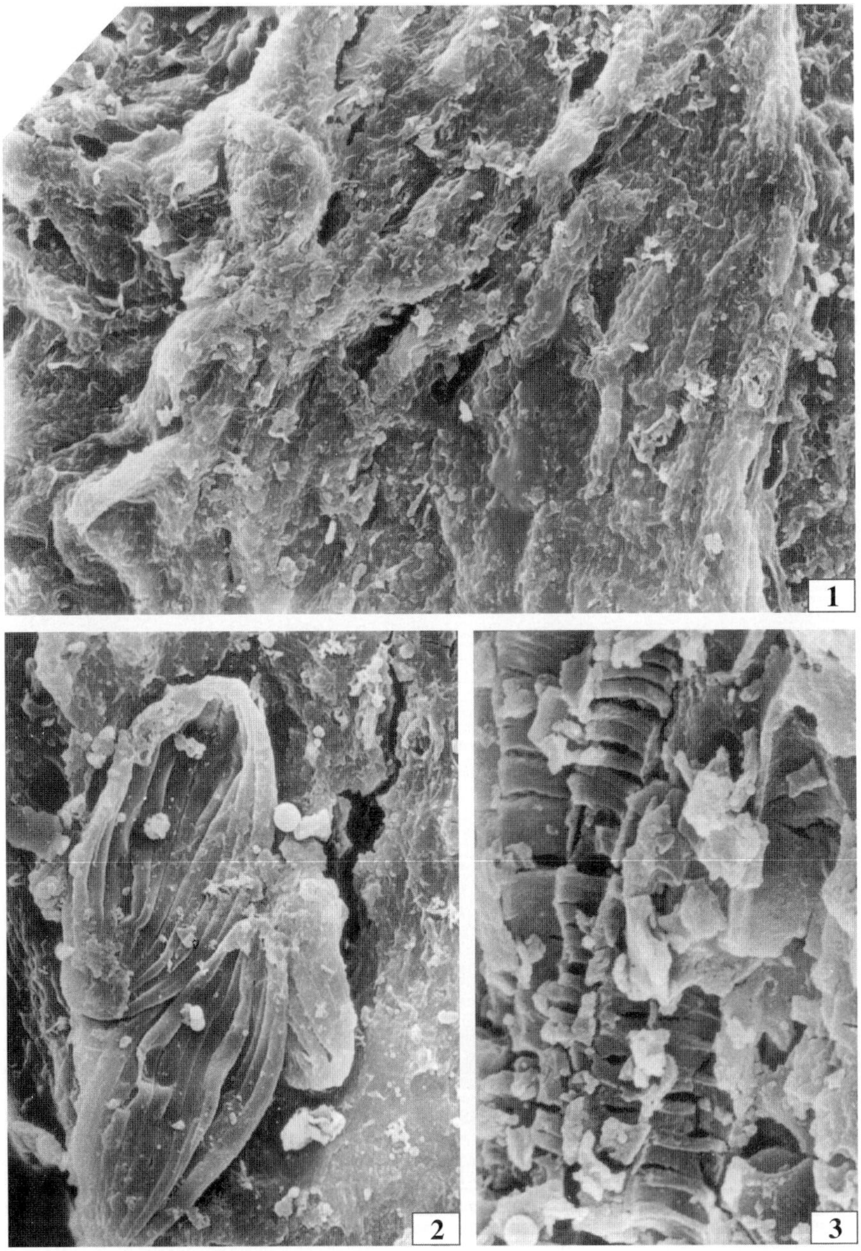

Plate 27

Vitimantha, preflower of the Baisian Assemblage, Early Cretaceous of Transbaikalia (Krassilov & Bugdaeva, 1997), SEM: (1) marginal papillate fringe of follicular organ (marked by arrow in Plate 25), x3 000, (2) clump of polyplicate pollen grains stuck to the marginal fringe showing sulcus and the uncleft interrib exine, x3 000.

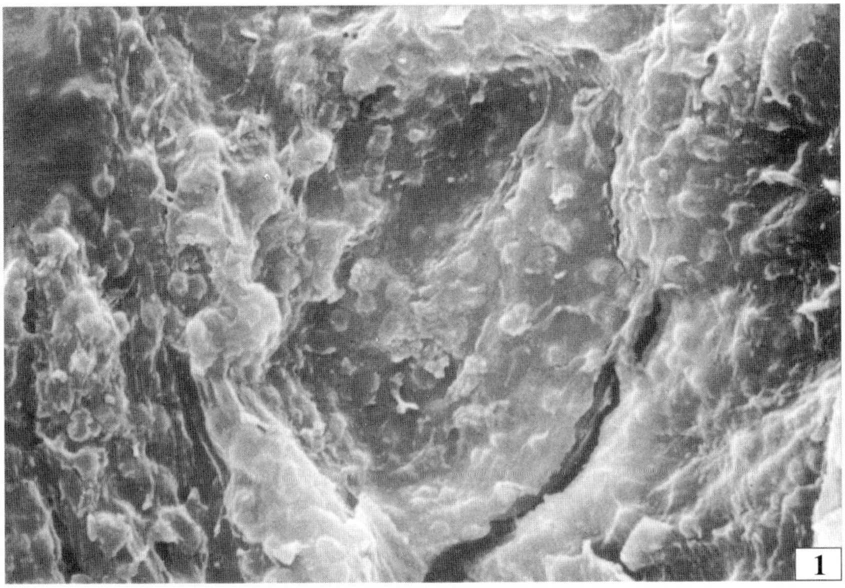

Plate 28

Caytonia sewardii Thomas, from the Jurassic of Yorkshire, England (Krassilov, 1978b, 1984): (1) partly cleared cupule showing the outlines of ovules, x15, (2) part of a nucellus (left) and the inner integument, same as in (4), x150, (3) canals leading from the exostome to the ovules, with a stuck pollen grain (arrow), x395, (4) nucellus with a partly preserved inner integument (outer integument removed), x70.

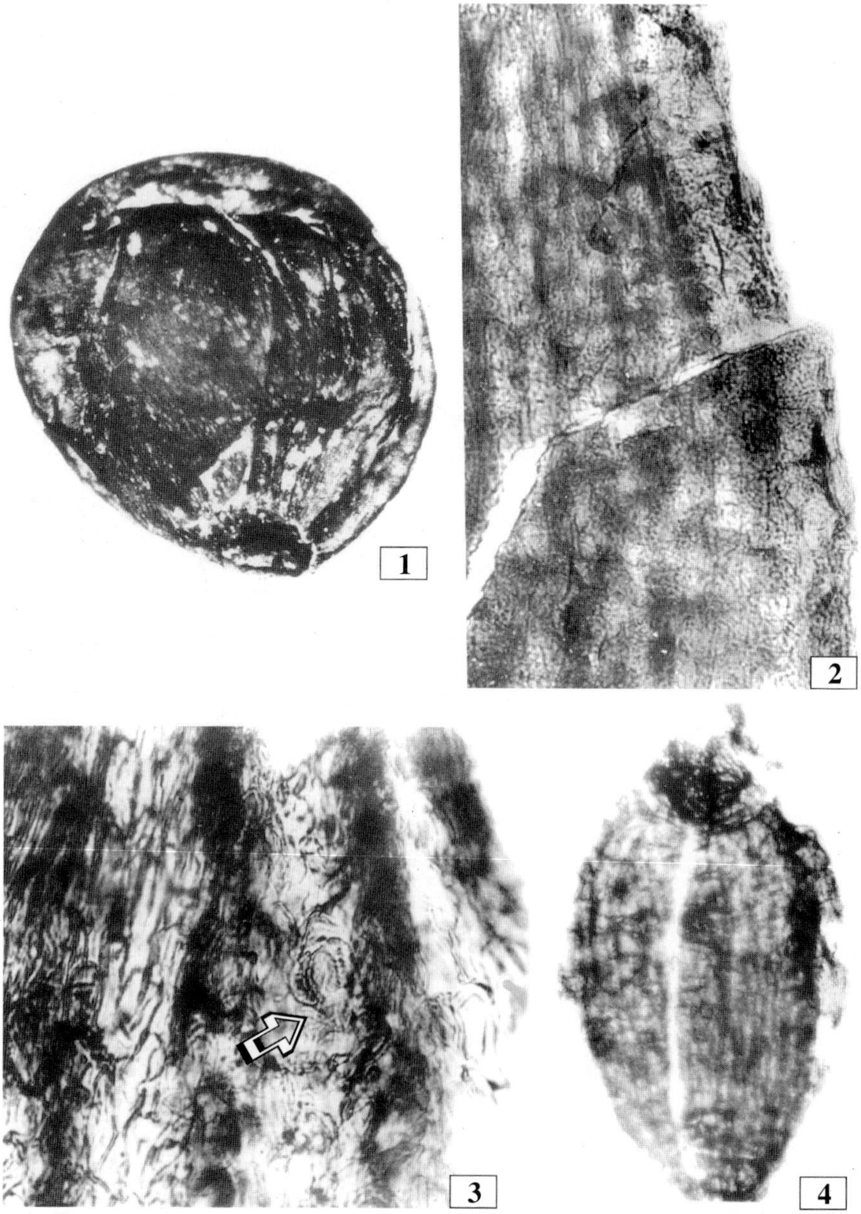

Plate 29

Caytonia sewardii Thomas, from the Jurassic of Yorkshire, England (Krassilov, 1978b, 1984): (1) nucellus showing a median ridge, x166, (2) seed showing coat cells, x100, (3) micropyles of two adpressed ovules showing bulging cells, x395.

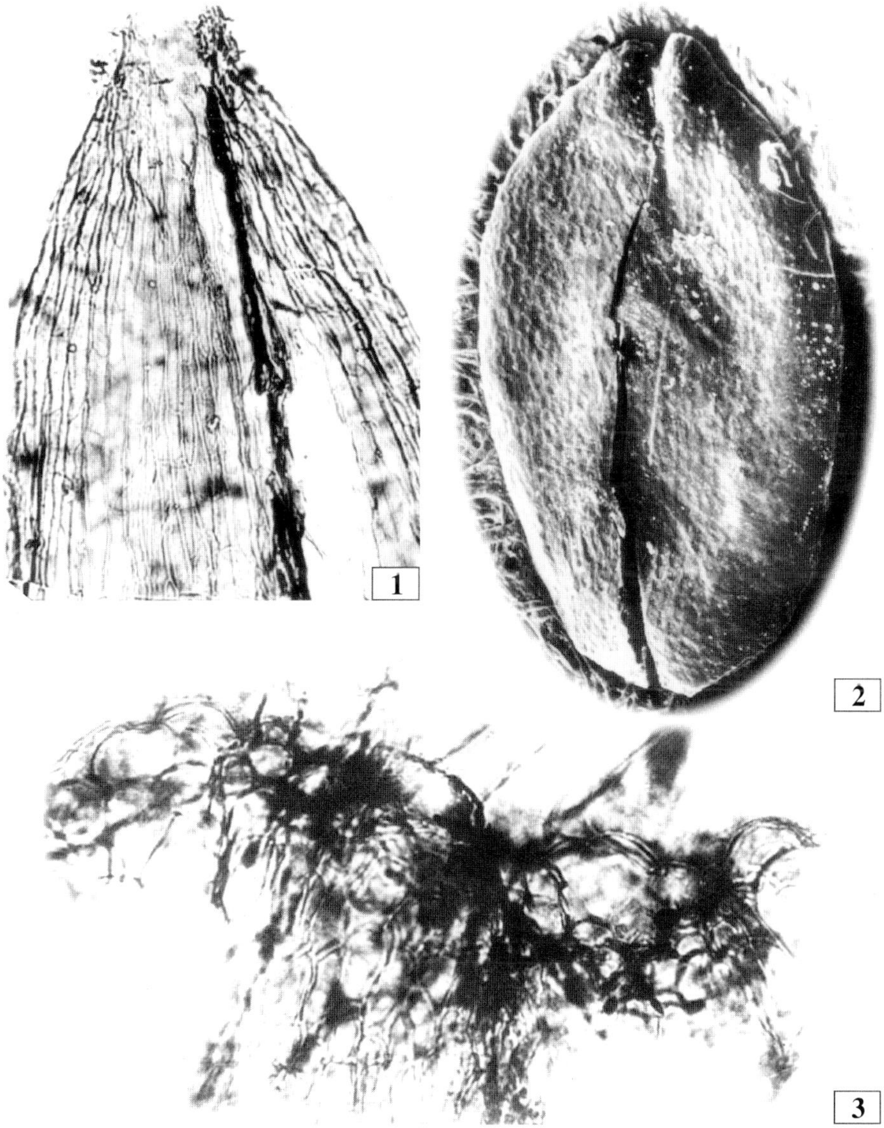

Plate 30

Caytonanthus tyrmensis Krassil., a caytonialean androclade from the basal Cretaceous of the Tyrma River, Far East, Russia (Krassilov, 1978b): (1, 2) androclade branches bearing synangia that are clustered on shorter branches (1) or turned to one side and pendent on longer branches (2), x8, (3) synangium, x16, (4) pollen grain, SEM, x3 500.

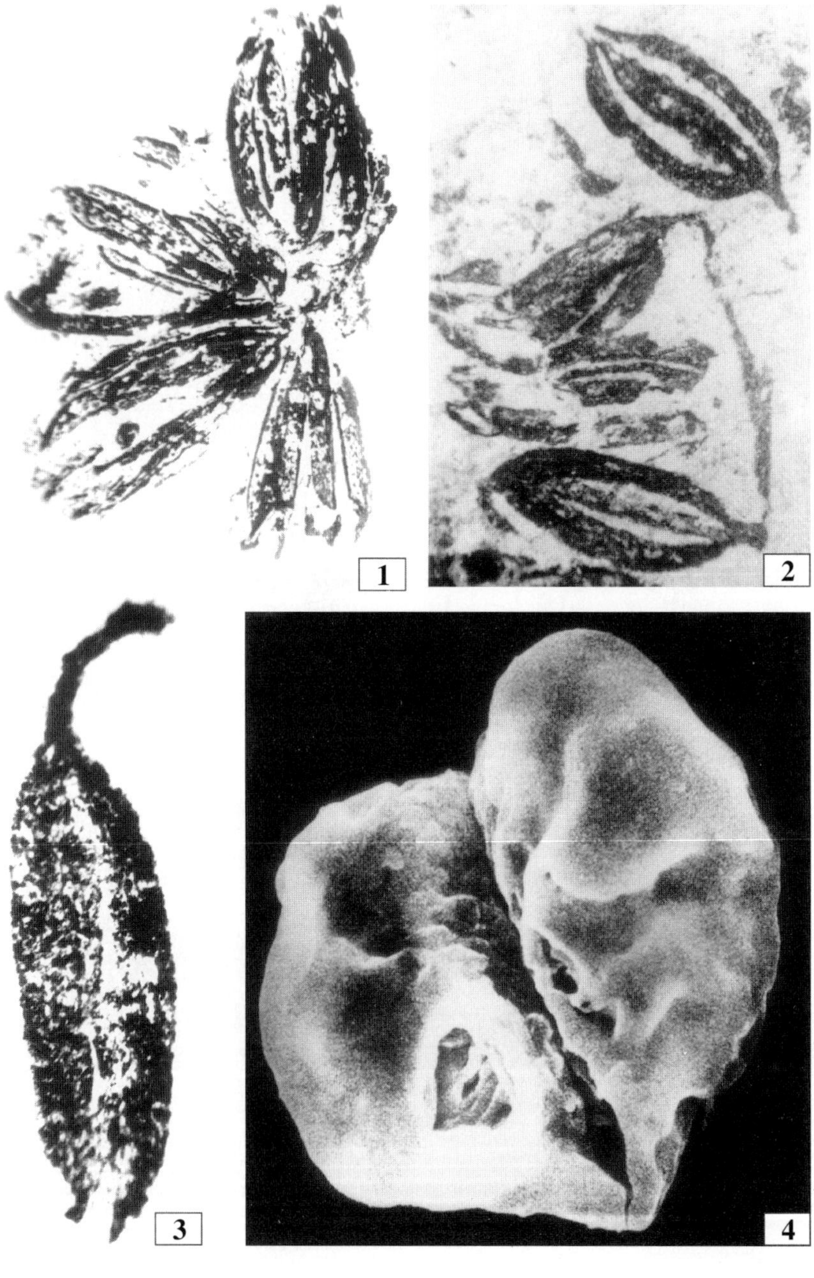

Plate 31

Czekanowskialean reproductive structures: (1) *Leptostrobus stigmatoideus* Krassil., from the Late Jurassic of Bureya Basin, Far East, Russia (Krassilov, 1969, 1972a), gynoclade with lateral cupules on variably developed stalks, some appearing as short-shoots, x1, (2, 3) cupules cleared from the rock and opened to show valves and the pitted papillate fringes, x3, (4) pits and papillae on marginal fringes, x117, (5) *Ixostrobus schmidtianus* (Heer) Krassil., pollen organ from the same locality, lateral sporangiophores with 4-lobed synangia, x8.

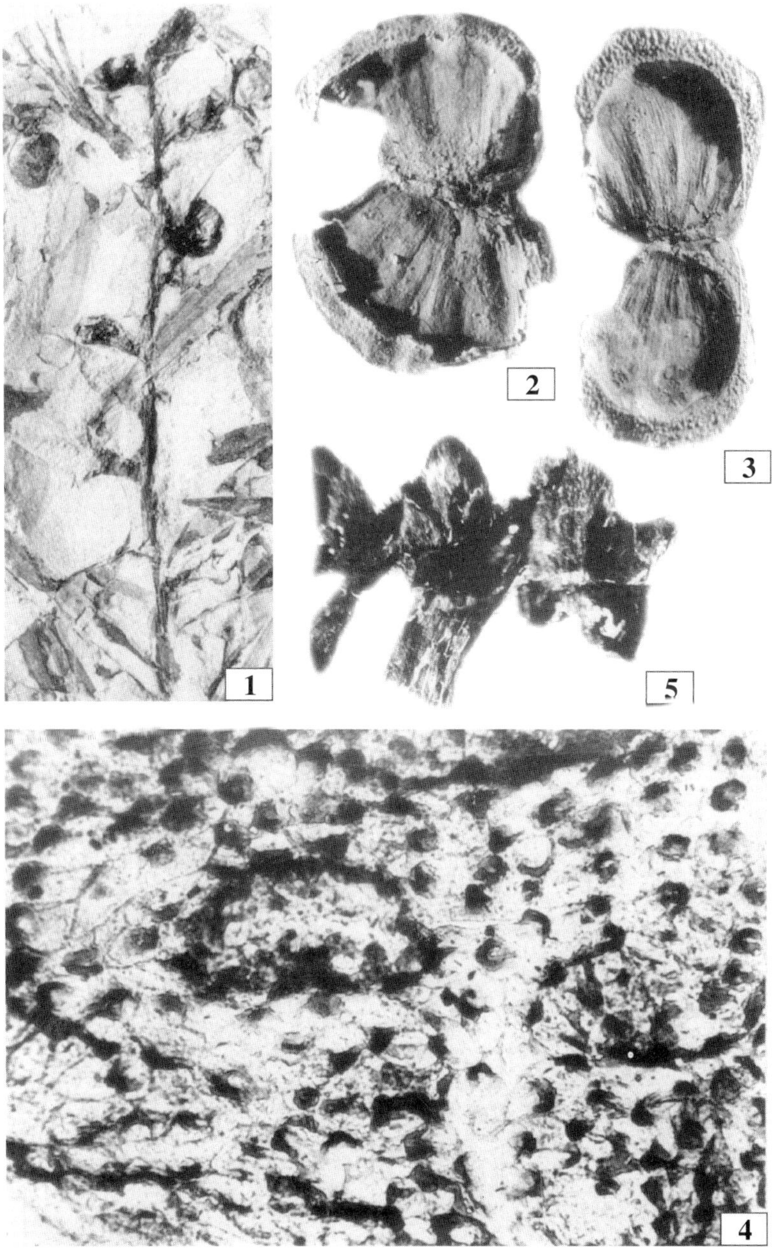

Plate 32

Leptostrobus stigmatoideus Krassil., czekanowskialwean reproductive structures from the Late Jurassic of Bureya Basin, Far East, Russia (Krassilov, 1969, 1972): (1, 2) cupule valves showing scars of the ovules, x6, (3) interior extension of the stigmatic marginal flange intruding the locule, SEM, x120.

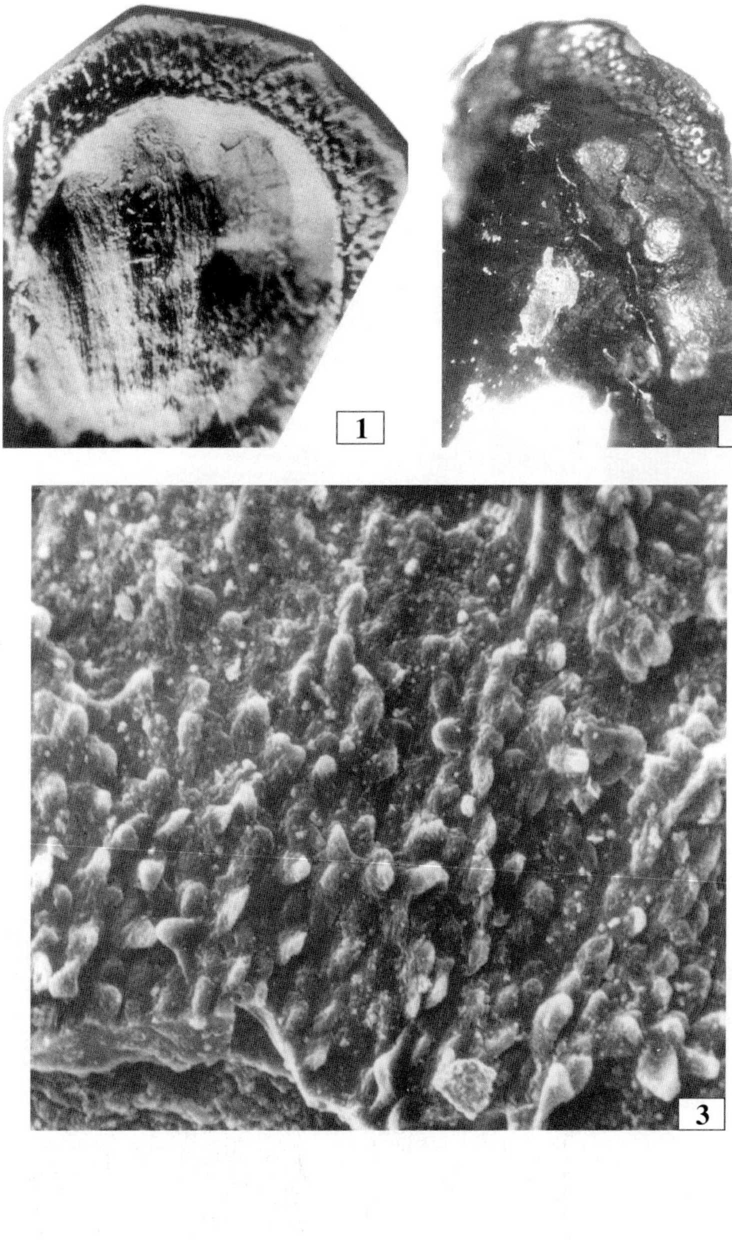

Plate 33

Reproductive structures of Cretaceous platanoids: (1, 2, 5) *Sarbaya radiata* Krassil. & Shilin, a staminate platanoid head from the Late Cretaceous (Cenomanian) of Kazakhstan (Krassilov & Shilin, 1995), stamen with a short filament, a recurved connective apophysis and narrow pollen sacs, SEM, x50 (1), staminate head of distinct wedge-shaped flowers, x10 (2), and flower of four stamens (three exposed) with short perianth lobes (left), SEM, x25 (5); (3, 4) axis bearing small pistillate heads, from the Late Cretaceous of the Amur River Valley (Krassilov, 1976), x1 and 7.

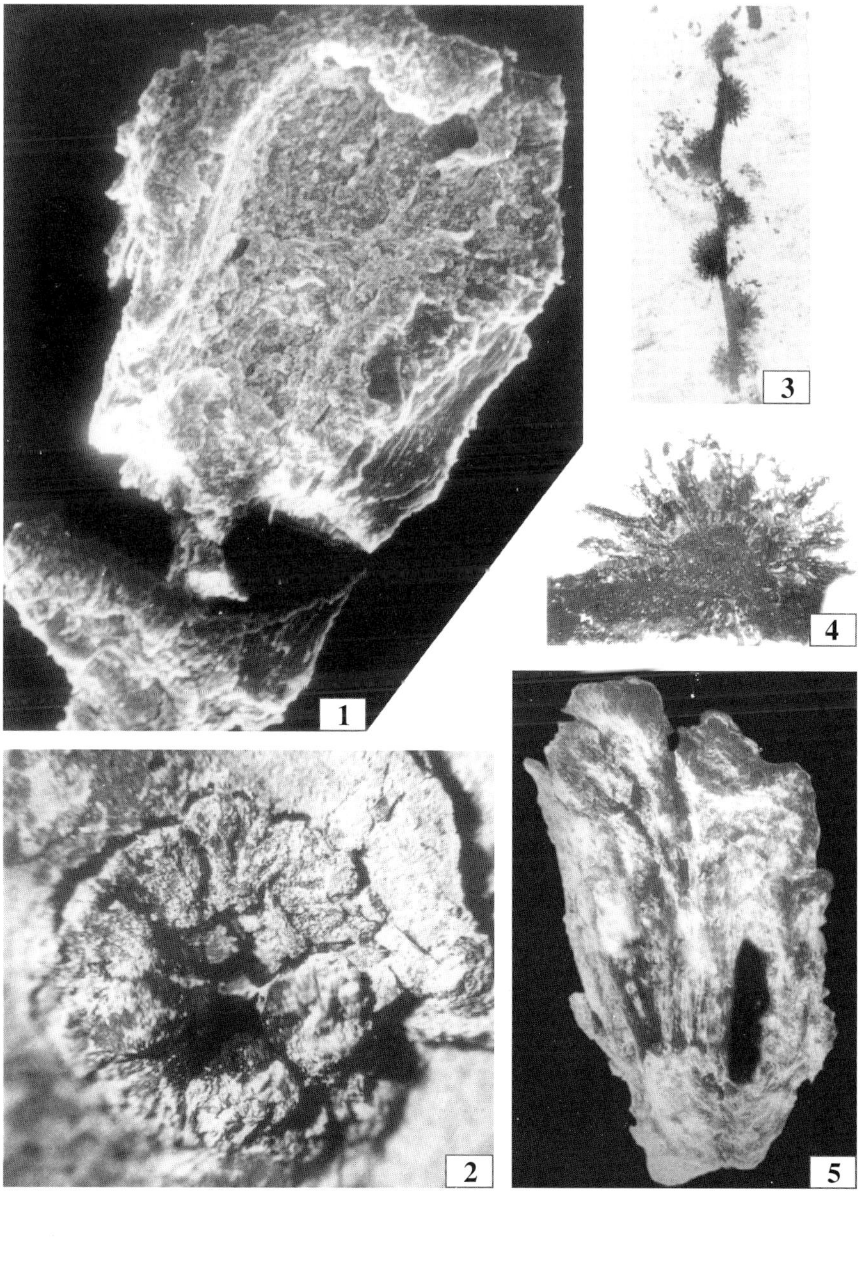

Plate 34

Sarbaya radiata Krassil. & Shilin, a staminate platanoid head from the Late Cretaceous (Cenomanian) of Kazakhstan (Krassilov & Shilin, 1995): (1) pollen grain showing an exocolpus with a granular membrane, slightly gaping in the middle over the porous mesoaperture, x7 000, (2) details of the exocolpus, SEM, x15 000, (3) oblique polar view of a pollen grain with narrow, almost straight colpi, SEM, x4 000, (4) pollen grains at different foci showing porous mesoapertures and lalongate endoapertures forming a thin equatorial girdle, or endocingulum, x2 000.

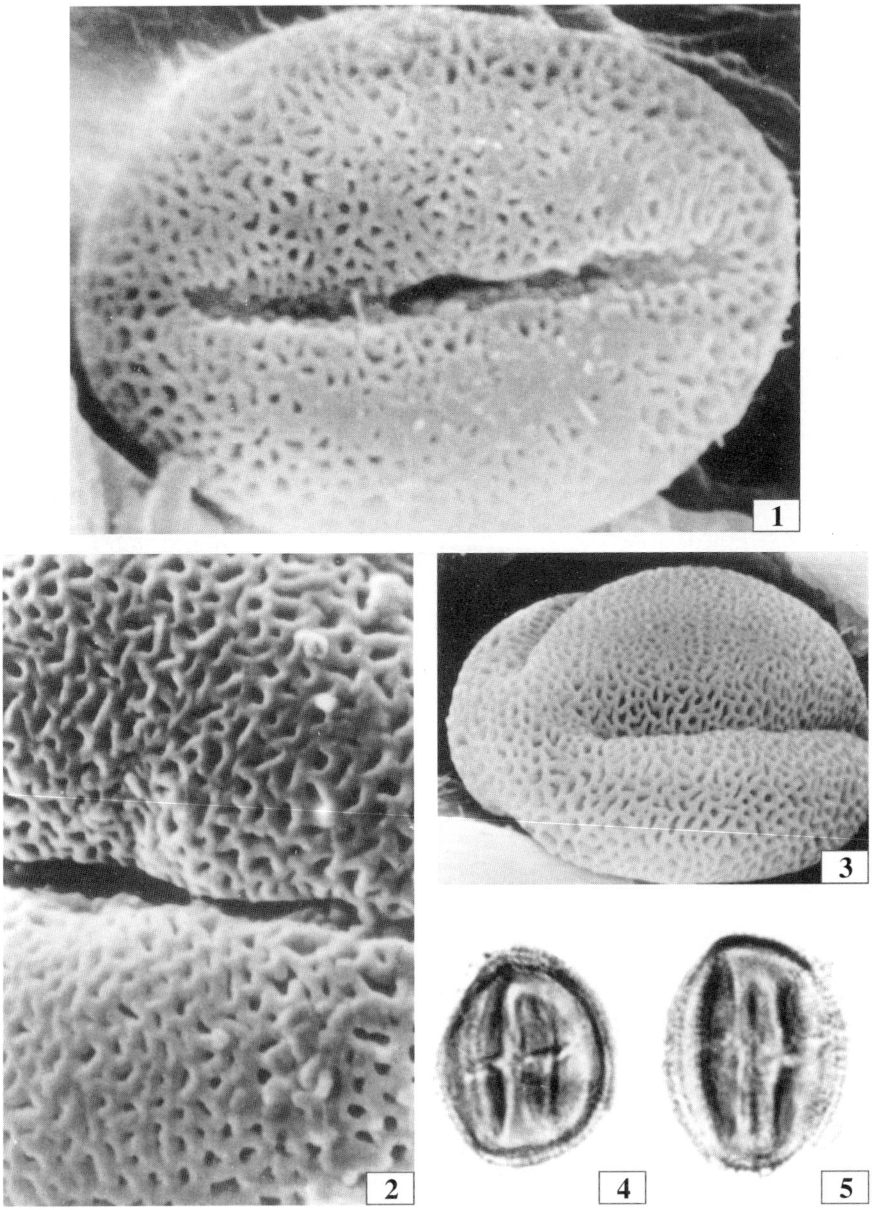

Plate 35

Caspiocarpus paniculiger Krassil. & Vachr., a ranunculoid angiosperm from the Early Cretaceous of Kazakhstan (Vakhrameev & Krassilov, 1979; Krassilov, 1984): (1) branching shoot with two panicles terminal on both central axis and lateral branch, the central axis bearing two opposite leaves (pale and poorly seen in the photograph, arrow on the leaf node) arising below the panicle, x1, (2) panicle with two head-like basal branches, x5, (3-5) panicle showing short lateral racemes, dense and overlapping, bearing small elliptical follicles, in (5) terminating in a looser raceme, x10.

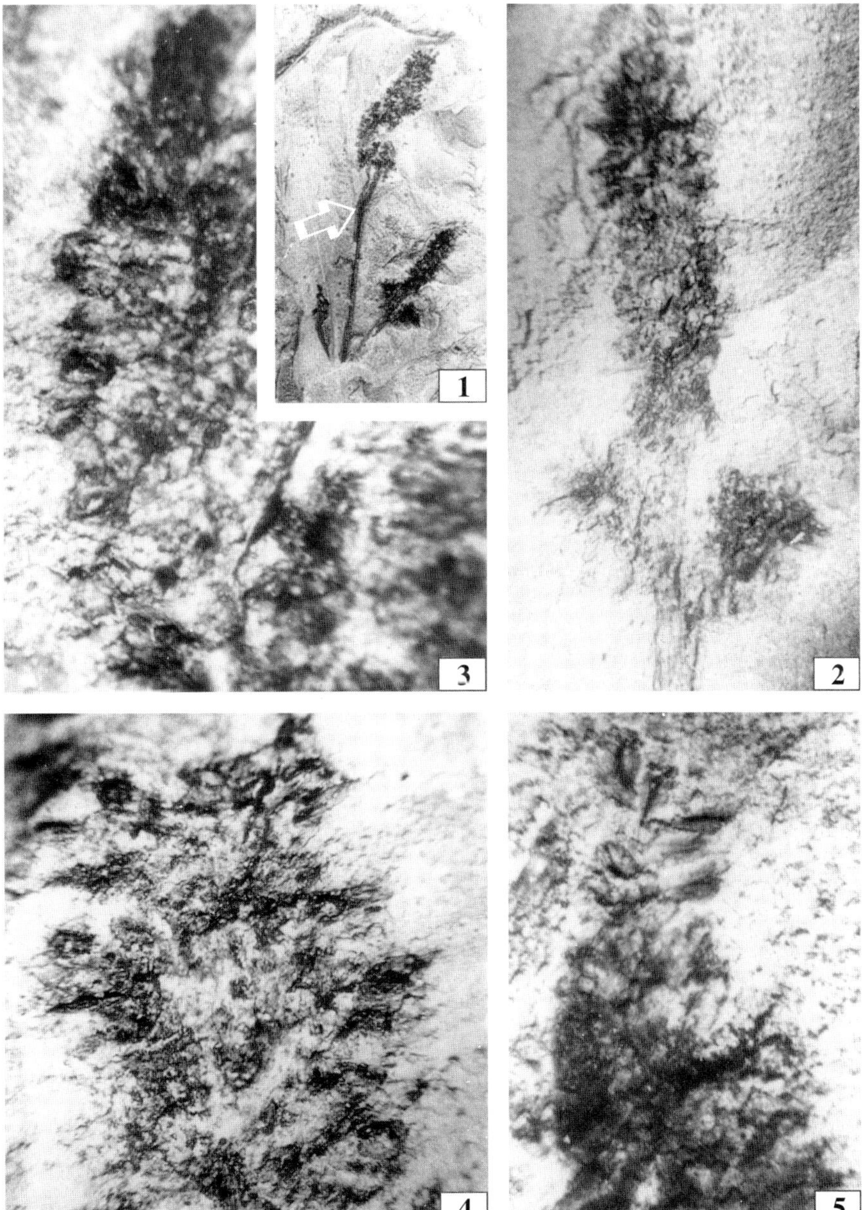

Plate 36

Caspiocarpus paniculiger Krassil. & Vachr., a ranunculoid angiosperm from the Early Cretaceous of Kazakhstan (Vakhrameev & Krassilov, 1979; Krassilov, 1984): (1) two ovules from a follicle, x100, (2) ovule showing two thin integuments (left), the inner one wedging out much below the micropyle, x200, (3) follicle with small appendages (tepals?) at base, x70, (4) follicle wall with hair bases, x2 000.

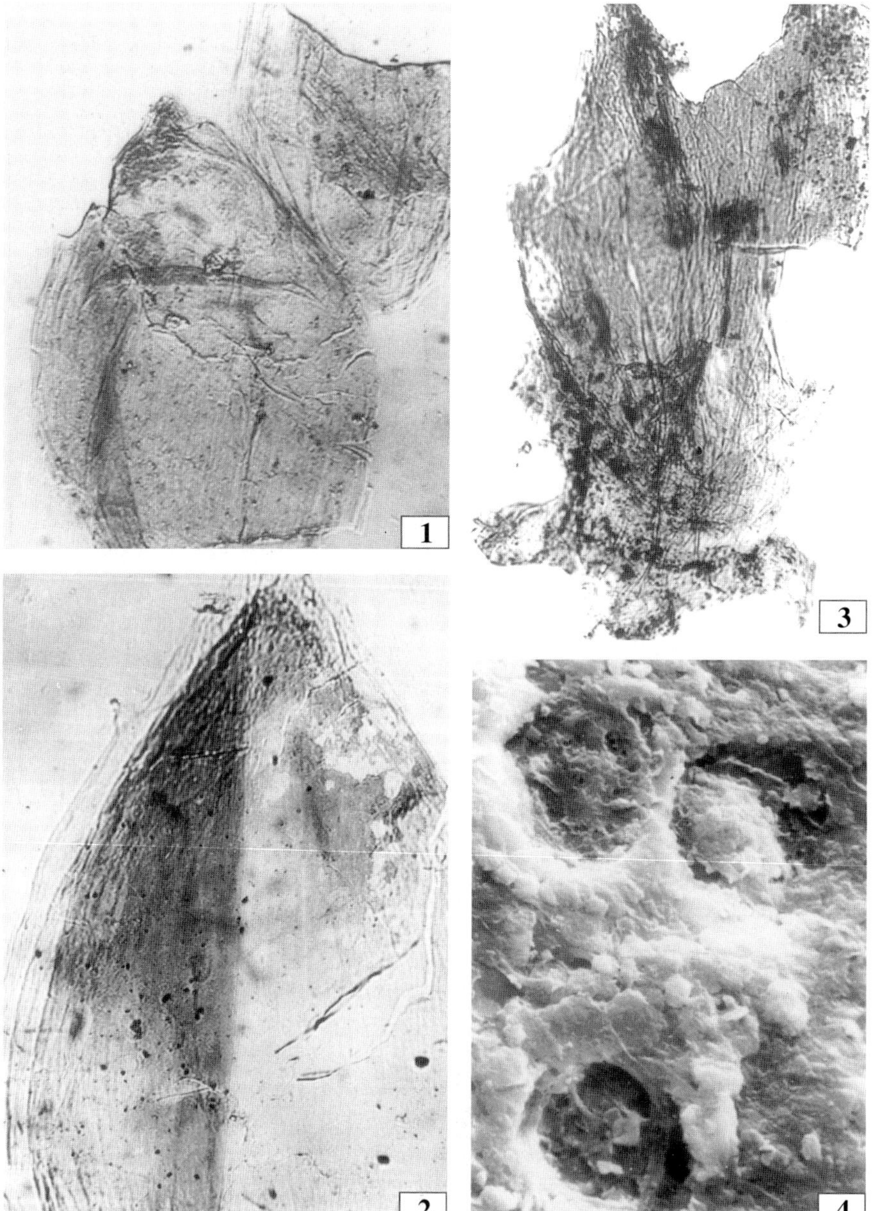

Plate 37

Hyrcantha karatscheensis (Vachr.) Krassil., a paeonialean inflorescence from the Early Cretaceous of Kazakhstan (Krassilov & al., 1983): (1) cymose inflorescence, x2, (2) flower showing filaments of poorly preserved stamens at the base of a median carpel, x7.

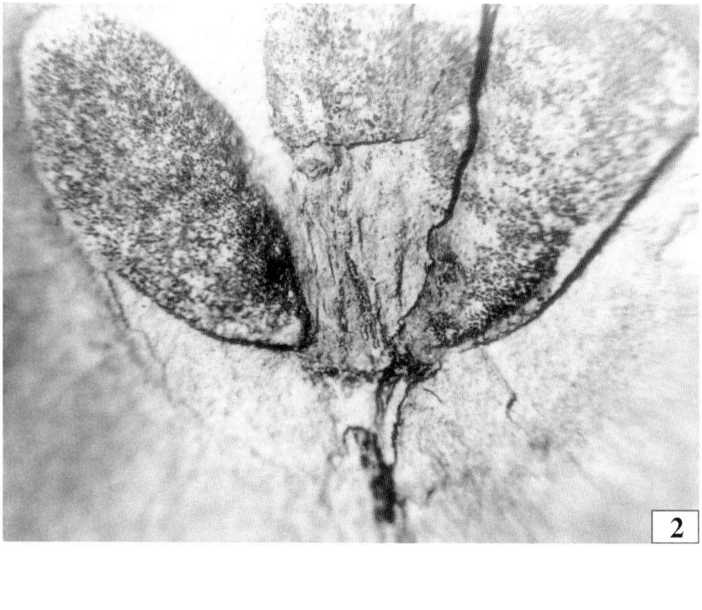

Plate 38

Hyrcantha karatscheensis (Vachr.) Krassil., a paeonialean inflorescence from the Early Cretaceous of Kazakhstan (Krassilov & al., 1983): (1) flower at a fruiting stage showing reflexed lobes of a persistent calyx, x7, (2) internal cast of a dehiscent carpel showing transverse striation of the locule, x10, (3) carpel showing a broad stigma, x7, (4) casts of stomata in the impression of *"Leguminosites" karatscheensis* Vachr., ternate compound leaves found with *Hyrcantha* flowers, SEM, x350.

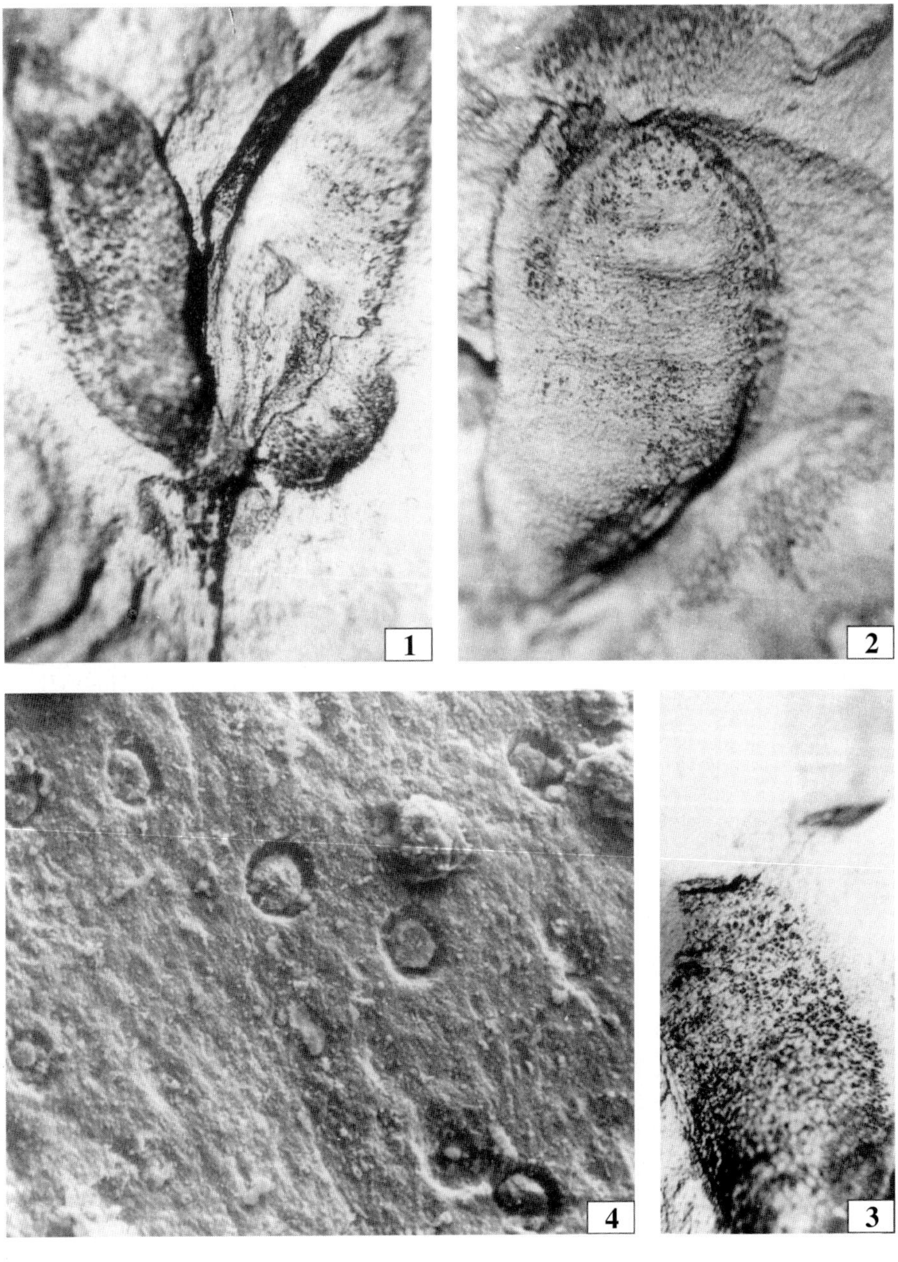

Plate 39

Asterocelastrus cretacea Velen. & Viniklar, a rosaceous fruit from the Late Cretaceous (Cenomanian) of Bohemia (Krassilov & Pacltová, 1989): (1) open pentamerous capsule, syncarpous and adnate to the calyx, lobes of the latter spreading star-like outside of the capsule, x3, (2-5) details of the capsule and anatropous ovules, x7 (2) and 15 (3-5).

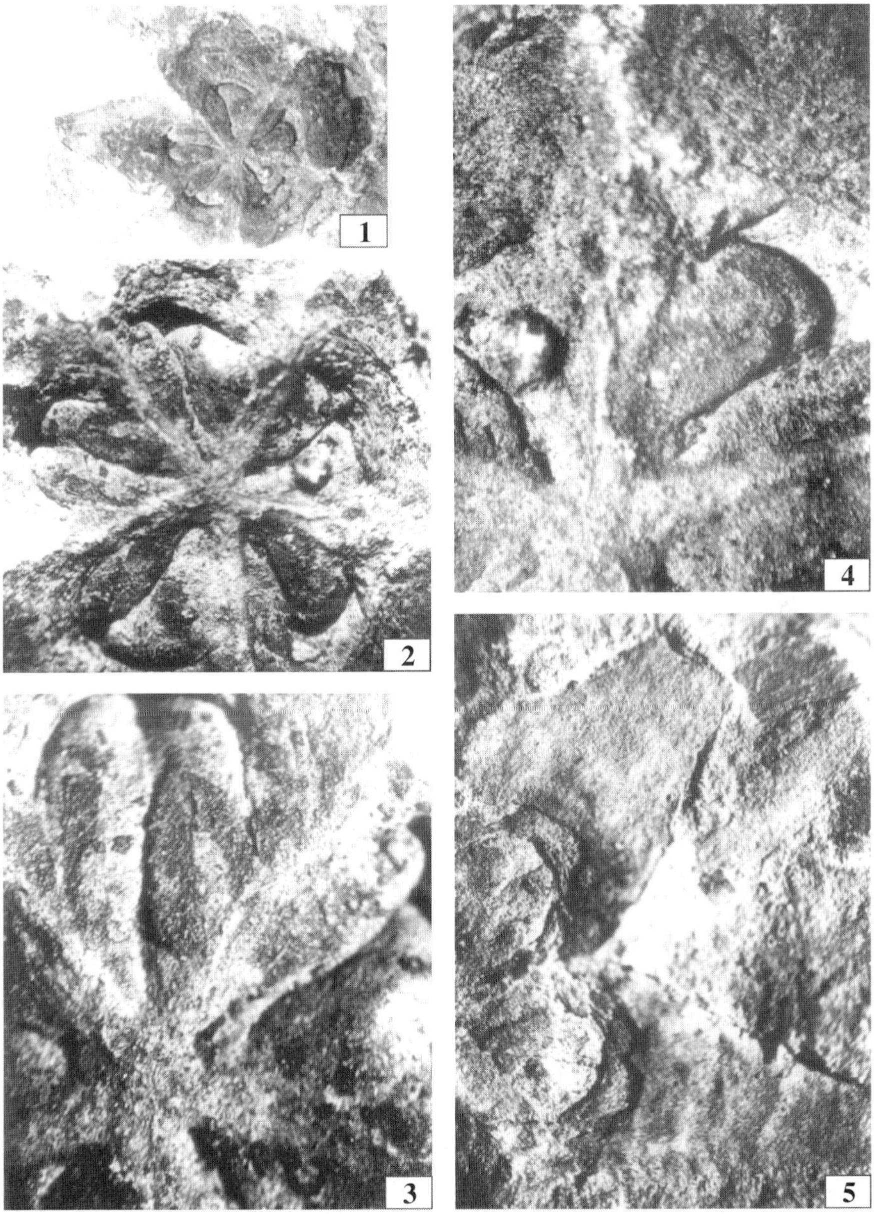

Plate 40

Sarysua pomona Krassil. & Shilin, flowers from the Late Cretaceous of Kazakhstan (Krassilov & al., 1983): (1) longitudinally split flower with the gynoecium much longer than the stamen (left); three of the carpels and their free styles are exposed, with the median carpel showing the outlines of a locule, x10, (2) styles, x15, (3) flower with a better preserved calyx, x10, (4) three flowers showing folds between the lobes of syncarpous ovaries, x3, (5) hairs on the ovary, x150; (6) hooked staminode and a clavate gland of the androecium, x250.

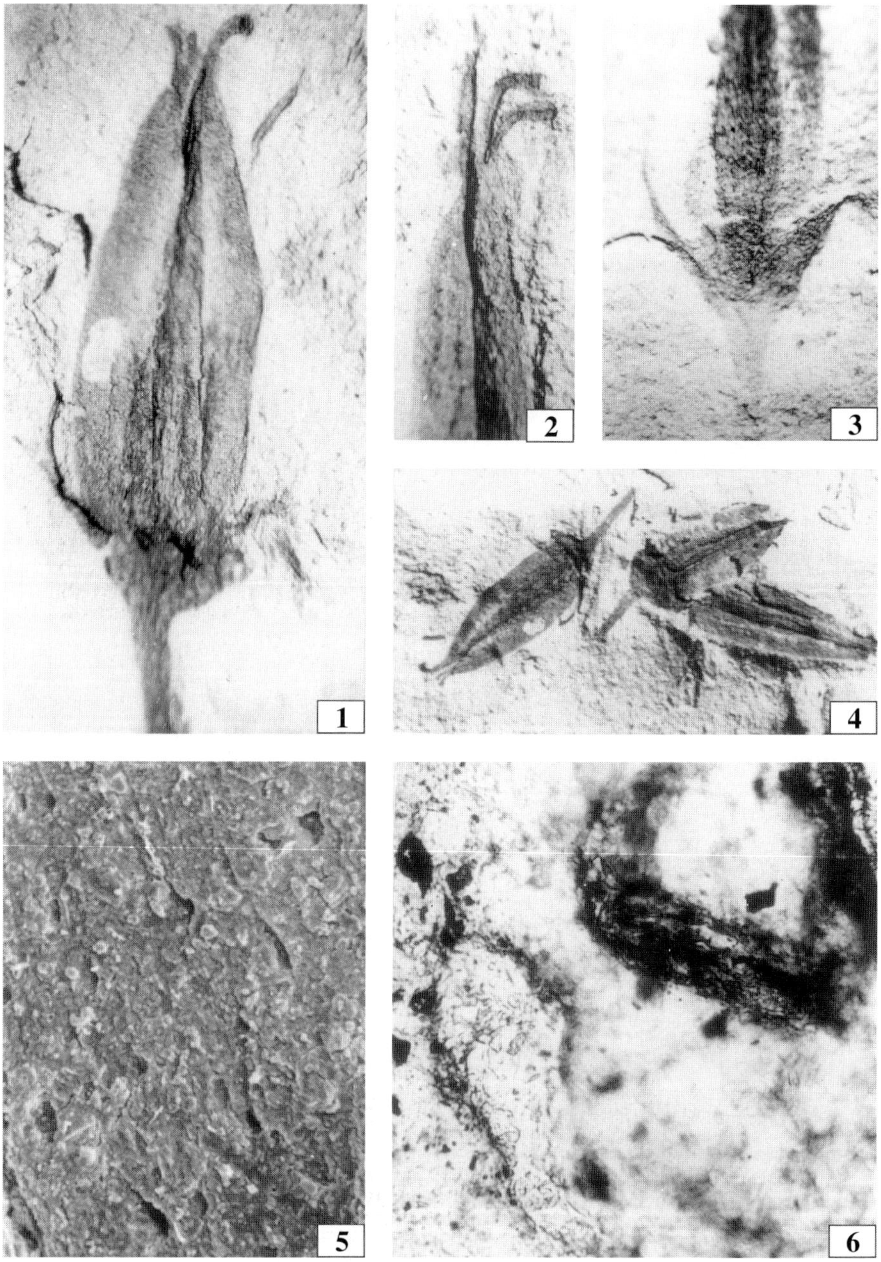

Plate 41

Ramonicarya nevoi Krassil. & Dobrusk., a samaroid fruit from the Early Cretaceous (Aptian) of Makhtesh Ramon, Israel (Krassilov & Dobruskina, 1995): (1) adaxial aspect (n - nutlet, s - stigma, w - winglet), x6, (2) bract venation, x9, (3) winglet and nutlet, stylar end with biarmed stigma, x15, (4) adaxial aspect with a partly preserved pedicel, x5.

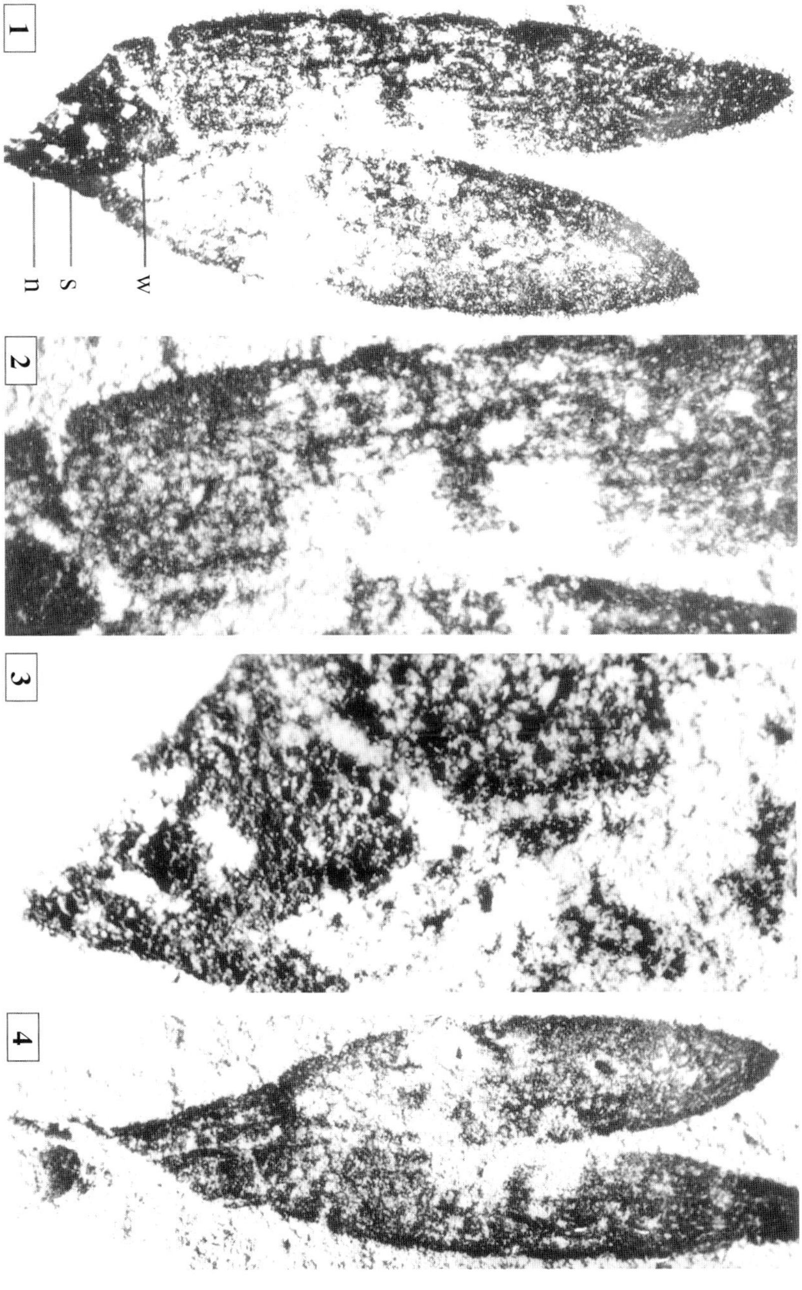

Plate 42

Taldysaja medusa Krassil. & Shilin, a monocot inflorescence from the Late Cretaceous of Kazakhstan (Krassilov & al., 1983): (1) fasciculate inflorescence on a long peduncle, x2, (2) flower showing the gynoecium with a bifid stigma, x10, (3, 4) fascicle rays showing areolate scars of shed flowers, x10, (5) distal parts of fascicle rays with solitary flowers, x10.

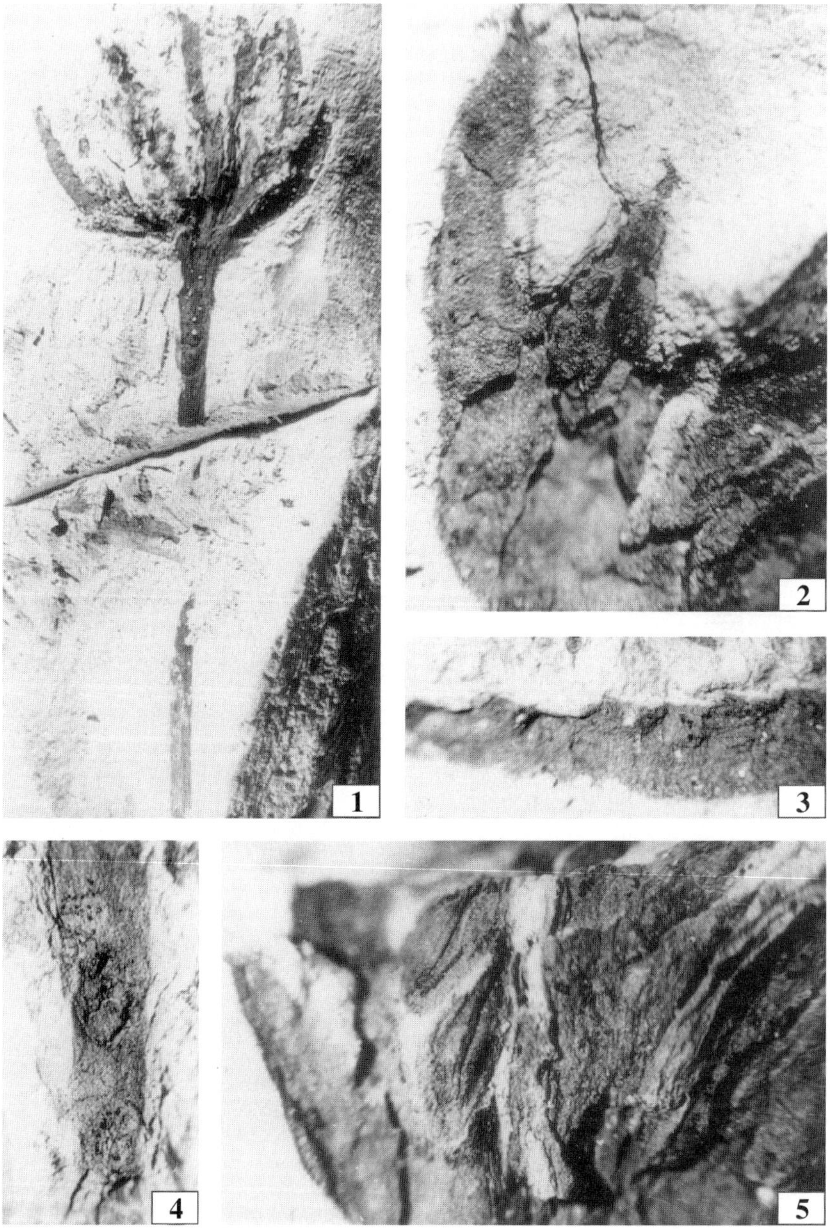

Plate. 43

Tyrmocarpus spinulosus Krassil., fruit-like structures from the terminal Jurassic - lowermost Cretaceous of the Tyrma River, Far East, Russia: (1, 2) external view with spinules, x7, (3) split fruit-like body showing a solitary ovule, x7, (4) fruit surface, x12, (5-7) fruit wall cuticle showing a spinule (5), a scar of detached spinule (6) and stoma (7), x146.

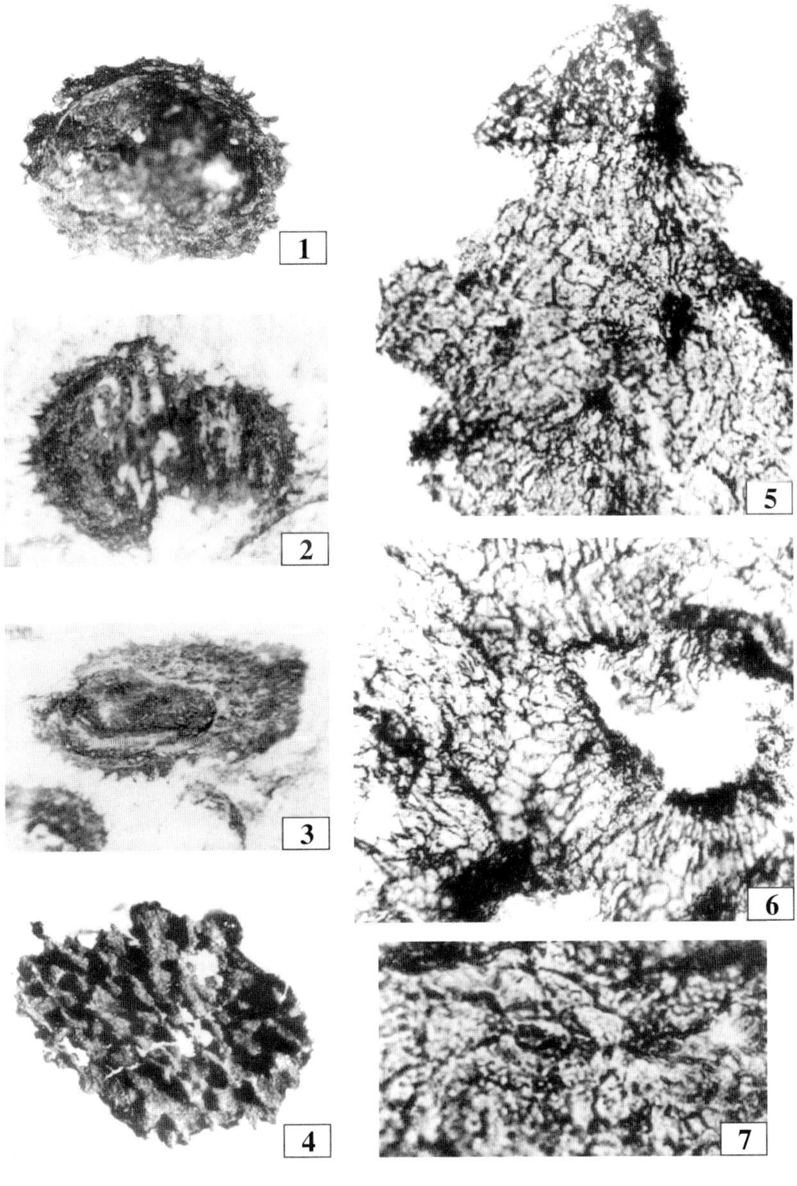

Plate 44

Dominant angiosperms of the post-crisis early Paleocene plant assemblages from the Tsagajan Formation of the Amur Province (Krassilov, 1976): (1-3) *Trochodendrocarpus arcticus* Krysht., panicles of paired carpels, x1, (4) *Trochodendroides*-type leaf of supposedly the same plant, x1, (5, 6) *Carinalaspermum bureicum* Krassil., liquidambroid seeds, x3, (7-9) *Tricolpopollianthus burejensis* Krassil., staminate head, x7, and pollen grains, x395 and 1 000.

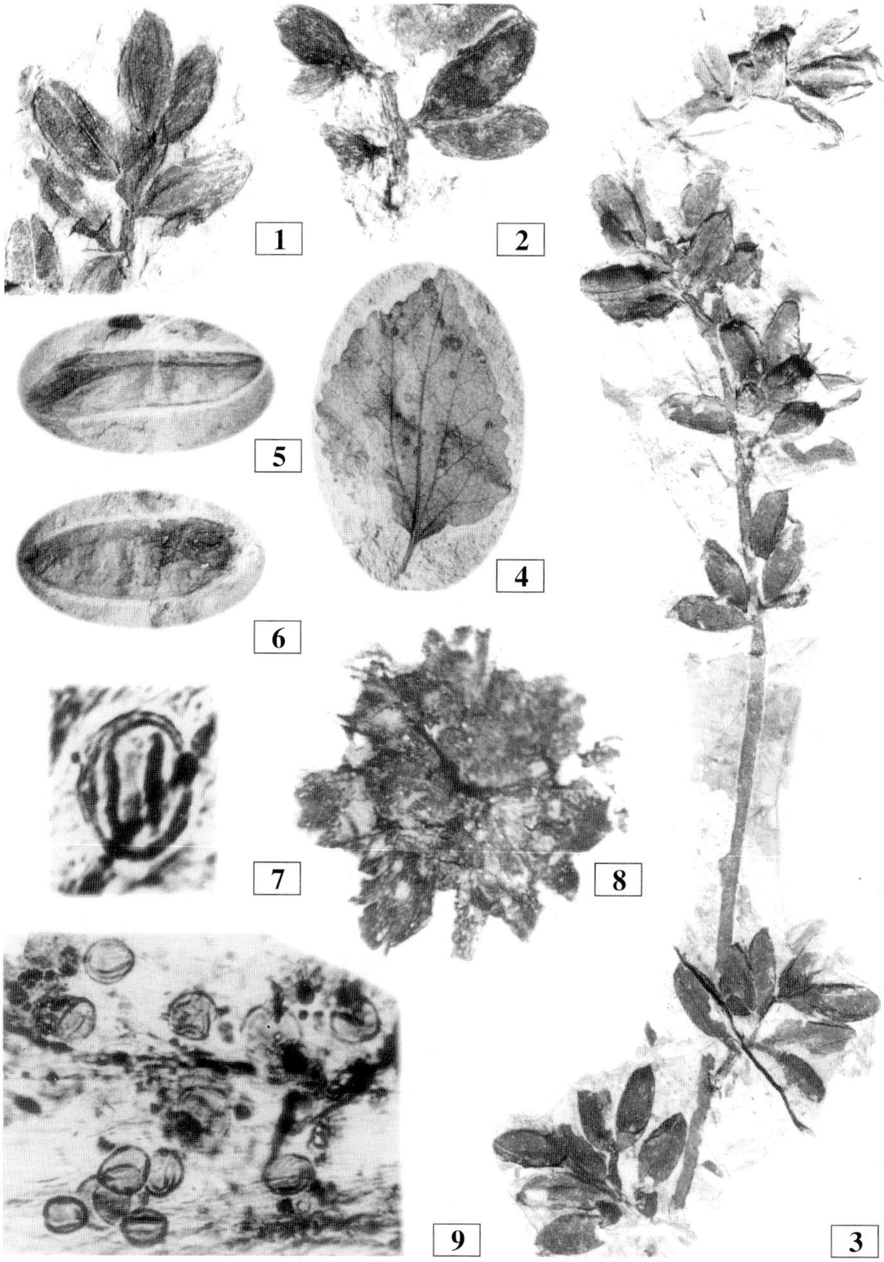

Plate 45

Leaf polymorphism in *Imania* from the Late Triassic of Primorye, Russian Far East (Krassilov & Shorokhova, 1970) showing transitions from pinnate to dichopodial to palmate morphologies (1-6), the latter similar to *Sagenopteris,* a caytonialean leaf genus with anostomosing veins (7), x8.

Plate 46

Meeusella proteiclada Krassil. & Bugd., a supposed prototype of fasciculate stamens from the Early Cretaceous of Transbaikalia (Krassilov & Bugdaeva, 1988b): (1) androclade bearing fertile and sterile branches, x3, (2) paraclade with two stalked sporangial heads lateral to the apex, x10, (3) paraclade with sporangial heads almost sessile and with the apex protruding between them, x10, (4) sterile paraclade, x10.

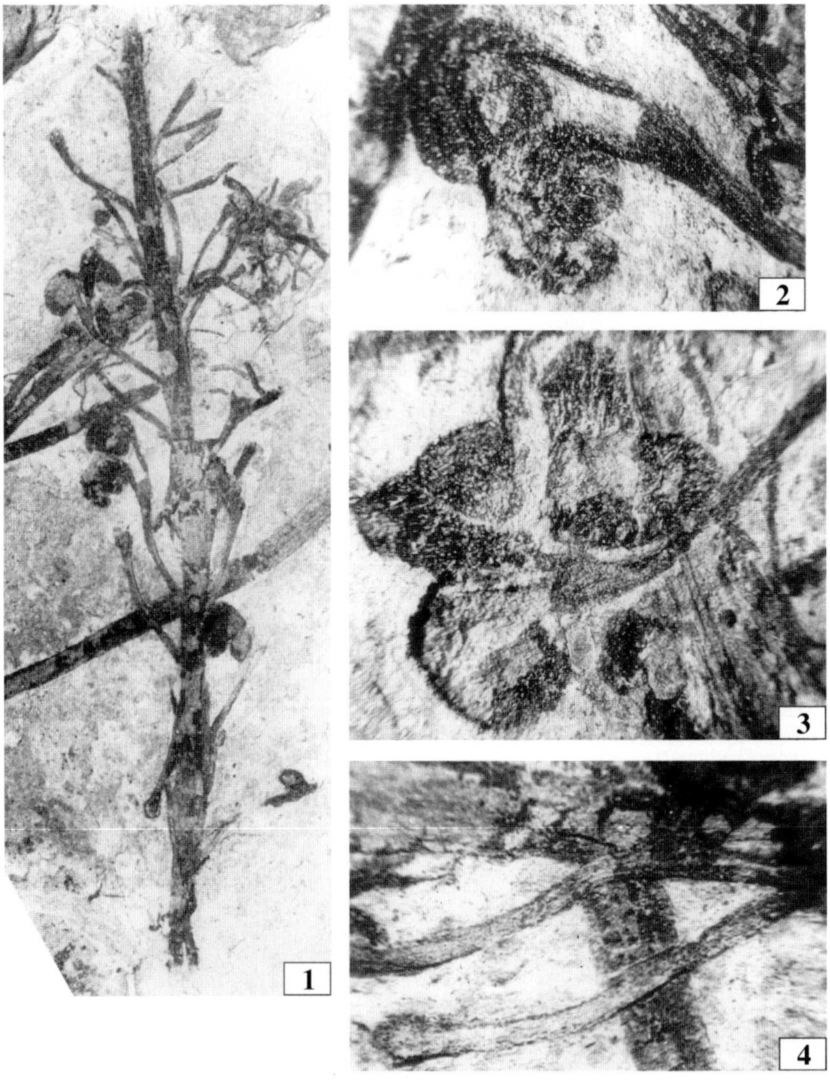

Plate 47

Lemnoid fruits from the Late Cretaceous of Mongolia (Krassilov & Makulbekov, 1995) supposedly retaining a primitive cupular structure of the utricle with an orthotropous ovule, SEM: (1) utricle, surface with lime-secreting glands, x70, (2) split utricle showing the ovule, x70, (3) ovule removed from the utricle, showing the funicle and operculum at the opposite ends but not quite in line with each other, x70, (4) proximal part of the ovule showing the funicle as a direct extension of the pedicel, x70.

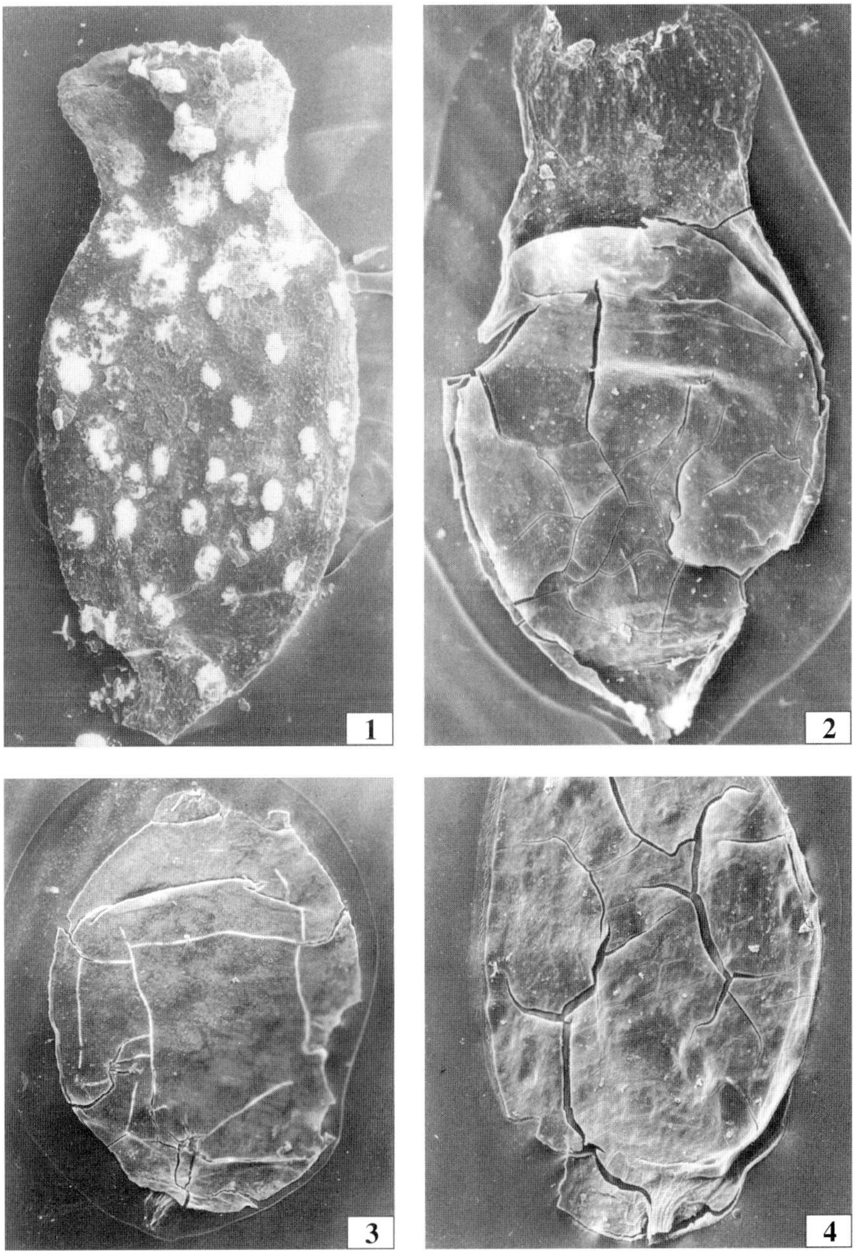

Plate 48

Lemnoid fruits from the Late Cretaceous of Mongolia (Krassilov & Makulbekov, 1995) supposedly retaining a primitive cupular structure of the utricle with an orthotropous ovule, SEM: (1) ampohoriform utricle with a perfect funnel-shaped stigma, x70, the latter enlarged in (2), x150, (3) operculum with an apical coruncle, x300.

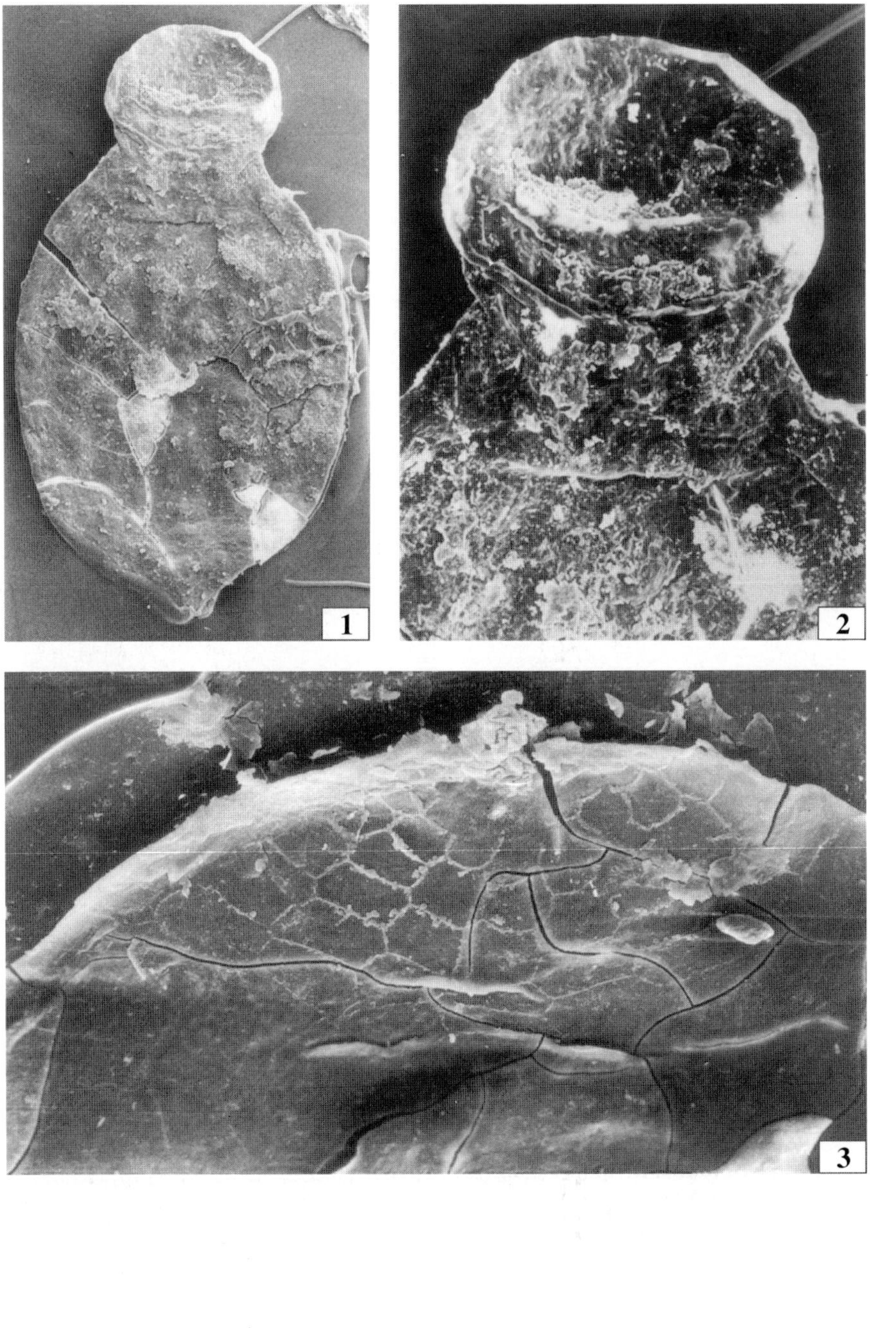

References

Agadjanian, A.K. 1996. Modes of phytophagy in mammals. - Paleont. Zh. (Moscow), No. 6, p. 723-729.

Allen, G.S. 1943. The embryogeny of *Pseudotsuga taxifolia* (Lamb.) Britt. - Amer. J. Bot., vol. 30, p. 655-661.

Alvin, K.L. 1982. Cheirolepidiaceae: biology, structure and paleoecology. - Rev. Palaeobot. Palynol., vol. 37, p. 71-98.

Alvin, K.L., Spicer, R.A., Watson, J.A. 1978. *Classopollis*-containing male cone associated with *Pseudofrenelopsis*. - Palaeontology, vol. 21, no. 4, p. 96-98.

Anderson, J.M., Anderson, H.M. 1983. Palaeoflora of Southern Africa. Molteno Formation (Triassic). Rotterdam: Balkema, vol. 1, 227 p.

Anderson, J.M., Anderson, H.M. 1985. Palaeoflora of Southern Africa. Rotterdam: Balkema, 423 p.

Antonov, A.S. 1995. Is Darwin's "abominable mystery" solved? - In: Evol. Biochem. & Related Areas of Physiochem. Biol., Moscow, p. 273-282 (in Russian).

Arber, A. 1918. The phyllode theory of monocotyledonous leaf, with special reference to anatomical evidence. - Ann. Bot. (Gr. Brit.), vol. 32, p. 465-501.

Arber, A. 1925. Monocotyledons, a morphological study. Cambridge: Univ. Press, 258 p.

Arber, A. 1950. The natural philosophy of plant form. Cambridge: Univ. Press, 247 p.

Arber, E.A.N., Parkin, J. 1907. On the origin of angiosperms. - Bot. J. Linn. Soc., vol. 38, p. 29-44.

Archangelsky, S. 1965. Fossil Ginkgoales from the Tico flora, Santa Cruz Province, Argentina. - Bull. Brit. Mus. (Nat. Hist.), Geol., vol. 10, no. 5, p. 121-137.

Archangelsky, S. 1968a. On the genus *Tomaxellia* (Coniferae) from the Lower Cretaceous of Patagonia and its male and female cones. - Bot. J. Linn. Soc., vol. 61, no. 384, p. 153-165.

Archangelsky, S. 1968b. Studies on Triassic fossil plants from Argentina. 4. The leaf genus *Dicroidium* and its possible relation to *Rhexoxylon* stems. - Palaeontology, vol. 11, no. 4, p. 500-512.

Archangelsky, S., Cuneo, R. 1987. Ferugliocladaceae, a new conifer family from the Permian of Gondwana. - Rev. Palaeobot. Palynol., vol. 51, nos. 1/3, p. 3-30.

Arnold, C. 1948. Classification of gymnosperms from the viewpoint of paleobotany. - Bot. Gaz., vol. 110, no. 1, p. 1-12.

Arnold, C.A., Lowther, J.S. 1955. A new Cretaceous conifer from northern Alaska. - Amer. J. Bot., vol. 42, p. 522-528.

Arnott, H. 1959. Anastomoses in the venation of *Ginkgo biloba*. - Amer. J. Bot., vol. 46, p. 405-411.

Asama, K. 1962. Evolution of Shansi flora and origin of simple leaf. - Sci. Rep. Tohoku Univ., Ser. 2, Spec. vol., no. 5, p. 247-273.

Asama, K. 1976. *Gigantopteris* flora in southeast Asia and its phytogeographic significance. - In: Geology and Palaeontology of Southeastern Asia. Tokyo: Univ. Press, p. 191-207.

Ash, S.R. 1970. *Dinophyton*, a problematical new plant genus from the Upper Triassic of the South-Western United States. - Palaeontology, vol. 13, no. 4, p. 646-663.

Ash, S. 1976. The systematic position of *Eoginkgoites*. - Amer. J. Bot., vol. 63, no. 10, p. 1321-1331.

Axelrod, D.I. 1952. A theory of angiosperm evolution. - Evolution, vol. 6, p. 29-60.

Axelrod, D.I. 1959. Poleward migration of early angiosperm flora. - Science, vol. 130, p. 203-207.

Axelrod, D.I. 1972. Edaphic aridity as a factor in angiosperm evolution. - Amer. Natur., vol. 106, no. 949, p. 311-320.

Bailey, J.W. 1956. Nodal anatomy in retrospect. - J. Arnold Arbor., vol. 37, p. 269-287.

Bailey, J.W., Swamy, B.G.L. 1951. The conduplicate carpel of dicotyledons and its initial trends of specialization. - Amer. J. Bot., vol. 38, p. 373-379.

Basinger, J.F. 1976. *Paleorosa similkameensis*, gen. et sp. nov., permineralized flowers (Rosaceae) from the Eocene of British Columbia. - Canad. J. Bot., vol. 54, no. 20, p. 2293-2305.

Basinger, J.F., Dilcher, D.L. 1984. Ancient bisexual flowers. - Science, vol. 224, no. 4648, p. 511-513.

Bate-Smith, E.C. 1972. Chemistry and phylogeny of the angiosperms. - Nature, vol. 236, p. 353-354.

Baum, H. 1949. Der einheitliche Bauplan der angiospermen Gynöcien. - Österr. bot. Ztschr., Bd. 96, S. 64-82.

Beck, C.B. 1960. The identity of *Archaeopteris* and *Callixylon*. - Brittonia, vol. 12, p. 351-368.

Behnke, H.-D. 1986. Sieve element characters and systematic position of *Austrobaileya*, Austrobaileyaceae - with comments to the distinction and definition of sieve cells and sieve-tube members. - Plant Syst. Evol., vol. 152, nos. 1/2, p. 101-121.

Béland, P., Russel, D.A. 1978. Palaeoecology of Dinosaur Provincial Park (Cretaceous), Alberta, interpreted from the distribution of articulated vertebrate remains. - Canad. J. Earth Sci., vol. 15, p. 1012-1024.

Bell, W.A. 1957. Flora of Upper Cretaceous group of Vancouver Island, British Columbia. - Mem. Geol. Surv. Canada, vol. 293, p. 1-84.

Bentham, G., Hooker, J.D. 1862-1883. Genera plantarum. London, vol. I-III.

Berry, E.W. 1911. Systematic paleontology of the Lower Cretaceous deposits of Maryland. Baltimore: Md. Geol. Surv., 597 p.

Berry, E.W. 1916. The Upper Cretaceous deposits of Maryland. Baltimore: Md. Geol. Surv., p. 757-901.

Bessey, C.E. 1897. Phylogeny and taxonomy of angiosperms. - Bot. Gaz., vol. 24, p. 145-178.

Bessey, C.E. 1915. The phylogenetic taxonomy of flowering plants. - Ann. Mo. Bot. Gard., vol. 2, p. 109-164.

Bierhorst, D.W. 1958. Vessels in *Equisetum*. - Amer. J. Bot., vol. 45, no. 7, p. 534-537.

Binda, P.L., Nambudiri, E.W.V. 1983. Fossil seed cuticles from the Upper Cretaceous Whitemud beds of Alberta and Saskatchewan, Canada. - Canad. J. Bot., vol. 61, p. 2717-2728.

Blackmore, S., van Helvoort, H.A.M., Punt, W. 1984. On the terminology, origins and functions of caveate pollen in Compositae. - Rev. Palaeobot. Palynol., vol. 43, p. 293-301.

Blagoveshchenskiy, A.V. 1975. Seed proteins and the phylogeny of Angiospermae. - Byull. Mosk. obshch. ispyt. prir., Biol., t. 80, no. 4, p. 87-92 (in Russian).

Blank-Louvel, C. 1984. Le genre "*Ranunculus* L." dans le Berriasien (Crétacé inf.) de la province de Lerida (Espagne). - La Revista Ilerida, t. 45, p. 83-92.

Blanc-Louvel, C., Barale, G. 1983. *Montsechia vidali* (Zeiller) Teixeira 1954. Nouvelles observations et réflexions sur son attribution systématique. - Ann. Paléontol., t. 69, p. 151-174.

Bogle, A.L. 1986. The floral morphology and vascular anatomy of the Hamamelidaceae: subfamily Liquidambroideae. - Ann. Mo. Bot. Gard., vol. 73, no. 2, p. 325-347.

Boltenhagen, B. 1982. Amendement de *Sofrepites legouxae* Jard. 1967 pollen Albien du Gabon. - Rev. Micropaléontol., t. 25, fasc. 3, p. 150-152.

Bose, M.N. 1966. A revision of *Rajmahalia*. - Palaeobotanist, vol. 14, nos. 1/2, p. 85-88.

Boulter, D., Bamshaw, G.A.M., Thompson, E.W. 1972. A phylogeny of higher plants based on the amino acid sequences of cytochrome "c" and its biological implications. - Proc. Roy. Soc. London, B, vol. 181, p. 441-455.

Bouman, F. 1984. The ovule. Embryology of angiosperms. Berlin etc.: Springer, p. 123-157.

Bouman, F., Calis, J.L.M. 1977. Integumentary shifting - a third way to unitegmy. - Berichte Dtsch. bot. Ges., Bd. 90, S. 15-28.

Brenner, G.J. 1967. Early angiosperm pollen differentiation in the Albian to Cenomanian deposits of Delaware (U.S.A.). - Rev. Palaeobot. Palynol., vol. 1, nos. 1/3, p. 219-227.

Brenner, G.J. 1976. Middle Cretaceous floral provinces and early migrations of angiosperms. Origin and early evolution of angiosperms. N.Y.: Columbia Univ. Press, p. 23-47.

Brown, R.W. 1956. Palmlike plants from the Dolores Formation (Triassic) in southwestern Colorado. - U.S. Geol. Surv. Profess. Pap., vol. 274H, p. 205-209.

Burger, W.C. 1977. The Piperales and the monocots. Alternate hypotheses for the origin of monocotyledonous flowers. - Bot. Rev., vol. 43, p. 345-393.

Cairns, H., Cairns, C. 1976. Psycholinguistics, a cognitive view of language. N.Y.: Holt.

Cahoon, E.J. 1972. *Paraphyllanthoxylon alabamense* - a new species of fossil dicotyledonous wood. - Amer. J. Bot., vol. 59, p. 5-11.

Canright, J.E. 1952. The comparative morphology and relationships of the Magnoliaceae. 1. Trend of specialization in the stamens. - Amer. J. Bot., vol. 39, no. 7, p. 484-497.

Carlquist, S. 1969. Toward acceptable evolutionary interpretation of floral anatomy. - Phytomorphology, vol. 19, p. 332-362.

Carlquist, S. 1987. Presence of vessels in wood of *Sarcandra* (Chloranthaceae): comments on vessel origins in angiosperms. - Amer. J. Bot., vol. 174, p. 1765-1771.

Chadefaud, M. 1966. Sur le sac embryonnaire des angiospermes. - Bull. Soc. bot. Fr., t. 113, p. 375-378.

Chakroun, S. 1983. Les formations secondaires des monocotylédons: conceptions actuelles. - Ann. Biol., t. 22, fasc. 2, p. 187-199.

Chaloner, W.G. 1976. The evolution of adaptive features of fossil exines. - In: Fergusson, I.K. & Muller, J. (eds.), The evolutionary significance of the exine. London: Academic Press, p. 1-14.

Chaloner, W.G., Hemsley, A.R. 1991. Heterospory: cul-de-sac or pathway to the seed? - In: Blackmore, S. & Barnes, S.H. (eds.), Pollen and spores: patterns of diversification. Oxford: Univ. Press, p. 151-167.

Chaloner, W.G., Hill, A.J., Lacey, W.S. 1977. First Devonian platyspermic seed and its implications in gymnosperm evolution. - Nature, vol. 265, p. 233-235.

Chamberlain, C.J. 1966a. Gymnosperms. Structure and evolution. N.Y.: Dover, 484 p.

Chamberlain, C.J. 1966b. The living cycads. N.Y.: Hafner, 172 p.

Chandler, M.E.J., Axelrod, D.J. 1961. An early angiosperm fruit from California. - Amer. J. Sci., vol. 259, p. 441-446.

Chen Zu-keng, Wang Fu-xiog. 1985. The ovule structure and development of male and female gametophytes in *Amentotaxus*. - Acta Bot. Sin., vol. 27, p. 19-25.

Chute, H.W. 1930. The morphology and anatomy of the achene. - Amer. J. Bot., vol. 17, no. 8, p. 703-720.

Clement-Westerhof, J.A. 1974. In situ pollen from gymnospermous cone of the Upper Permian of the Italian Alps - a preliminary account. - Rev. Palaeobot. Palynol., vol. 17, p. 65-73.

Clement-Westerhof, J.A. 1984. Aspects of Permian palaeobotany and palynology, 4. - Rev. Palaeobot. Palynol., vol. 41, p. 51-166.

Clement-Westerhof, J.A. 1987. Aspects of Permian palaeobotany and palynology, 7. - Rev. Palaeobot. Palynol., vol. 52, p. 375-402.

Clement-Westerhof, J.A., van Konijnenburg-van Cittert, J.H.A. 1991. *Hirmeriella muensteri*: new data on the fertile organs leading to a revised concept of the Cheirolepidiaceae. - Rev. Palaeobot. Palynol., vol. 68, p. 147-179.

Cocucci, A.E. 1973. Some suggestions on the evolution of gametophytes of higher plants. - Phytomorphology, vol. 23, nos. 1/2, p. 109-124.

Cocucci, A.E., Astegiano, M.F. 1978. Interpretacion del embrion de las Poaceas. - Kurtziana, t. 11, p. 41-54.

Cohen, L.I. 1970. Development of the pistillate flower in the dwarf mistletoes, *Arceuthobium*. - Amer. J. Bot., vol. 57, no. 4, p. 477-485.

Colin, J.-P. 1973. Microfossiles végétaux dans le Cénomanien et le Turonien de Dordogne (S.O. France). - Palaeontographica B, Bd. 143, S. 106-119.

Conover, M.H. 1982. The vegetative morphology of the reticulate-veined Liliiflorae. - Telopea, vol. 2, no. 4, p. 401-412.

Cornet, B. 1979. Angiosperm-like pollen with tectate-columellate wall structure from the Upper Triassic (and Jurassic) of the Newark Supergroup, U.S.A. - Palynology, vol. 3, p. 281-282.

Cornet, B. 1986. The leaf venation and reproductive structures of a Late Triassic angiosperm, *Sanmiguellia lewisii*. - Evol. Theory, vol. 7, no. 5, p. 231-309.

Couper, R.A. 1958. British Mesozoic microspores and pollen grains. - Palaeontographica B, Bd. 103, S. 75-179.

Crane, P.R., Dilcher, D.L. 1984. *Lesqueria:* an early angiosperm fruiting axis from the mid-Cretaceous. - Ann. Mo. Bot. Gard., vol. 71, no. 2, p. 384-402.

Crane, P.R., Friis, E.M., Pedersen, K.R. 1986. Lower Cretaceous angiosperm flowers: fossil evidence on early radiation of dicotyledons. - Science, vol. 232, p. 852-854.

Crane, P.R., Friis, E.M., Pedersen, K.R. 1989. Reproductive structure and function in Cretaceous Chloranthaceae. - Plant Syst. Evol., vol. 165, p. 211-226.

Crane, P.R., Stockey, R.A. 1985. Growth and reproductive biology of *Joffrea speirsii* gen. et sp. nov., a *Cercidiphyllum*-like plant from the Late Paleocene of Alberta, Canada. - Canad. J. Bot., vol. 63, p. 340-364.

Crane, P.R., Stockey, R.A. 1986. Morphology and development of pistillate inflorescences in extant and fossil Cercidiphyllaceae. - Ann. Mo. Bot. Gard., vol. 73, p. 382-393.

Crane, P.R., Upchurch, R. (Jr.) 1987. *Drewria potomacensis* gen. et sp. nov., an early Cretaceous member of Gnetales from the Potomac Group of Virginia. - Amer. J. Bot., vol. 74, p. 1722-1736.

Cranwell, L.M. 1961. Coniferous pollen types of the Southern Hemisphere. 1. Aberration in *Acmopyle* and *Podacarpus dacrydioides*. - J. Arnold Arbor., vol. 42, no. 4, p. 416-423.

Crepet, W.L. 1974. Investigations of North American cycadeoids: the reproductive biology of *Cycadeoidea*. - Palaeontographica B, Bd. 148, S. 144-169.

Crepet, W.L., Delevoryas, T. 1972. Investigations of North American cycadeoids: early ovule ontogeny. - Amer. J. Bot., vol. 59, p. 209-215.

Crepet, W.l., Nixon. K.C. 1994. Flowers of Turonian Magnoliidae and their implications. - Plant Syst. Evol., Suppl., vol. 8, p. 73-92.

Cridland, A.A., Morris, J.E. 1960. *Spermopteris*, a new genus of pteridosperms from the Upper Pennsylvanian Series of Kansas. - Amer. J. Bot., vol. 47, p. 855-859.

Croizat, L. 1964. Thoughts on high systematics, phylogeny and floral morphogeny with a note on the origin of angiosperms. - Candollea, vol. 19, p. 18-96.

Cronquist, A. 1968. The evolution and classification of the flowering plants. Boston: Houghton Mifflin Co., 396 p.

Cronquist, A. 1974. Thought of an origin of the monocotyledons. - B. Sahni Inst. Palaeobot. Spec. Publ., no. 1, p. 19-24.

Cronquist, A. 1977. On the taxonomical significance of secondary metabolites in angiosperms. - Plant Syst. Evol., Suppl., vol. 1, p. 179-189.

Cronquist, A. 1981. An integrated system of classification of flowering plants. N.Y.: Columbia Univ. Press, 1262 p.

Dahlgren, R. 1983. General aspects of angiosperm evolution and macrosystematics. - Nord. J. Bot., vol. 3, p. 119-149.

Daghlian, C.P. 1981. A review of the fossil record of monocotyledons. - Bot. Rev., vol. 47, p. 517-555.

Darwin, Ch. R. 1859. On the origin of species. London: Murray, 490 p.

Davis, G.L. 1966. Systematic embryology of the angiosperms. N. Y.: Wiley.

De Boer, R., Bouman, F. 1972. Integumentary studies in the Polycarpicae. 2. *Magnolia stellata* and *Magnolia virginiana*. - Acta Bot. Neerl., vol. 21, no. 6, p. 617-629.

De Candolle, A.P. 1827. Organographie végétale. Paris: Detreville, t. I, 558 p., t. II, 304 p.

Delevoryas, T. 1955. The Medullosae - structure and relationships. - Palaeontographica B, Bd. 97, S. 114-167.

Delevoryas, T. 1964. A probable pteridosperm microsporangiate fructification from the Pennsylvanian of Illinois. - Paleontology, vol. 7, p. 60-63.

Delevoryas, T. 1965 (1966). Investigations of North American cycadeoids: microsporangiate structures and phylogenetic implications. - Palaeobotanist, vol. 14, nos. 1/3, p. 89-99.

Delevoryas, T. 1968. Some aspects of cycadeoid evolution. - Bot. J. Linn. Soc., vol. 61, p. 137-146.

Dettmann, M.E. 1963. Upper Mesozoic microfloras from southern Australia. - Proc. Roy. Soc. Victoria, vol. 77, p. 1-148.

Dettmann, M.E. 1973. Angiospermous pollen from Albian to Turonian sediments of eastern Australia. - Spec. Publ. Geol. Soc. Austral., vol. 4, p. 3-34.

Diggle, P.K., de Mason, D.A. 1983. The relationship between the primary thickening meristem and the secondary thickening meristem in *Yucca whipplei* Torr., 2. - Amer. J. Bot., vol. 70, no. 8, p. 1205-1216.

Dilcher, D.L. 1974. Approaches to the identification of angiosperm leaf remains. - Bot. Rev., vol. 40, p. 1-157.

Dilcher, D.L. 1979. Early angiosperm reproduction: an introductory report. - Rev. Palaeobot. Palynol., vol. 27, p. 291-328.

Dilcher, D.L., Crane, P.R. 1984. *Archaeanthus*: an early angiosperm from the Cenomanian of the western interior of North America. - Ann. Mo. Bot. Gard., vol. 71, no. 2, p. 351-383.

Dilcher, D.L., Crepet, W.L., Beeker, C.D., Reynolds, H. 1976. Reproductive and vegetative morphology of a Cretaceous angiosperm. - Science, vol. 27, p. 854-856.

Dilcher, D.L., Kovach, W.L. 1986. Early angiosperm reproduction: *Caloda delevoryana* gen. et sp. nov., a new fructification from the Dakota Formation (Cenomanian) of Kansas. - Amer. J. Bot., vol. 73, p. 1230-1237.

Dilcher, D.L., Krassilov, V.A., Douglas, J.G. 1996. Angiosperm evolution: fruits with affinities to Ceratophyllales from the Lower Cretaceous. - Abstr. 5th Conf. Int. Org. Paleobot., Santa Barbara, Calif., p. 23.

Donoghue, M.J., Doyle, J.A. 1989. Phylogenetic analysis of angiosperms and the relationships of Hamamelididae. - In: Crane, P. R. & Blackmore, S. (eds.), Evolution, systematics and fossil history of the Hamamelididae, vol. 1. Oxford: Clarendon, p. 17-45.

Dorf, E. 1942. Flora of the Lance Formation in its type locality, Niobrara County, Wyoming. - Carnegie Inst. Wash. Publ., no. 580, p. 79-159.

Douglas, G.E. 1944. The inferior ovary. - Bot. Rev., vol. 10, p. 125-186.

Douglas, J.G. 1969. The Mesozoic floras of Victoria, 1-2. - Mem. Geol. Surv. Vic., vol. 28, p. 1-310.

Doyle, J.A. 1973. Fossil evidence on early evolution of the monocotyledon. - Quart. Rev. Biol., vol. 48, p. 399-413.

Doyle, J.A., Donoghue, M.J. 1986. Seed plant phylogeny and the origin of angiosperms: an experimental cladistic approach. - Bot. Rev., vol. 52, no. 4, p. 321-431.

Doyle, J.A., Hickey, L.J. 1972. Coordinated evolution in Potomac Group angiosperm pollen and leaves. - Amer. J. Bot., vol. 59, p. 660.

Doyle, J.A., Van Campo, M., Lugardon, B. 1975. Observations on exine structure of *Eucommiidites* and lower Cretaceous angiosperm pollen. - Pollen et Spores, vol. 17, p. 429-486.

Drinnan, A.N., Chambers, T.C. 1986. Flora of the Lower Cretaceous Koonwarra Fossil Bed (Korumburra Group), South Gippsland, Victoria. - Mem. Assoc. Austral. Palaeont., vol. 3, p. 1-77.

Drinnan, A.N., Crane, P.R., Friis, E.M., Pedersen, K.R. 1990. Lauraceous flowers from the Potomac Group (mid-Cretaceous) of eastern North America. - Bot. Gaz., vol. 151, no. 3, p. 370-384.

Eames, A.J. 1913. The morphology of *Agathis australis*. - Ann. Bot. (Gr. Brit.), vol. 27, p. 1-38.

Eames, A.J. 1961. Morphology of the angiosperms. N.Y.: McGraw Hill, 518 p.

Eiten, L.T. 1976. Inflorescence units in the Cyperaceae. - Ann. Mo. Bot. Gard., vol. 63, p. 81-112.

Emberger, L. 1954. Sur les Ginkgoales et quelques rapprochements avec d'autres groupes systématiques. - Svensk. Bot. Tidskr., Bd. 48, H. 2, S. 361-367.

Emberger, L. 1960. Les végétaux vasculaires. Traité de botanique systématique. Paris: Mason, 1539 p.

Endress, P.K. 1983. The early floral development of *Austrobaileya*. - Bot. Jb. (Syst.), Bd. 103, H. 4, S. 481-497.

Endress, P.K. 1986a. Reproductive structures and phylogenetic significance of extant primitive angiosperms. - Plant Syst. Evol., vol. 152, nos. 1/2, p. 1-28.

Endress, P.K. 1986b. Floral structure, systematics, and phylogeny in Trochodendrales. - Ann. Mo. Bot. Gard., vol. 73, no. 2, p. 297-324.

Endress, P.K. 1987. The Chloranthaceae: reproductive structures and phylogenetic position. - Bot. Jb. (Syst.), Bd. 109, S. 153-226.

Endress, P.K., 1978. Blütenontogenese, Blütenabgrenzung und systematische Stellung der perianthlosen Hamamelidoideae. - Bot. Jb. (Syst.), Bd. 100, S. 249-317.

Engler, A., Prantl, K. (eds.) 1897-1915. Die natürlichen Pflanzenfamilien. Leipzig: Engelmann.

Ettingshausen, C. 1868. Die fossile Flora der alteren Braunkohlenformation der Wetterau. - Sitz.-Ber. Kgl. Akad. Wiss. Wien, Bd. 57, S. 807-893.

Eyde, R.H. 1971. Evolutionary morphology: distinguishing ancestral structure from derived structure in flowering plants. - Taxon, vol. 20, no. 1, p. 63-73.

Faegri, K., van der Pijl, L. 1966. The principles of pollination ecology. Pergamon Press, 248 p.

Farris, J.S. 1989. The retention index and homoplasy excess ratio. - Syst. Zool., vol. 38, p. 406-407.

Favre-Duchartre, M. 1956. Contribution à l'étude de la reproduction chez le *Ginkgo biloba*. - Rev. Cytol. Biol. Végét., t. 17, fasc. 1/2, p. 1-214.

Favre-Duchartre, M. 1965. A propos des gametophytes femelles et des archégones des plantes ovulées. - Ann. Sci. Nat., Bot. & Biol. Végét., t. 6, fasc. 1, p. 157-182.

Favre-Duchartre, M. 1979. Time relations and sexual reproduction in *Cichorium* and other angiosperms as compared with Archegoniates. - Phytomorphology, vol. 29, no. 2, p. 166-178.

Fiordi, A.C. 1984. Fine structure of the stigmatic micropyle of *Larix leptolepis*. - Phytomorphology, vol. 34, nos. 1/4, p. 110-117.

Florin, R. 1931. Untersuchungen zur Stammesgeschichte der Coniferales und Cordaitales, I. - Kgl. Svensk. Vetensk.-Akad. Handl., Bd. 10, S. 1-588.

Florin, R. 1936. Die fossilen Ginkgophyten von Franz-Joseph Land nebst Erörterungen über vermeintliche Cordaitales mesozoischen Alters. 1, 2. - Palaeontographica B, Bd. 81, S. 71-713; Bd. 82, S. 1-72.

Florin, R. 1938-1945. Die Koniferen des Oberkarbons und des unteren Perms, 1-8. - Palaeontographica B, Bd. 85, S. 1-729.

Fontaine, W.M. 1899. The Potomac or younger Mesozoic flora. - U.S. Geol. Surv. Monogr., vol. 15, p. 1-377.

Foster, A.S., Gifford, E.M. (Jr.) 1974. Comparative morphology of vascular plants. San Francisco: Freeman, 751 p.

Francis, J.E. 1983. The dominant conifer of the Jurassic Purbeck Formation, England. - Palaeontology, vol. 26, p. 277-294.

Friis, E.M. 1983. Upper Cretaceous (Senonian) floral structures of Juglandalean affinity containing *Normapolles* pollen. - Rev. Palaeobot. Palynol., vol. 39, p. 161-188.

Friis, E.M. 1985. Structure and function in Late Cretaceous angiosperm flowers. - Kgl. Dansk. Videnskab. Selskab. Biol. Skr., vol. 25, p. 1-37.

Friis, E.M., Crane, P.R., Pedersen, K.R. 1986. Floral evidence for Cretaceous chloranthoid angiosperms. - Nature, vol. 320, no. 6058, p. 163-164.

Friis, E.M., Crane, P.R., Pedersen, K.R. 1988. Reproductive structures of Cretaceous Platanaceae. - Biol. Skr., vol. 31, p. 1-53.

Friis, E.M., Pedersen, K.R., Crane, P.R. 1994. Angiosperm floral structures from the Early Cretaceous of Portugal. - Plant Syst. Evol., Suppl., vol. 8, p. 31-50.

Friis, E.M., Pedersen, K.R., Crane, P.R. 1995. *Apomattoxia ancistrophora*gen. et sp. nov., a new Early Cretaceous plant with similarities to *Circaeaster* and extant Magnoliidae. - Amer. J. Bot., vol. 82, no. 7, p. 933-943.

Friis, E.M., Skarby, A. 1982. *Scandianthus* gen. nov., angiosperm flowers of Saxifragalean affinity from the Upper Cretaceous of Southern Sweden. - Ann. Bot. (Gr. Brit.), N.S., vol. 50, p. 569-583.

Fu Li-kuo (ed.) 1992. China Plant Red Data Book. Rare and endangered plants, vol. 1. Beijing, New York: Science Press, 741 p.

Gao Rui-qi 1982. Evolutionary trends of Cretaceous angiospermous pollen from the Songliao Basin, NE China. - Acta Palaeontol. Sin., vol. 21, no. 2, p. 217-224.

Gensel, P.G., Skog, J.E. 1977. Two early Mississippian seeds from the Price Formation of southwestern Virginia. - Brittonia, vol. 29, no. 3, p. 332-351.

Gillespie, W.H., Pfefferkorn, H.W. 1986. Taeniopteroid lamina on *Physmatocycas* megasporophylls (Cycadales) from the Lower Permian of Kansas, U.S.A. - Rev. Palaeobot. Palynol. vol. 49, nos. 1/2, p. 99-116.

Gillespie, W.H., Rothwell, G.W., Scheckler, S.E. 1981. The earliest seeds. - Nature, vol. 293, no. 5832, p. 462-464.

Goethe, J.W. von. 1790. Versuch die Metamorphose der Pflanzen zu Erklären. Gotha, 86 S.

Georgiev, G.N. 1985. Morphogenic and cytoembryological studies in *Taxus baccata* L. 1. Development of male reproductive structures. - Annu. Sofia Univ., 1981-1982, vol. 75, p. 21-30 (in Bulgarian).

Gerasimova-Navashina, E.N. 1968. Double fertilization in angiosperms and some theoretical aspects. - Materials of All-Union Conf. on Plant Embryol., Kiev: Naukova Dumka, p. 44-49 (in Russian).

Goldberg, A. 1986. Classification, evolution and phylogeny of the families of Dicotyledons. - Smiths. Contrib. Bot., no. 58, p. 1-314.

Golovneva, L.B. 1995. Environmental change and the patterns of floral evolution during the Cretaceous-Tertiary transition in northeastern Asia. - Paleonotol. zhurn. (Moscow), vol. 29, no. 2A, p. 36-49.

Gottlieb, L.D. 1984. Genetics and morphological evolution in plants. - Amer. Natur., vol. 123, no. 5, p. 681-709.

Gould, R.E., Delevoryas, T. 1977. The biology of *Glossopteris*: evidence from pertrified seed-bearing and pollen-bearing organs. - Alcheringa, vol. 1, p. 387-399.

Grant, V. 1975. Genetics of flowering plants. N.Y., London: Columbia Univ. Press, 514 p.

Grauvogel-Stamm, L. 1978. La flore du ges à *Voltzia* (Bundsandstein supérieur) des Vosges du nord (France) - Mém. Univ. L. Pasteur Strasb., Geol. Sci., no. 50, p. 1-226.

Grimsky, N., Hohn, T., Davies, J.W., Hohn, B. 1987. *Agrobacterium*-mediated delivery of infectious maize streak virus into maize plants. - Nature, vol. 325, no. 7000, p. 177-179.

Grove, G.C., Rothwell, G.W. 1980. *Mitrospermum vinculum* sp. nov., a cardiocarpalean ovule from the Upper Pennsylvanian of Ohio. - Amer. J. Bot., vol. 67, p. 1051-1058.

Guédès, M., Dupuy, P. 1983. From decussation to distichy, with some comments on current theories of phyllotaxy. - Bot. J. Linn. Soc., vol. 87, no. 1, p. 1-12.

Guignard, J.-L. 1984. The development of cotyledon and shoot apex in monocotyledons. - Canad. J. Bot., vol. 62, p. 1316-1318.

Gupta, K.M. 1958. *Williamsonia* fructifications from the Jurassic of Rajmahal Hills - their preservation and plan of construction. - J. Palaeont. Soc. Ind., vol. 3, p. 230-232.

Hall, J.M. 1963. Megaspores and other fossils in the Dakota Formation (Cenomanian) of Iowa (U.S.A.). - Pollen et Spores, vol. 5, p. 425-443.

Hallier, H. 1912. L'origine et la système phylétique des angiospermes. - Arch. Néer. Sci., vol. 1, p. 146-234.

Harris, T.M. 1932. The fossil flora of Scoresby Sound, East Greenland. Caytoniales and Bennettitales. - Medd. Groenl., vol. 85, no. 5, p. 1033.

Harris, T.M. 1946. A coprolite of *Caytonia* pollen. - Ann. Mag. Nat. Hist., Ser., vol. 12, p. 357-378.

Harris, T.M. 1951a. The relationships of the Caytoniales. - Phytomorphology, vol. 1, nos. 1/2, p. 29-39.

Harris, T.M. 1951b. The fructification of *Czekanowskia* and its allies. - Phil. Trans. Roy. Soc. London, vol. 235B, p. 483-508.

Harris, T.M. 1953. Conifers of the Taxodiaceae from the Wealden Formation of Belgium. - Mém. Inst. Roy. Sci. Nat. Belg., t. 126.

Harris, T.M. 1964. The Yorkshire Jurassic flora. 2. Caytoniales, Cycadales and Pterridosperms. London: Brit. Mus. (Nat. Hist.), 191 p.

Harris, T.M. 1969. The Yorkshire Jurassic flora. 3. Bennettitales. London: Brit. Mus. (Nat. Hist.), 186 p.

Harris, T.M. 1979. The Yorkshire Jurassic flora. 5. Coniferales. London: Brit. Mus. (Nat. Hist.), 166 p.

Harris, T.M. 1983. Fossils from New Zealand ascribed to the *Pentoxylon* plant. - Phyta, D.D. Pant Commemorial vol., p. 91-103.

Harris, T.M., Millington, W., Miller, J. 1974. The Yorkshire Jurassic flora. 4. Ginkgoales, Czekanowskiales. London: Brit. Mus. (Nat. Hist.), 150 p.

Hecker, R.F. 1948. The Karatau locality of fauna and flora of Jurassic age. - Trudy Paleont. inst. Akad. nauk SSSR (Moscow), t. 15 (1) (in Russian).

Heel, W.A. von. 1981. A SEM investigation of the development of free carpels. - Blumea, vol. 27, p.499-522.

Heer, O. 1869. Contributions to the fossil flora of North Greenland. - Phil. Trans. Roy. Soc. London, vol. 2B, p. 39-56.

Heer, O. 1876. Beiträge zur Jura-Flora Ostsibiriens und des Amurlandes. - Mém. Acad. Imp. Sci. St.-Pétersbourg, t. 22, p. 1-122.

Heer, O. 1878. Beiträge zur fossilen Flora Sibiriens und des Amurlandes. - Mém. Acad. Imp. Sci. St.-Pétersbourg, sér. 25, t. 7, p. 1-58.

Heer, O. 1880. Nachträge zur fossilien Flora Grönlands. - Flora fossilis arctica, Zürich, Bd. 6. S. 1-224.

Heinsbroek, P. G., van Heel, W.A. 1969. Note on the bearing of the pattern of vascular bundles on the morphology of the stamens of *Victoria amazonica* (Poepp.) Sowerb. - Proc. Kon. Nederl. Akad. Wetensch., C., vol. 72, p. 431-444.

Hennig, W. 1950. Grundzüge einer Theorie der phylogenetischen Systematik. Berlin: Deutscherzentralverlag (English edition: 1966, Univ. Illinis Press).

Henslow, G. 1911. The origin of monocotyledons from dicotyledons through self-adaptation to a moist or aquatic habitat. - Ann. Bot. (Gr. Brit.), vol. 26, p. 717-744.

Herendeen, P.S., Crepet, W.L., Nixon, K.C. 1993. *Chloranthus*-like stamen from the Upper Cretaceous of New Jersey. - Amer. J. Bot., vol. 80, p. 865-871.

Herngreen, G.E.W., Chlonova, A.F. 1981. Cretaceous microfloral provinces. - Pollen et Spores, vol. 23, p. 441-555.

Herzfeld, S. 1922. *Ephedra campylopoda* Mey. Morphologie der weiblichen Blüte und Befruchtungsvorgang. - Denkschr. Akad. Wiss. Wien, Bd. 98, S. 1-26.

Herzfeld, S. 1927. Beiträge zur Kenntnis von *Ginkgo*. - Jb. Wiss. Bot., Bd. 66, H. 5, S. 814-818.

Heslop-Harrison, J. 1976. The adaptive significance of the exine. Linn. Soc. Symp. Ser., London: Academic Press, no. 1, p. 27-37.

Heslop-Harrison, Y., Heslop-Harrison, J., Reger, B.J. 1985. The pollen-stigma interaction in the grasses. 7. Pollen-tube guidance and the regulation of tube number in *Zea mays* L. - Acta Bot. Neerl., vol. 34, no. 2, p. 193-211.

Hesse, M. 1984. Pollenkitt is lacking in Gnetatae: *Ephedra* and *Welwitschia*, further proof for its restriction to the angiosperms. - Plant Syst. Evol., vol. 144, no. 1, p. 9-16.

Hickey, L.J. 1973. Classification of the architecture of dicotyledonous leaves. - Amer. J. Bot., vol. 60, p. 17-33.

Hickey, L.J., Doyle J.A. 1977. Early Cretaceous fossil evidence for angiosperm evolution. - Bot. Rev., vol. 43, no. 1, p.3-104.

Hill R. 1980. Three new Eocene cycads from eastern Australia. - Austral. J. Bot. vol. 28, p. 105-122.

Hochuli, P.A. 1979. Ursprung und Verbreitung der Restionaceen. - Vierteiljahrschr. naturf. Ges. Zürich, Bd. 124, S. 109-131.

Holmes, W.B. 1974. On some fructifications of the Glossopteridales from the Upper Permian of N.S.W. - Proc. Linn. Soc. N.S.W., vol. 98, no. 3, p. 131-141.

Hörhammer, L. 1933. Über die Coniferen-Gattung *Cheirolepis* Schimper und *Hirmeriella* nov. gen. aus dem Rhät - Lias von Franken. - Bibl. Bot., Bd. 27, S. 1-33.

Howard, R.A. 1974. The stem-leaf continuum of the Dicotyledoneae. - J. Arnold Arbor., vol. 55, no. 2, p. 125-181.

Hu Zheng-hai, Tian Lan-xin. 1985. Studies on morphology of *Kingdonia unifora* F. Balfour et W.W. Smith: 3. The morphology and anatomy of flowers, fruits and seeds. - Acta Phytotaxon. Sin., vol. 23, p. 170-178.

Hughes, N.F. 1961a. Fossil evidence and angiosperm ancestry. - Sci. Progr., vol. 49, p. 84-102.

Hughes, N.F. 1961b. Further interpretation of *Eucommiidites* Erdtman, 1948. - Palaeontology, vol. 4, p. 292-299.

Hughes, N.F. 1976. Palaeobiology of angiosperm origins. Cambridge: Univ. Press, 242 p.

Hughes, N.F. 1994. The enigma of angiosperm origins. Cambridge: Univ. Press, 303 p.

Hughes, N.F., McDougall, A.B. 1987. Record of angiospermid pollen entry into the English Early Cretaceous succession. - Rev. Palaeobot. Palynol., vol. 50, no. 3, p. 255-272.

Hutchinson, J. 1959. The families of flowering plants. 2nd edition. Oxford: Clarendon Press, 903 p.

Hutchinson, J. 1969. Evolution and phylogeny of flowering plants. Facts and theory. London, N.Y.: Academic Press, 717 p.

Ibrahim, M.I.A. 1996. Aptian-Turonian palynology of the Ghazalat-I Well (GTX-I), Qattara Depression, Egypt. - Rev. Paleobot Palynol, vol. 94, p. 137-168.

Inamdar, J.A., Mohan, J.S.S., Bagavathi, S.R. 1986. Stomatal classification - a review. - Feddes Rep., vol. 97, nos. 3/4, p. 147-160.

Johnson, R.W., Riding, R.T. 1981. Structure and ontogeny of the stomatal complex in *Pinus strobus* L. and *P. banksiana* Lamb. - Amer. J. Bot., vol. 68, no. 2, p. 260-268.

Johri, B.M. (ed.).1984. Embryology of angiosperms. Berlin etc.: Springer, 830 p.

Johri, B.M., Ambegaokar, K.B. 1984. Some unusual features in the embryology of angiosperms. - Proc. Ind. Acad. Sci., Plant. Sci., vol. 93, no. 3, p. 413-427.

Jones, J.H., Dilcher, D.L. 1986. A study of the *"Dryophyllum"* leaf forms from the Paleogene of southeastern North America. - Palaeontographica B, Bd. 208, S. 53-80.

Jung, W. 1967. Eine neue Rekonstruktion der Fruchtzapfens von *Cheirolepis muensteri* Schimper. - Neues Jb. Geol. Paläont. Mh., H. 2, S. 111-114.

Kanta, K. 1962. Morphology and embryology of *Piper nigrum* L. - Phytomorphology, vol. 12, no. 3, p. 207-221.

Kaplan, D.R. 1970. Comparative foliar histogenesis in *Acorus calamus* and its bearing on the phyllode theory of monocotyledonous leaves. - Amer. J. Bot., vol. 57, p. 331-361.

Kedves, M.J. 1982. Ultrastructural investigation of the early *Normapolles* taxa *Complexiopollis* and *Limaipollenites*. - Palynology, vol. 6, p. 149-159.

Kedves, M. 1983. La stratification de l'exine et la morphologie des *Normapolles*. - Physio-Géo, t. 6, p. 53-67.

Kedves, M.J., Párdutz, Á. 1983a. Scanning electron microscopy of some selected recent amentiflorae pollens. 2. - Acta Biol. Szeged., t. 29, fasc. 1/4, p. 67-76.

Kedves, M.J., Párdutz, Á. 1983b. Electron microscope investigations of the early *Normapolles* pollen genus *Atlantopollis*. - Palynology, vol. 7, p. 153-169.

Keng, H. 1965. Report on an abnormal ovulate strobilus of *Gnetum gnemon* L. - Gard. Bull. Singapore, vol. 21, no. 2, p. 213-214.

Kerp, J.H.F. 1983. Aspects of Permian palaeobotany and palynology. 1, *Sobernheimia Jonkeri* nov. gen. sp., a new fossil plant of cycadalean affinity from the Waderner Gruppe of Sobernheim. - Rev. Palaeobot. Palynol., vol. 38, nos. 3/4, p. 173-184.

Kerp, J.H.F. 1986. On some interesting fructifications from the Lower Permian of the Saar-Nahe Area (Western Germany). - Cour. Forsch.-Inst. Senckenberg, Bd. 86, S. 73-87.

Kerp, H. 1996. Post-Variscan Late Paleozoic Northern Hemisphere gymnosperms: the onset to the Mesozoic. - Rev. Palaeobot. Palynol., vol. 90. p. 263-285.

Khan, R. 1940. A note on "double fertilization" in *Ephedra foliata*. - Curr. Sci., vol. 9, p. 323-324.

Khlonova, A.F. 1961. Spores and pollen from the upper part of the Late Cretaceous in the eastern part of the West-Siberian Plain. Novosibirsk: Siberian Branch, Akad. nauk SSSR, 138 p. (in Russian).

Khlonova, A.F. 1974. Palynology of the Cretaceous in Siberia and the Far East. Novosibirsk: Nauka, 166 p. (in Russian).

Kimura, T., Sekido, S. 1975. *Nilssoniocladus* n. gen. (Nilssoniaceae n. fam.) newly found from the early Lower Cretaceous of Japan. - Palaeontographica B, Bd. 153, H. 1/3, S. 111-118.

Kirchheimer, F. 1957. Die Laubgewächse der Braunkohlenzeit. Halle; Knapp, 672 S.

Klaus, W. 1979. Zur entwicklungsgeschichtlichen Bedeutung triadischer angiospermider Pollenapertur und Strukturanlagen. - Beitr. Paläontol. Österr., Bd. 6, S. 135-177.

Knobloch, E. 1960. Paleogenni flora z Ceskeho Chloumku u Karlových Var. - Sborn. ùstr. úst. Geol., sv. 27, s. 101-155.

Knobloch, E. 1978. On some primitive angiosperm leaves from the Upper Cretaceous of the Bohemian Massif. - Palaeontographica B, Bd. 166, S. 83-98.

Knobloch, E. 1983. Die Gattung *Costatheca* Hall in der mitteleuropäischen Kreide. - Paleontol. Sborn. Geol. Ved, sv. 24, s. 95-105.

Knobloch, E., Mai, D.H. 1986. Monographie der Früchte und Samen in der Kreide von Mitteleuropa. - Rozpr. ùstred. ústavu Geol., sv. 47, s. 1-219.

Knoll, F. 1948. Bau, Entwicklung und morphologische Bedeutung unifazialer Verläuferspitzen an Monokotylenblättern. - Österr. Bot. Ztschr., Bd. 95, S. 163-193.

Knowlton, F.H. 1930. The flora of the Denver and associated formations of Colorado. - U.S. Geol. Surv. Profess. Pap., vol. 155, p. 1-142.

Kohn, J.R., Waser, N.M. 1985. The effect of *Delphinium nelsonii* pollen on seed set in *Ipomopsis aggregata*, a competitor for hummingbird pollination. - Amer. J. Bot., vol. 72, no. 7, p. 1144-1148.

Komar, G.A. 1972. On the nature of the lower ovary. - Bot. Zhurn. (Leningrad), vol. 57, p. 595-604 (in Russian).

Konar, R., Oberoi, Y.P. 1969. Recent work on reproductive structure of living conifers and taxads - a review. - Bot. Rev., vol. 35, no. 2, p. 89-117.

Krassilov, V.A. 1967. Early Cretaceous flora of Southern Primorye and its stratigraphic significance. Moscow: Nauka, 264 p. (in Russian).

Krassilov, V.A. 1969. Approach to the classification of Mesozoic "Ginkgoalean" plants from Siberia. - Palaeobotanist, vol. 18, p. 12-19.

Krassilov, V.A. 1972a. Phytogeographical classification of Mesozoic floras and their bearing on continental drift. - Nature, vol. 237, p. 49.

Krassilov, V.A. 1972b. Mesozoic flora of Bureya River. Ginkgoales and Czekanowskiales. Moscow, Nauka, 152 p. (in Russian).

Krassilov, V.A. 1972c. Paleoecology of terrestrial plants. Vladivostok: Far East Branch, Akad. nauk SSSR, 208 p. (in Russian; English translation: 1975, N.Y.: Wiley).

Krassilov, V.A. 1973a. Mesozoic plants and the problem of angiosperm ancestry. - Lethaia, vol. 6, p. 163-178.

Krassilov, V.A. 1973b. The Jurassic disseminules with pappus and their bearing on the problem of angiosperm ancestry. - Geophytology, vol. 3, p. 1-4.

Krassilov, V.A. 1973c. Cuticular structures of Cretaceous angiosperms from the Far East of the USSR. - Palaeontographica B, Bd. 142, S. 105-116.

Krassilov, V.A. 1973d. Climatic changes in Eastern Asia as indicated by fossil floras. 1. Early Cretaceous. - Palaeogeogr., Palaeoclimatol., Palaeoecol., vol. 13, p. 261-273.

Krassilov, V.A. 1975a. Dirhopalostachyaceae - a new family of proangiosperms and its bearing on the problem of angiosperm ancestry. - Palaeontographica B, Bd. 153, S. 100-110.

Krassilov, V.A. 1975b. Climatic changes in Eastern Asia as indicated by fossil floras. 2. Late Cretaceous and Danian. - Palaeogeogr., Palaeoclimatol., Palaeoecol., vol. 17, p. 157-172.

Krassilov, V.A. 1976. The Tsagajan flora of the Amur Province. Moscow: Nauka, 92 p. (in Russian).

Krassilov, V.A. 1977a. The origin of angiosperms. - Bot. Rev., vol. 43, no. 1, p. 143-176.

Krassilov, V.A. 1977b. Contributions to the knowledge of the Caytoniales. - Rev. Palaeobot. Palynol., vol. 24, p. 155-178.

Krassilov, V.A. 1978. Bennettitalean stomata. - Palaeobotanist, vol. 25, p. 179-184.

Krassilov, V.A. 1979. Cretaceous flora of Sakhalin. Moscow: Nauka, 182 p. (in Russian).

Krassilov, V.A. 1981. Changes of Mesozoic vegetation and the extinction of dinosaurs. - Palaeogeogr., Palaeoclimatol., Palaeoecol., vol. 34, p. 207-224.

Krassilov, V.A. 1982a. On the ovuliferous organ of *Hirmerella*. - Phyta, D.D. Pant Commemorial vol., p. 141-144.

Krassilov, V.A. 1982b. Early Cretaceous flora of Mongolia. - Palaeontographica B, Bd. 181, S. 1-43.

Krassilov, V.A. 1984a. New paleobotanical data on origin and early evolution of angiospermy. - Ann. Mo. Bot. Gard., vol. 71, p. 577-592.

Krassilov, V.A. 1984b. Albian-Cenomanian flora of the Kacha-Bodrak Intershed, Crimea. - Byull. Mosk. obshch. ispyt. prir., Geol., t. 59, no. 4, p. 104-112 (in Russian).

Krassilov, V.A. 1985. Cretaceous Period. Evolution of earth crust and biosphere. Moscow: Nauka, 240 p. (in Russian).

Krassilov, V.A., 1986. New floral structure from the Lower Cretaceous of Lake Baikal area. - Rev. Palaeobot. Palynol., vol. 47, p. 9-16.

Krassilov, V. A. 1989a. Origin and early evolution of angiosperms. Moscow: Nauka, 263 p. (in Russian).

Krassilov, V.A. 1989b. Correlation of the Cretaceous volcanogenic strata in the Soviet Far East - problems and perspectives. - In: Krassilov, V.A. (ed.), Volcanic Cretaceous of the Far East. Vladivostok: Far East Branch, Akad. nauk SSSR, p. 6-13 (in Russian).

Krassilov, V. A. 1990. Fossil links reconsidered. - Proc. 3rd IOP Conf., Melbourne, p. 7-10.

Krassilov, V.A. 1991. The origin of angiosperms: new and old problems. - TREE, vol. 6, p. 215-220.

Krassilov, V.A. 1994. Global climatic changes as a factor of biosphere evolution. - In: Ecosystem restructurings and the evolution of the biosphere, 1. Moscow: Nedra, p. 285-294 (in Russian).

Krassilov, V.A. 1995. *Scytophyllum* and the origin of angiosperm leaf characters. - Paleont. Zh. (Moscow), vol. 29, No 1A, p. 63-73

Krassilov, V.A., Ash, S. 1988. On *Dinophyton* - protognetalean Mesozoic plant. - Palaeontographica B, Bd. 208, S. 33-38.

Krassilov, V.A., Blokhina, N.I., Markevich, V.S. 1987. The Cretaceous-Paleogene of the Lesser Kuril Islands. Vladivostok: Far East Branch, Akad. nauk SSSR, 140 p. (in Russian).

Krassilov, V.A., Bugdaeva, E.V. 1982. Achene-like fossils from the Lower Cretaceous of the Lake Baikal area. - Rev. Palaeobot. Palynol., vol. 36, p. 279-295.

Krassilov, V.A., Bugdaeva, E.V. 1988a. Gnetalean plants from the Jurassic of Ust-Balej, East Siberia. - Rev. Palaeobot. Palynol., vol. 53, p. 359-374.

Krassilov, V.A., Bugdaeva, E.V. 1988b. *Meeusella* and the origin of stamens. - Lethaia, vol. 21, p. 425-431.

Krassilov, V.A., Bugdaeva, E.V. 1988c. Protocycadopsid pteridosperms from the Lower Cretaceous of Transbaikalia and the origin of cycads. - Palaeontographica B, Bd. 208, S. 27-32.

Krassilov, V.A., Bugdaeva, E.V. 1997. Baisian assemblage of proangiosperms. - Palaeontographica B (in press).

Krassilov, V.A., Burago, V.I. 1981. New interpretation of *Gaussia* (Vojnovskyales). - Rev. Palaeobot. Palynol., vol. 32, p. 227-237.

Krassilov, V.A., Delle, G.V., Vladimirova, H.V. 1996. A new Jurassic pollen cone from Georgia and its bearing on cycad phylogeny. - Palaeontographica B, 1996, Bd. 238, S. 71-75.

Krassilov, V.A., Dilcher, D.L., Douglas, J.G., 1996. Ephedroid plant from the Lower Cretaceous of Koonwarra, Australia. - Abstr. 5th Conf. Int. Org. Paleobot., Santa Barbara, Calif., p. 54.

Krassilov, V.A., Dilcher, D.L., Douglas, J.G. 1997. New ephedroid plant from the Lower Cretaceous of Koonwarra, Australia. - Alcheringa (in press).

Krassilov, V.A., Dobruskina, I.A. 1995. Angiosperm fruit from the Lower Cretaceous of Israel and origins in rift valleys. - Paleont. Zh. (Moscow), vol. 29, No. 1A, p. 110-115.

Krassilov. V.A., Fotyanova, L.I. 1995. A new hamamelid fructification from the Paleogene of Kamchatka, with comments on *Trochodendrocarpus* and *Nordenskioldia*. - Paleonotol. Zh. (Moscow), vol. 29, no. 1A, p. 87-95.

Krassilov, V.A., Golovneva, L.B., Nessov, L.A. 1990. A cycadophyte from the Late Cretaceous dinosaur locality in northern Koryakia. - In: Krassilov, V.A. (ed.), Continental Cretaceous of the USSR. Vladivostok: Far East Branch, Akad. nauk SSSR, p. 213-215 (in Russian).

Krassilov, V.A., Makulbekov, N.M. 1995. Maastrichtian aquatic plants from Mongolia. - Paleonotol. Zh. (Moscow), vol. 29, no. 2A, p. 119-140.

Krassilov, V.A., Martinson, H.H. 1982. Fruits from the Late Cretaceous of Mongolia. - Paleontol Zh. (Moscow), no. 1, p. 113-121 (in Russian).

Krassilov, V.A., Pacltová, B. 1989. *Asterocelastrus cretacea*, a mid-Cretaceous angiosperm from Bohemia. - Rev. Paleobot. Palynol., vol. 60, p. 1-6.

Krassilov, V.A., Rasnitsyn, A.P. 1982. Unique find: pollen in the guts of Early Cretaceous xyelotomid insects. - Paleont. Zh. (Moscow), no. 4, p. 83 -96.

Krassilov, V.A., Rasnitsyn, A.P. 1997. Pollen in the guts of Permian insects: first evidence of pollinivory and its evolutionary significance. - Lethaia (in press).

Krassilov, V.A., Raskatova, M.G., Istschenko, A.A. 1987. A new Archaeopteridalean plant from the Devonian of Pavlovsk, U.S.S.R. - Rev. Paleobot. Palynol., vol. 53, p. 163-173.

Krassilov, V.A., Shilin, P.V. 1995. New staminoid heads from the mid-Cretaceous of Kazakhstan. - Rev. Palaeobot. Palynol., vol. 85, p. 207-211.

Krassilov, V.A., Shilin, P.V., Vachrameev, V.A. 1983. Cretaceous flowers from Kazakhstan. - Rev. Palaeobot. Palynol., vol. 40, p. 91-113.

Krassilov. V.A., Shorokhova, S.A. 1970. New Triassic plants from the Iman River basin (Primorye) and some problems of Mesozoic pteridophyll morphology. - In: Triassic invertebrates and plants of the eastern USSR. Vladivistok: Far East Branch, Akad. nauk SSSR, p. 98-121 (in Russian).

Krassilov, V.A., Shorokhova, S.A. 1988. Phytostratigraphy of the Alchan Formation in northern Primorye. In: Krassilov, V.A. (ed.), Volcanic Cretaceous of the Far East. Vladivostok: Far East Branch, Akad. nauk SSSR, p. 43-49 (in Russian).

Krassilov, V.A., Zakharova, T.V. 1995. *Moresnetia*-like plants from the Upper Devonian of Minusinsk Basin, Siberia. - Paleonotol. zhurn. (Moscow), vol. 29, no. 1A, p. 35-43.

Krassilov. V.A., Zherikhin, V.V., Rasnitsyn, A.P. 1997. *Classopollis* in the guts of Jurassic insects. - Palaeontology, (in press).

Kräusel, R. 1956. Zur Geschichte der Angiospermen. - Bot. Mag. (Tokyo), vol. 69, no. 822, p. 537-544.

Kryshtofovich, A.N. 1958. Fossil floras of Penzhin Gulf, Tastakh Lake, and Rarytkin Range. - Trudy Bot. inst. Akad. nauk SSSR, ser. 8, Paleobotany, t. 3, p. 73-121 (in Russian).

Kubitzki, K. 1973. Probleme der Großsystematik der Blütenpflanzen. - Berichte Dtsch. bot. Ges., Bd. 85, S. 259-277.

Kubitzki, K., Gottlieb, O.R. 1984a. Micromolecular patterns and the evolution and major classification of angiosperms. - Taxon, vol. 33, no. 3, p. 375-391.

Kubitzki, K., Gottlieb, O.R. 1984b. Phytochemical aspects of angiosperm origin and evolution. - Acta Bot. Neerl., vol. 33, no. 4, p. 457-468.

Kügler, H. 1970. Blütenökologie. Jena: G. Fischer, 345 S.

Kupriyanova, L.A. 1976. Pollen morphology in the genus *Nymphaea* in the European part of the USSR. - Bot. Zhurn. (Leningrad), vol. 61, p. 1558-1563 (in Russian).

Kuznetsov, N.I. 1936. Introduction to the systematics of flowering plants, 2nd edition. Leningrad: Akad nauk SSSR, 456 p. (in Russian).

Kvaček, Z., Knobloch, E. 1967. Zur Nomenklatur der Gattung *Daphnogene* Ung. und die neue Art *Daphnogene pannonica* sp. n. - Vestn. Ústred. geol., sv. 42, s. 201-210.

Lam, H.J. 1950. Stachyospory and phyllospory as factors in the natural system of the Cormophyta. - Svenk. Bot. Tidskr., Bd. 44, H. 4, S. 517-534.

Lang, W.H. 1900. Studies in the development and morphology of Cycadean sporangia. 2. The ovule of *Stangeria paradoxa*. - Ann. Bot. (Gr. Brit.), vol. 14, p. 281-306.

Lanyon, S.M. 1994. Polyphyly of the blackbird genus *Agelaius* and the importance for assumptions of monophyly in comparative studies. - Evolution, vol. 48, no. 3, p. 679-693.

Laubenfels, D.J. de. 1953. The external morphology of coniferous leaves. - Phytomorphology, vol. 3, nos. 1/2, p. 1-19.

Lee, C.L. 1955. Fertilization in *Ginkgo biloba*. - Bot. Gaz., vol. 117, no. 2, p. 79-100.

Leinfellner, W. 1950. Der Bauplan des synkarpen Gynözeums. - Österr. bot. Ztschr., Bd. 97, S. 403-436.

Leins, P., Erbar, C. 1985. Ein Beitrag zur Blütenentwicklung der Aristolochiaceen, einer Vermittelgruppe zu den Monokotylen. - Bot. Jb. (Syst.), Bd. 107, S. 343-368.

Leroy, J.-F. 1983. The origin of angiopsperms: an unrecognized ancestral dicotyledon, *Hedyosmum* (Chloranthales), with a strobiloid flower is living today. - Taxon, vol. 32, no. 2, p. 169-175.

Lesquereux, L. 1874. Contributions to the fossil flora of the Western Territories, I. The Cretaceous flora. - U.S. Geol. Geogr. Surv. Territ. Rept., vol. 6, p. 1-136.

Lesquereux, L. 1876. Palaeontology. - U.S. Geol. Geogr. Surv. Territ. Rept., vol. 8, p. 271-366.

Lesquereux, L. 1883. Contributions to the fossil flora of the Western Territories, 3. The Cretaceous and Tertiary floras. - U.S. Geol. Geogr. Surv. Territ. Rept., vol. 8, p. 1-283.

Li, H., Tian, B, Taylor, E.L., Taylor, T.N. 1994. Foliar anatomy of *Gigantonoclea guizouensis* (Gigantopteridales) from the Upper Permian of Guizhou Province, China. - Amer. J. Bot., vol. 21, no. 6, p. 678-689.

Li Xingxue, Yao Zhaoqi 1983. Fructifications of gigantopterids from South China. - Palaeontographica B, Bd. 185, nos. 1/3, S. 11-26.

Lillegraven, J.A., Kielan-Jaworowska, Z., Clemens, W.A. (Jr.) 1979. Mesozoic mammals: the first two-thirds of mammalian history. Berkeley: Univ. California Press, 311 p.

Linnaei, C. 1751. Philosophia botanica. Stockholm.

Linnel, T. 1933. Zur Morphologie und Systematik triassischer Cycadophyten. 2. Über *Scytophyllum* Bornem., eine wenig bekannte Cycadophytengattung aus dem Keuper. - Svensk. Bot. Tidskr., Bd. 27, H. 3, S. 310-331.

Long, A.G. 1966. Some Lower Carboniferous fructifications from Berwickshire, together with a theoretical account of the evolution of ovules, cupules, and carpels. - Trans. Roy. Soc. Edinburgh, vol. 66, p. 345-375.

Long, A.G. 1969. *Eurystoma trigona* sp. nov., a pteridosperm ovule borne on a frond of *Alcicarnopteris* Kidston. - Trans. Roy. Soc. Edinburgh, vol. 68, p. 171-182.

Long, A.G. 1977a. Some Lower Carboniferous pteridpsperm cupules bearing ovules and microsporangia. - Trans. Roy. Soc. Edinburgh, vol. 70, nos. 1/3, p. 1-11.

Long, A.G. 1977b. Lower Carboniferous pteridosperm cupules and the origin of angiosperms. - Trans. Roy. Soc. Edinburgh, vol. 70, nos. 1/3, p. 13-35.

Long, A.G. 1986. The cupule-carpel theory. A defence. - Trans. Roy. Soc. Edinburgh, vol. 44, no. 2, p. 281-285.

Lucke, E. 1984. Samenstruktur und Samenkeimung europäischer Orchideen nach VEYRET sowie weitere Untersuchungen. - Orchidee, Bd. 35, H. 4, S. 153-158.

Macdonald, A.D. 1977. Myricaceae: floral hypothesis for *Gale* and *Comptonia*. - Canad. J. Bot., vol. 55, p. 2636-2651.

Mädel, E. 1962. Die fossilen Euphorbiaceen-Holzer mit besondere Berücksichtigung neuer Funde aus der Oberkreide Süd-Afrikas. - Senckenberg. lethaea, Bd. 43, H. 4, S. 283-321.

Maheshwari, S. C. 1958. *Spirodela polyrhiza*, the link between the aroids and the duckweeds. - Nature, vol. 181, p. 1745-1746.

Maheshwari, P., Singh, H. 1967. The female gametophyte of gymnosperms. - Biol. Rev. Cambridge Phil. Soc., vol. 42, no. 1, p. 88-130.

Maheshwari, P., Vasil, V. 1961. *Gnetum*. New Delhi: Counc. Sci. & Industr. Res., 142 p.

Maley, J. 1996. The African rain forest - main characteristics of changes in vegetation and climate from the Upper Cretaceous to the Quaternary. - In: Alexander, I.J., Swaine, M.D. & Watling, R. (eds.), Essays on the ecology of the Guinea-Congo rain forest. - Proc. Roy. Soc. Edinburgh, B (Biol. Sci.), vol. 104, p. 31-74.

Mamay, S.H. 1966. *Tinsleya*, a new genus of seedbearing callipterid plants from the Permian of North-Central Texas. - U.S. Geol. Surv. Profess. Pap., vol. 523, pt. E, p. 1-15.

Mamay, S.H. 1968. *Russelites*, new genus, a problematical plant from the Lower Permian of Texas. - U.S. Geol. Surv. Profess. Pap., vol. 593, pt. I, P, p. 1-15.

Mamay, S.H. 1976. Paleozoic origin of cycads. - U.S. Geol. Surv. Profess. Pap., vol. 934, p. 1-48.

Mamay, S.H., Watt, A.D. 1971. An ovuliferous callipteroid plant from the Hermit Shale (Lower Permian) of the Grand Canyon, Arizona. - U.S. Geol. Surv. Profess. Pap., vol. 750, pt. C, p. 48-51.

Manchester, S.R. 1986. Vegetative and reproductive morphology of an extinct plane tree (Platanaceae) from the Eocene of western North America. - Bot. Gaz., vol. 147, p. 200-226.

Mapes, G., Rothwell, G.W. 1980. *Questora amplecta* gen. et sp. n., a structurally simple medullosan stem from the Upper Mississippian of Arkansas. - Amer. J. Bot., vol. 67., p. 636-647.

Mapes, G., Rothwell, G.W. 1984. Permineralized ovulate cones of *Lebachia* from late Palaeozoic limestones of Kansas. - Palaeontology, vol. 27, p. 69-94.

Martens, P. 1971. Les gnétophytes. - Handb. Pflanzenanat., Bd. 12, S. 1-295.

Martin, P.G., Dowd, J.M. 1984. The study of plant phylogeny using amino acid sequences of ribulose-1,5-bisphosphate carboxylasae. 5. Magnoliaceae, Polygonaceae and the concept of primitiveness. - Austral. J. Bot., vol. 32, p. 301-309.

Martin, P.G., Dowd, J.M. 1986. A phylogenetic tree for some monocotyledons and gymnosperms derived from protein sequences. - Taxon, vol. 35, no. 3, p. 469-475.

Maslova, N.P., Krassilov, V.A. 1997. New hamamelid infructescence from the Paleocene of Western Kamchatka, Russia. - Rev. Palaeobot. Palynol. (in press).

Matten, L.C., Lacey, W.s., Lyces, R.C. 1980. Studies on the cupulate seed genus *Hydrosperma* Long from Berwickshire and east Lothian in Scotland and County Kerry in Ireland. - Bot. J. Linn. Soc., vol. 81, p. 249-273.

Médus, J. 1983. Morphologie, fonction et évolution de quelques grains de pollen *Normapolles* du Crétacée supérieur. - Lethaia, vol. 16, p. 265-271.

Meeuse, A.D.J. 1963. The so-called "megasporophyll" of *Cycas* - a morphological misconception. Its bearing on the phylogeny and the classification of the Cycadophyta. - Acta Bot. Neerl., vol. 12, no. 1, p. 119-128.

Meeuse, A.D.J., Bouman, F. 1974. The inner integument - its probable origin and homology. - Acta Bot. Neerl., vol. 24, p. 237-249.

Mehra, P.N. 1950. Occurrence of hermaphrodite flowers and the development of female gametophyte in *Ephedra intermedia* Shrenk et May. - Ann. Bot. (Gr. Brit.), N.S., vol. 14, p. 165-180.

Melville, R. 1963. A new theory of the angiosperm flower. - Kew Bull., vol. 17, no. 1, p. 1-63.

Melville, R. 1976. The terminology of leaf architecture. - Taxon, vol. 25, nos. 5/6, p. 549-561.

Merril, E.K. 1986. Heteroblastic seedling of green ash. 2. Early development of simple and compound leaves. - Canad. J. Bot., vol. 64, p. 2650-2601.

Mestre, J.-C., Guédès, M. 1983. Nature et signification phylogénétique du cotyledon. - Bull. Soc. bot. Fr., t. 130, p. 7-22.

Meyen, S. 1981. Some true and alleged Permotriassic conifers of Siberia and the Russian platform and their alliance. - Palaeobotanist, vol. 28/29, p. 161-176.

Meyen, S.V. 1982. Reproductive organs of gymnosperms and their evolution. - Zhurn. obshch. biol. (Moscow), vol. 43, p. 303-323 (in Russian).

Meyen, S.V. 1984. Basic features of gymnosperm systematics and phylogeny as evidenced by fossil record. - Bot. Rev., vol. 50, no. 1, p. 1-111.

Meyen, S.V. 1987. Fundamentals of palaeobotany. London, N.Y.: Chapman & Hall, 432 p.

Meyer, N.R. 1980. Probable directions of evolution in the sporoderm of gymnosperms and some angiosperms. - In: Systematics and evolution of higher plants. Leningrad: Nauka, p. 82-92 (in Russian).

Millay, M.A., Taylor, T.N. 1979. Paleozoic seed fern pollen organs. - Bot. Rev., vol. 45, p. 301-375.

Moeliono, B.M. 1970. Cauline or carpellary placentation among dicotyledons. 1. The cauline ovules of Centrosperms. Assen: Van Gorcum-Prakke, 292 p.

Mosley, M.F., Uhl, N.W. 1985. Morphological studies of the Nymphaeaceae sensu lato. 15. The anatomy of the flower of *Nelumbo*. - Bot. Jb. (Syst.), Bd. 106, S. 61-98.

Moussel, B. 1978. Double fertilization in the genus *Ephedra*. - Phytomorphology, vol. 28, no. 3, p. 336-344.

Muhammad, A.F., Sattler, R. 1982. Vessel structure of *Gnetum* and the origin of angiosperms. - Amer. J. Bot., vol. 69, no. 6, p. 1004-1021.

Muller, H. 1966. Palynological investigations of Cretaceous sediments in northeastern Brazil. - Proc. 2nd West African Micropaleontol. Colloq., Leiden, p. 123-136.

Muller, J. 1970. Palynological evidence on early differentiation of angiosperms. - Biol. Rev. Cambrige Phil. Soc., vol. 45, p. 417-450.

Muller, J. 1974. A comparison of southeast Asian with European fossil angiosperm pollen floras. - Symp. Orig. & Phytogeogr. Angiosperms, Lucknow: B. Sahni Inst. Palaeobot., no. 1, p. 49-56.

Muller, J. 1984. Significance of fossil pollen for angiosperm history. - Ann. Mo. Bot. Gard., vol., 71, no. 2, p. 419-443.

Nambudiri, K.K., Beck, C.B. 1968. A comparative study of the primary vascular system of conifers. - Amer. J. Bot., vol. 55, no. 4, p. 447-472.

Negbi, M., Koller, D. 1962. Homologies in the grass embryo. A reevaluation. - Phytomorphology, vol. 12, no. 12, p. 289-296.

Neuburg, M.F. 1963. Family Vojnovskyaceae Neuburg. - In: Fundamentals of paleontology. Gymnosperms and angiosperms. Moscow: Gostekhizdat, 301 p. (in Russian).

Neumayer, H. 1924. Die Geschichte der Blüte. - Abh. zool.-bot. Ges. Wien, Bd. 14, S. 1-112.

Newberry, J.S. 1895. The flora of the Amboy Clays. - U.S. Geol. Surv. Monogr., no. 26, p. 1-137.

Nishida, H. 1985. A structurally preserved magnolialean fructification from mid-Cretaceous of Japan. - Nature, vol. 318, no. 6041, p. 58-59.

Nishida, H. 1994. *Elsemaria,* a new Late Cretaceous angiosperm fructification from Hokkaido, Japan. - Plant Syst. Evol., Suppl., vol. 8, p. 123-136.

Nishida, H., Nishida, M. 1988. *Protomonimia kasai-nakajhongii* gen. et sp. nov.: a permineralized magnolialean fructification from the mid-Cretaceous of Japan. - Bot. Mag. (Tokyo), vol. 101, p. 397-426.

Nishida, M. 1974. Diagnostic characters between vesselless angiospermous woods and cycadean woods. - Symp. Morphol. & Stratigr. Palaeobot., Lucknow: B. Sahni Inst. Palaeobot., no. 2, p. 47-49.

Nishida, M., Nishida, H. 1985. Petrified woods from the Upper Cretaceous of Quiriquina Island. A report of the botanical survey to Bolivia and southern Chile. Chiba: Chiba Univ., p. 27-34.

Nowicke, J.W., Skvarla, J.J. 1979. Pollen morphology: the potential influence in higher order systematics. - Ann. Mo. Bot. Gard., vol. 66, no. 4, p. 633-700.

Ohana, T., Kimura, T. 1987. Preliminary notes on the multicarpelous female flower with conduplicate carpels from the Upper Cretaceous of Hokkaido, Japan. - Proc. Jap. Acad., vol. 63B, p. 175-178.

Ohri, D., Khoshoo, T.N. 1986. Genome size in gymnosperms. - Plant Syst. Evol., vol. 153, p. 119-132.

Oishi, S. 1940. The Mesozoic floras of Japan. - J. Fac. Sci. Hokkaido Univ., vol. 5, p. 123-480.

Oldham, T.C.B. 1976. Flora of the Wealden plant debris beds of England. - Palaeontology, vol. 19, p. 437-502.

Orel, L.I., Kupriyanova, L.A., Golubeva, E.A. 1987. Ultrastructure of acetyl-resistant cell walls of tapetal cells and pollen grains of *Gnetum africanum* (Gnetaceae). - Bot. Zhurn. (Leningrad), vol. 71, p. 750-759 (in Russian).

Osborn, J.M., Taylor, T.N. 1993. Pollen morphology and ultrastructure of the Corystospermales: permineralized in situ grains from the Triassic of Antarctica. - Rev. Paleobot. Palynol., vol. 79, p. 205-220.

Osborn, J.M., Taylor, T.N., de Lima, M.R. 1993. The ultrastructure of ephedroid pollen with gnetalean affinities from the Lower Cretaceous of Brazil. - Rev. Paleobot. Palynol., vol. 77, p. 171-184.

Osborn, J.M., Taylor, T.N., Crane, P.R. 1991. The ultrastructure of *Sahnia* pollen (Pentoxylales). - Amer. J. Bot., vol. 78, no. 11, p. 1560-1569.

Owens, J.N., Blake, M.D. 1984. The pollination mechanism of Sitka spruce (*Picea sitchensis*). - Canad. J. Bot., vol. 62, no. 6, p. 1136-1148.

Oyama, T., Matsuo, H. 1964. Notes on palmaean leaf from the Oarai flora (Upper Cretaceous), Oarai Machi, Ibaraki Prefecture, Japan. - Trans. Palaeont. Soc. Jap., N.S., vol. 55, p. 241-246.

Pacltová, B. 1971. Palynological study of Angiospermae from the Peruc Formation (?Albian-Lower Cenomanian) of Bohemia. - Sborn. Geol. Ved, sv. 13, s. 105-141.

Pacltová, B. 1977. Cretaceous angiosperms of Bohemia - central Europe. - Bot. Rev., vol. 43, no. 1, p. 128-142.

Pankow, H., Sothmann, E. 1967. Histogenetische Untersuchungen an den weiblichen Blüten von *Ginkgo biloba* L. - Berichte Dtsch. bot. Ges., Bd. 80, H. 4, S. 265-272.

Pant, D.D. 1959. The classification of gymnospermous plants. - Palaeobotanist, vol. 6, p. 65-70.

Pant, D.D. 1968. On the stem and attachment of *Glossopteris* leaves. - Phytomorphology, vol. 17, p. 351-359.

Pant, D.D. 1977. The plant of *Glossopteris*. - J. Ind. Bot. Soc., vol. 56A, p. 1-23.

Pant, D.D. 1982. The lower Gondwana gymnosperms and their relationships. - Rev. Palaeobot. Palynol., vol. 37, p. 55-70.

Pant, D.D., Kidwai, P.F. 1978. The origin and evolution of flowering plants. - J. Ind. Bot. Soc., vol. 56A, p. 242-274.

Pant, D.D., Mehra, B. 1963. On a cycadophyte leaf, *Pteronilssonia gopalii* gen. et sp. nov. from the Lower Gondwana of India. - Palaeontographica B, Bd. 113, S. 126-134.

Pant, D.D., Nautiyal, D.D. 1963. Cuticle and epidermis of recent Cycadales leaves, sporangia and seeds. - Senckenberg. biol., Bd. 44, H. 4, S. 257-336.

Pant, D.D., Nautiyal, D.D. 1987. *Diphyllopteris verticillata* Srivastava, the probable seedling of *Glossopteris* from the Paleozoic of India. - Rev. Palaeobot. Palynol., vol. 51, nos. 1/3, p. 31-36.

Pauzé, F., Sattler, R. 1978. L'androcée centripète d'*Ochna atropurpurea*. - Canad. J. Bot., vol. 56, no. 20, p. 2500-2511.

Pauzé, F., Sattler, R. 1979. La placentation axillaire chez *Ochna atropurpurea*. - Canad. J. Bot., vol. 57, no. 2, p. 100-107.

Pearson, H.H.W. 1929. Gnetales. Cambridge: Univ. Press, 194 p.

Pedersen, K.R., Friis, E.M. 1986. *Caytonanthus* pollen from the Lower and Middle Jurassic. - Geoskrifter, vol. 24, p. 255-266.

Pedersen, K.R., Friis, E.M., Crane, P.R. 1988. *Eucommiidites* pollen in situ in Cretaceous reproductive structures. - Abstr. 7th Intern. Palynol. Congr., Brisbane, p. 128.

Pellmyr, O., Thien, L.B. 1986. Insect reproduction and floral fragrances: keys to the evolution of the angiosperms? - Taxon, vol. 35, no. 1, p. 76-85.

Pennell, R.I., Bell, P.R. 1986. Microsporogenesis in *Taxus baccata* L. The formation of the tetrad and development of microspores. - Ann. Bot. (Gr. Brit.), vol. 57, p. 545-555.

Penny, J.H.J. 1991. Early Cretaceous angiosperm pollen from the borehole Mersa Matruch I, Northwest desert, Egypt. - Palaeontographica B, Bd. 222, S. 31-88.

Periasamy, K., Muraganathan, E.A. 1985. Ontogeny of palmately compound leaves in angiosperms: 1. *Tebebuia pentaphlla* Hense. - Proc. Ind. Acad. Sci., Plant Sci., vol. 95, no. 6, p. 429-436.

Pettitt, J.M. 1969. Pteridophytic features in some Lower Carboniferous seed megaspores. - Bot. J. Linn. Soc., vol. 62, p. 233-239.

Pettitt, J.M., Chaloner, W.G. 1964. The ultrastructure of Mesozoic pollen *Classopollis*. - Pollen et Spores, vol. 6, p. 611-620.

Philipson, W.R. 1985. Is the grass gynoecium monocarpellary? - Amer. J. Bot., vol. 72, no. 2, p. 1954-1961.

Philomen, P.A., Shah, C.K. 1985. Unusual germination and seedling development in two monocotyledonous dicotyledons. - Proc. Ind. Acad. Sci., Plant. Sci., vol. 95, no. 4, p. 221-225.

Pigg, K.B., Stockey, R.A. 1991. Platanaceous plants from the Paleocene of Alberta, Canada. - Rev. Palaeobot. Palynol., vol. 70, p. 125-146.

Playford, G., Rigby, J.F., Archibald, D.C. 1982. Middle Triassic flora from the Moolayember Formation, Bowen Basin, Queensland. - Geol. Surv. Queensland Publ., vol. 380, p. 1-52.

Pocock, S.A.J., Vasanthy, G. 1988. *Cornetipollis reticulata*, a new pollen with angiospermid features from upper Triassic (Carnian) sediments of Arizona (U.S.A.) with notes on *Equisetosporites*. - Rev. Palaeobot. Palynol., vol. 55, p. 337-356.

Pocock, S.A.J., Vasanthy, G., Venkatachala, B.S. 1990. Pollen of *Circumpolles* - an enigma of morphotrends showing evolutionary adaptation. - Rev. Palaeobot. Palynol., vol. 65, p. 179-193.

Popova, I.S. 1971. Morphological types of pollen grains in tetraploid forms of winter wheat. - Izvestiya Sibirsk. otd. Akad. nauk SSSR, t. 3, no. 15, p. 62-64 (in Russian).

Popper, K. 1972. Objective knowledge: an evolutionary approach. Oxford: Univ. Press, 380 p.

Potonié, R., Schweitzer, H.I. 1960. Der Pollen von *Ulmannia frumentaria*. - Paläont. Ztschr., Bd. 34, H. 1, S. 27-39.

Prakash, U. 1965. A survey of fossil dicotyledonous woods from India and the Far East. - J. Paleont., vol. 39, no. 5, p. 815-827.

Quinn, C.J. 1966. Gametophyte development and embryogeny in the Podocarpaceae. 3. *Dacrydium bidwillii*. - Phytomorphology, vol. 16, no. 1, p. 81-91.

Rao, A.R. 1935. On an abnormal flower of *Convolvulus pluricaulis* Chois. - Curr. Sci., vol. 4, p. 414-415.

Rao, L.N. 1974. *Cycas beddomei* Dyer. - Proc. Ind. Acad. Sci., vol. 79, p. 59-66.

Rasmussen, H. 1983. Stomatal development in families of Liliales. - Bot. Jb. (Syst.), Bd. 104, H. 2, S. 261-287.

Rasnitsyn, A.P., Krassilov, V.A. 1996. Pollen in the gut compressions of fossil insects as evidence of coevolution. - Paleont. Zh. (Moscow), no. 6, p. 716-122.

Ratter, J.A., Milne, C. 1973 Chromosome numbers of some primitive angiosperms. - Notes Roy. Bot. Gard. Edinburgh, vol. 32, no. 3, p. 423-428.

Reid, E., Chandler, M.E.J. 1933. The London Clay flora. London: Brit. Mus. (Nat. Hist.), 561 p.

Reihman, M.A., Schabilon, J.T. 1985. Stomatal structure of *Alethopteris ullivantii* and *Neuropteris scheuchzeri*, Pennsylvanian pteridosperm foliage. - Amer. J. Bot., vol. 72, no. 9, p. 1392-1396.

Remy, W., Remy, R. 1980. Devonian gametophytes with anatomically preserved gametangia. - Science, vol. 208, p. 295-296.

Remy, W., Remy, R., Hass, H., Schultka, S., Franzmayer, F. 1980. *Sciadophyton* Steinmann - ein Gametophyt aus den Siegen. - Argumenta Palaeobotanica, H. 5, S. 73-94.

Retallack, G., Dilcher, D.L. 1981a. Arguments for a glossopterid ancestry of angiosperms. - Paleobiology, vol. 7, no. 1, p. 54-67.

Retallack, G., Dilcher, D.L. 1981b. A coastal hypothesis for the dispersal and rise to dominance of flowering plants. - Paleobotany, paleoecology and evolution. N.Y.: Praeger, vol. 2, p. 27-77.

Retallack, G., Dilcher, D.L. 1981c. Early angiosperm reproduction: *Prisca reynoldsii* gen. et sp. nov. from mid-Cretaceous coastal deposits, Kansas, U.S.A. - Palaeontographica B, Bd. 179, S. 103-137.

Retisma, T. 1966. Pollen morphology of some European Rosaceae. - Acta Bot. Neerl., vol. 15, p. 290-307.

Reymanówna, M. 1968. On seeds containing *Eucommiidites troedsonii* pollen from the Jurassic of Groec, Poland. - Bot. J. Linn. Soc., vol. 61, no. 384, p. 147-152.

Reymanówna, M. 1974. On anatomy and morphology of *Caytonia*. - Symp. Morphol. Stratigr. & Palaeobot., Lucknow: B. Sahni Inst. Palaeobot., no. 2, p. 50-57.

Rigby, J.E. 1962. Occurrence of Glossopteridae fructifications at Baralaba, Queensland and their evolution. - Nature, vol. 195, no. 4837, p. 196-198.

Rigby, J.E. 1978. Permian glossopterid and other cycadopsid fructifications from Queensland. - Geol. Surv. Queensl. Publ., vol. 367, p. 3-27.

Rodin, R.J. 1966. Leaf structure and evolution in American species of *Gnetum*. - Phytomorphology, vol. 16, p. 56-58.

Rohr, R. 1977. Etude comparée de la formation de l'exine au cours de la microsporogénèse chez une gymnosperme (*Taxus baccata*) et une prephanerogame (*Ginkgo biloba*). - Cytologia, vol. 42, p. 157-167.

Rohweder, O. 1967. Karpellbau und Synkarpie bei Ranunculaceen. - Berichte Schweiz. bot. Ges., Bd. 77, S. 376-425.

Romero, E.J., Archangelsky, S. 1986. Early Cretaceous angiosperm leaves from southern South America. - Science, vol. 234, no. 4783, p. 1580-1582.

Rothwell, G.W. 1980. The Callistophytaceae (Pteridospermopsida). 2. Reproductive features. - Palaeontographica B, Bd. 173, S. 85-106.

Rothwell, G.W. 1982a. *Cordaianthus duquesnensis* sp. n., anatomically preserved ovulate cone from the Upper Pennsylvanian of Ohio. - Amer. j. Bot., vol. 69, no. 2, p. 239-247.

Rothwell, G.W. 1982b. New interpretation of the earliest conifers. - Rev. Palaeobot. Palynol., vol. 37, p. 7-28.

Rothwell, G.W., Erwin, D.M. 1987. Origin of seed plants: an aneurophyte/seed fern link elaborated. - Amer. J. Bot., vol. 74, no. 6, p. 970-973.

Rothwell, G.W., Warner, S. 1984. *Cordaixylon dumusum* n. sp. (Cordaites). 1. Vegetative structures. - Bot. Gaz., vol. 145, no. 2, p. 275-291.

Roy Chowdhury, C. 1962. The embryogeny of conifers: A review. - Phytomorphology, vol. 12, p. 313-338.

Rüffle, L. 1980a. Wachstum-Modus und Blatt-Morphologie bei altertümlichen Fagales und Hamamelidales der Kreide und Gegenwart. 100 Jahre Arboretum, 1879-1979. Berlin, S. 329-341.

Rüffle, L. 1980b. Merkmals-Enkapsis und Homologie-Kriterien bei älteren Angiospermen und ihre Herkunft. - Schr. geol. Wiss. (Berlin), Bd. 16, S. 417-439.

Rugenstein, S.R., Lersten, N.R. 1981. Stomata on seeds and fruits of *Bauhinia* (Leguminosae: Caesalpinioideae). - Amer. J. Bot., vol. 68, no. 6, p. 873-876.

Rutishauser, R., Sattler, R. 1985. Complementarity and heuristic value of contrasting models in structural botany. 1. General considerations. - Bot. Jb. (Syst.), Bd. 107, H. 1/4, S. 415-455.

Rutishauser, R., Sattler, R. 1986. Architecture and development of the phyllode-stipule whorls of *Acacia longipedunculata*: controversial interpretation and continuum approach. - Canad. J. Bot., vol. 64, p. 1987-2019.

Sabato, S., De Luca, P. 1985. Evolutionary trends in *Dion* (Zamiaceae). - Amer. J. Bot., vol. 72, no. 9, p. 1353-1363.

Sanwal, M. 1962. Morphology and embryogeny of *Gnetum gnemon* L. - Phytomorphology, vol. 12, p. 243-264.

Sattler, R. 1974a. A new approach to gynoecial morphology. - Phytomorphology, vol. 24, nos. 1/2, p. 23-34.

Sattler, R.A. 1974b. A new conception of the shoot of higher plants. - J. Theor. Biol., vol. 47, p. 367-382.

Sattler, R. 1978. "Fusion" and "continuity" in floral morphology. - Notes Roy. Bot. Gard. Edinburgh, vol. 36, p. 397-405.

Sattler, R., Maier, U. 1977. Development of the epiphyllous appendages of *Begonia hispida* var. *cucullifera*: implications for comparative morphology. - Canad. J. Bot., vol. 55, no. 4, p. 411-425.

Schenk, A. 1867. Die fossile Flora der Grenzschichten des Keupers und Lias Frankens 5-9. Wiesbaden, 231 S.

Schmid, R. 1986. On Cornerian and other terminology of angiospermous and gymnospermous seed coats: historical perspectives and terminological recommendations. - Taxon, vol. 35, p. 476-491.

Schopf, J.M. 1976. Morphologic interpretation of fertile structures of glossopetrid gymnosperms. - Rev. Palaeobot. Palynol., vol. 21, p. 25-64.

Schopf, J.M. 1982. Forms and facies of *Vertebraria* in relation to Gondwana coal. - Antarct. Res. Ser., vol. 36, no. 3, p. 37-62.

Schrank, F. Scanning electron and light microscope investigations of angiosperm pollen from the Lower Cretaceous of Egypt. - Pollen et Spores, vol. 25, p. 212-242.

Schultze-Motel, W. 1959. Entwicklungsgeschichtliche und vergleichend-morphologische Untersuchungen im Blütenbereich der Cyperaceae. - Bot. Jb. (Syst.), Bd. 78, S. 129-170.

Schweitzer H.-J. 1963. Der weibliche Zapfen von *Pseudovoltzia liebeana* und seine Bedeutung für die Phylogenie der Koniferen. - Palaeontographica B, Bd. 113, S. 1-29.

Schweitzer H.-J. 1977. Die Rätische Zwitterblüte *Irania hermaphroditica* nov. spec. und ihre Bedeutung für die Phylogenie der Angiospermen. - Palaeontographica B, Bd. 161, S. 98-145.

Scott, A.C., Taylor, T.N. 1983. Plant/animal interactions during the Upper Carboniferous. - Bot. Rev., vol. 49, p. 259-307.

Scott, R.A., Barghoorn, E.S. 1958. *Phytocrene microcarpa* - a new species of Icacinaceae based on Cretaceous fruits from Kreischerville, New York. - Palaeobotanist, vol. 6, p. 25-28.

Scott, R.A., Barghoorn, E.S., Leopold, E. 1960. How old are angiosperms? - Amer. J. Sci., vol. 258A, p. 284-299.

Sellwood, B.W., Price, G.D., Valdes, P.J. 1994. Colder estimates of Cretaceous temperatures. - Nature, vol. 370, pp. 453-456.

Serlin, B.S. 1982a. On the structure of *Cardiocarpus tritolopus*, a cordaite seed from the Lower-Middle Pennsylvanian of Kentucky. - Rev. Palaeobot. Palynol., vol. 36, p. 297-304.

Serlin, B.S. 1982b. An Early Cretaceous fossil flora from northwest Texas: its composition and implications. - Palaeontographica B, Bd. 182, H. 1/3, S. 52-86.

Seward, A.C. 1894-1895. Catalogue of the Mesozoic plants in the Department of Geology, British Museum Natural History. The Wealden flora. London: Brit. Mus. (Nat. Hist.), pt. 1, 384 p.; pt. 2, 259 p.

Seward, A.C. 1917. Fossil plants. 3. Pteridospermeae, Cycadofilices, Cordaitales, Cycadophyta. Cambridge: Univ. Press, 656 p.

Seward, A.C. 1926. The Cretaceous plant-bearing rocks of western Greenland. - Phil. Trans. Roy. Soc. London, vol. 215B, p. 57-174.

Seward, A.C., Conway, V.M. 1935. Additional Cretaceous plants from western Greenland. - Kgl. Svenk. Vetensk.-Akad. Handl., Bd. 15, S. 1-51.

Sharma, B.D. 1970. On the structure of seeds of *Williamsonia* collected from the Middle Jurassic rocks of Amarjola in the Rajmahal Hills, India. - Ann. Bot. (Gr. Brit.), vol. 34, p. 1071-1078.

Shashi, N., Ueno, J. 1986. Pollen morphology of *Ginkgo biloba* and *Cycas revoluta*. - Canad. J. Bot., vol. 64, p. 3075-3078.

Shimakura, M. 1937. Studies on the fossil woods from Japan and adjacent lands, 2. - Sci. Rep. Tohoku Imp. Univ. Sendai. Ser. 2, vol. 19, no. 1, p. 1-73.

Singh, H., Maheshwari, K. 1962. A contribution to the embryology of *Ephedra gerardiana* Well. - Phytomorphology, vol. 12, no. 4, p. 361-372.

Singh, H., Oberoi, Y.P. 1962. A contribution to the life history of *Biota orientalis* Endl. - Phytomorphology, vol. 12, no. 4, p. 373-393.

Sinnott, E.W. 1913. The morphology and reproductive structures in the Podocarpaceae. - Ann. Bot. (Gr. Brit.), vol. 27, p. 39-82.

Sinnott, E.W., Bailey, I.W. 1914. Investigations on the phylogeny of the angiosperms. 4. The origin and dispersal of herbaceous angiosperms. - Ann. Bot. (Gr. Brit.), vol. 28, p. 547-600.

Sitholey, R.V. 1954. The Mesozoic and Tertiary floras of India - a review. - Palaeobotanist, vol. 3, p. 55-69.

Skipworth, J.P. 1970. Development of floral vasculature in the Magnoliaceae. - Phytomorphology, vol. 20, p. 228-236.

Smith, A.W. 1994. Systematics and the fossil record: documenting evolutionary patterns. Blackwell, 223 p.

Smith, F.G. 1907. Morphology of the trunk and development of the microsporangium of cycads. - Bot. Gaz., vol. 43, no. 3, p. 187-204.

Smoot, E.L., Taylor, T.N. 1983/1984. Studies on Carboniferous seeds: the relationship of the integument and nucellus in *Pachytesta gigantea*. - Rev. Palaeobot. Palynol., vol. 40, p. 165-176.

Smoot, E.L., Taylor, T.N., Delevoryas, T. 1985. Structurally preserved fossil plants from Antarctica. 1. *Antarcticycas*, gen. nov., a Triassic cycad stem from the Beardmore Glacier area. - Amer. J. Bot., vol. 72, p. 1410-1423.

Sporne, K.R. 1974. The morphology of angiosperms. London: Hutchinson, 207 p.

Srivastava, S.K. 1981. Evolution of Upper Cretaceous phytogeoprovinces and their pollen flora. - Rev. Palaeobot. Palynol., vol. 35, p. 155-173.

Stebbins, G.L. 1974. Flowering plants. Evolution above species level. Cambridge (Mass.): Harvard Univ. Press, 397 p.

Sterling, C. 1965. Comparative morphology of the carpel in the Rosaceae, IV. - Amer. J. Bot., vol. 52, p. 938-946.

Sterling, C. 1973. Comparative morphology of the carpel in the Liliaceae: Colchicae (*Androcymbium*). - Bot. J. Linn. Soc., vol. 67, p. 149-156.

Stevenson, D.W. 1980a. Radial growth in *Beaucarnea recurvata*. - Amer. J. Bot., vol. 67, no. 4, p. 476-489.

Stevenson, D.W. 1980b. Radial growth in the Cycadales. - Amer. J. Bot., vol. 67, no. 4, p. 465-475.

Stevenson, D.W. 1981. Observations on ptyxis, phenology, and trichomes in the Cycadales. - Amer. J. Bot., vol. 68, no. 8, p. 1104-1114.

Stewart, W., Delevoryas, T. 1952. Bases for determining relationships among the Medullosaceae. - Amer. J. Bot., vol. 39, no. 7, p. 505-516.

Stidd, B.M., Hall, J.W. 1970. *Callandrium callistophytoides*, gen. et. sp. nov., the probable pollen-bearing organ of the seed fern, *Callistophyton*. - Amer. J. Bot., vol. 57, no. 4, p. 394-403.

Stopes, M.C. 1912. Petrifications of the earliest European angiosperms. - Phil. Trans. Roy. Soc. London, vol. 203B, p. 75-100.

Stopes, M.C. 1915. Catalogue of the Mesozoic plants in the British Museum. The Cretaceous flora. London: Brit. Mus. (Nat. Hist.), pt. 2: Lower Greensand (Aptian) plants of Britain, 360 p.

Stopes, M.C., Fujii, K. 1911. Studies on the structure and affinities of Cretaceous plants. - Phil. Trans. Roy. Soc. London, vol. 201B, p. 1-90.

Stubblefield, S.P., Rothwell, G.W., Taylor, T.N. 1984. *Conostoma williamsonii* n. sp., a lagenostomalean ovule from the Pennsylvanian of the Illinois Basin. - Canad. J. Bot., vol. 62, no. 1, p. 96-101.

Stubblefield, S.P., Taylor, T.N., Daghlian, C.P. 1982. Compressed plants from the Lower Pennsylvanian of Kentucky (U.S.A.). 1. *Crossotheca kentuckiensis* n. sp. - Rev. Palaeobot. Palynol., vol. 36, p. 197-204.

Surange, K.R., Chandra, S. 1975. Morphology of the gymnospermous fructifications of the *Glossopteris* flora and their relationships. - Palaeontographica B, Bd. 149, S. 153-180.

Surova, T.D., Kavadze, E.V. 1988. Sporoderm ultrastructure in some gymnosperms (*Metasequoia, Cunninghamia, Sciadopitys*). - Bot. Zhurn. (Leningrad), vol.73, p. 34-44 (in Russian).

Swamy, B.G.L., Periasamy, K. 1964. The concept of conduplicate carpel. - Phytomorphology, vol. 14, p. 319-327.

Syvanen, M. 1994. Horizontal gene transfer: evidence and possible consequences. - Annu. Rev. Genet., vol. 28, p. 237-261.

Takaso, T., Bouman, F. 1986. Ovule and seed ontogeny in *Gnetum gnemon* L. - Bot. Mag. (Tokyo), vol. 99, p. 241-266.

Takeda, H. 1913. Some points in the anatomy of the leaf of *Welwitschia mirabilis*. - Ann. Bot. (Gr. Brit.), vol. 27, p. 347-357.

Takhtajan, A.L. 1966. System and phylogeny of flowering plants. Moscow, Leningrad: Nauka, 611 p. (in Russian).

Takhtajan, A.L. 1970. Origins and spread of flowering plants. Leningrad: Nauka, 145 p. (in Russian).

Takhtajan, A.L. 1980. Outline of the classification of flowering plants. - Bot. Rev., vol. 46, p. 225-359.

Tamura, M. 1963. Morphology, ecology and phylogeny of the Ranunculaceae. 1. - Sci. Rep. Osaka Univ., vol. 11, p. 115-126.

Tang, W. 1987. Insect pollination in the cycad *Zamia pumila* (Zamiaceae). - Amer. J. Bot., vol. 74, no. 1, p. 90-99.

Taylor, D.W., Hickey, L.J. 1990. An Aptian plant with attached leaves and flowers: implications for angiosperm origin. - Science, vol. 247, p. 702-704.

Taylor, T.N. 1965. Paleozoic seed studies: a monograph of the American species of *Pachytesta*. - Palaeontographica B, Bd. 117, S. 1-46.

Taylor, T.N. 1973. A consideration of the morphology, ultrastructure and multicellular microgametophyte of *Cycadeoidea dacotensis* pollen. - Rev. Palaeobot. Palynol., vol. 16, p. 157-165.

Taylor, T.N. 1978. The ultrastructure and reproductive significance of *Monoletes* (Pteridospermales) pollen. - Canad. J. Bot., vol. 56, no. 24, p.3105-3118.

Taylor, T.N. 1981. Palaeobotany. An introduction to fossil plant biology. N.Y. etc.: McGraw-Hill, 589 p.

Taylor, T.N. 1982. Ultrastructural studies of Paleozoic seed fern pollen: sporoderm development. - Rev. Palaeobot. Palynol., vol. 37, p. 29-53.

Taylor, T.N., Alvin, K.L. 1984. Ultrastructure and development of Mesozoic pollen: *Classopollis*. - Amer. J. Bot., vol. 71, no. 4, p. 575-587.

Taylor, T.N., Archangelsky, S. 1985. The Cretaceous pteridosperms *Ruflorinia* and *Ktalenia* and implications on cupule and carpel evolution. - Amer. J. Bot., vol. 72, p. 1842-1853.

Taylor, T.N., Daghlian, C.P. 1980. The morphology and ultrastructure of *Gothania* (Cordaitales) pollen. - Amer. J. Bot., vol. 29, p. 1-14.

Taylor, T.N., Millay, M.A. 1981. Morphological variability of Pennsylvanian lyginopteroid seed ferns. - Rev. Palaeobot. Palynol., vol. 32, p. 27-62.

Teixeira, C. 1948. Flora mesozóica portuguesa. - Port. Serv. Geol., pt. 1, p. 1-118.

Terabayashi, S. 1985. The comparative floral anatomy and systematics of the Berberidaceae. - Mem. Fac. Sci. Kyoto Univ., Ser. Biol., vol. 10, p. 73-90.

Thayn, G.F., Tidwell, W.D. 1984. A review of the genus *Paraphyllanthoxylon*. - Rev. Palaeobot. Palynol., vol. 43, p. 321-335.

Thayn, G.F., Tidwell, W.D., Stokes, W.L. 1983. Flora of the Lower Cretaceous Cedar Mountain Formation of Utah and Colorado. 1. *Paraphyllantoxylon utahense*. - Gr. Basin Natur., vol. 43, no. 3, p. 394-402.

Thayn, G.F., Tidwell, W.D., Stokes, W.L. 1985. Flora of the Lower Cretaceous Cedar Mountain Formation of Utah and Colorado. 3, *Icacinoxylon pittiense* n. sp. - Amer. J. Bot., vol. 72, no. 2, p. 175-180.

Thien, L.B., Bernhardt, P., Pellmyr, O. et al. 1985. The pollination of *Zygogynum* (Winteraceae) by a moth, *Sabatinca* (Micropterigidae): an ancient association? - Science, vol. 227, no. 4686, p. 540-542.

Thoday, M.G., Berridge, E.M. 1912. The anatomy and morphology of inflorescences and flowers in *Ephedra*. - Ann. Bot. (Gr. Brit.), vol. 26, p. 953-979.

Thomas, H.H. 1925. The Caytoniales, a new group of angiospermous plants from the Jurassic rocks of Yorkshire. - Phil. Trans. Roy. Soc. London, vol. 213B, p. 299-363.

Thomas, H.H. 1931. The early evolution of the angiosperms. - Ann. Bot. (Gr. Brit.), vol. 45, no. 180.

Thomas, H.H. 1932/1933. The old morphology and the new. - Proc. Linn. Soc. London, pt. 1, p. 17-32.

Thomas, H.H. 1933. On some pteridospermous plants from the Mesozoic rocks of South Africa. - Phil. Trans. Roy. Soc. London, vol. 222B, p. 193-265.

Thomas, H.H. 1952. *Glossopteris* with whorled leaves. - Palaebotanist, vol. 1, p. 435-438.

Thomas, H.H. 1955. Mesozoic pteridosperms. - Phytomorphology, vol. 5, p. 177-185.

Thomas, H.H., Bancroft, N. 1913. On the cuticles of some recent and fossil cycadean fronds. - Trans. Linn. Soc. London, Bot., vol. 8, p. 155-204.

Thorne, R.F. 1973. The "Amentiferae" or Hamamelidae as an artificial group: a summary statement. - Brittonia, vol. 25, no. 4, p. 395-405.

Thorne, R.F. 1976. A phylogenetic classification of angiosperms. - Evolutionary Biol., N.Y.: Plenum Press, vol. 9, p. 35-100.

Tiffney, B.H. 1977. Dicotyledonous angiosperm flower from the Upper Cretaceous of Martha's Vineyard, Massachusetts. - Nature, vol. 265, p. 136-137.

Tiffney, B.H. 1984. Seed size, dispersal syndromes, and the rise of angiosperms: evidence and hypothesis. - Ann. Mo. Bot. Gard., vol. 71, no. 2, p. 551-576.

Tilton, V.R. 1980. The nucellar epidermis and micropyle of *Ornithogalum caudatum* (Liliaceae) with a review of these structures in other taxa. - Canad. J. Bot., vol. 58, no. 17, p. 1872-1884.

Tomlinson, P.B. 1974. Development of stomatal complex as taxonomic character in the monocotyledons. - Taxon, vol. 23, p. 109-128.

Tomlinson, P.B., Wilder, G.J. 1984. Systematic anatomy of Cyclanthaceae (Monocotyledoneae) - an overview. - Bot. Gaz., vol. 145, no. 4, p. 535-549.

Tosh, K.J., Powell, G.R. 1986. Proliferated, bisporangiate, and other atypical cones occurring on young plantation growth of *Larix laricina*. - Canad. J. Bot., vol. 64, no. 3, p. 469-475.

Townrow, J.A. 1965. A new member of Corystospermaceae Thomas. - Ann. Bot. (Gr. Brit.), vol. 29, no. 115, p. 495-511.

Townrow, J.A. 1967. On *Rissikia* and *Matoia*, podocarpaceous conifers from the Lower Mesozoic of southern lands. - Pap. Roy. Soc. Tasmania, vol. 101, p. 103-125.

Trevisan, L. 1980. Ultrastructural notes and considerations on *Ephedripites*, *Eucommiidites* and *Monosulcites* pollen grains from Lower Cretaceous sediments of southern Tuscany (Italy). - Pollen et Spores, vol. 22, no. 1, p. 85-132.

Trivett, M.L., Rothwell, G.W. 1985. Morphology, systematics and paleoecology of Paleozoic fossil plants: *Mesoxylon priapi*, sp. nov. (Cordaitales). - Syst. Bot., vol. 10, no. 2, p. 205-223.

Troll, W. 1932. Morphologie der schildförmigen Blätter. - Planta, Bd. 17, S. 153-314.

Troll, W. 1939. Die morphologische Natur der Karpelle. - Chronica Bot., Bd. 5, S. 38-41.

Tsvelev, N.N. 1977. On the primary life form of angiosperms. - Byull. Mosk. obshch. ispyt. prir., Biol., t. 82, p. 79-88 (in Russian).

Tucker, S.C. 1960. Ontogeny of the floral apex of *Michelia fuscata*. - Amer. J. Bot., vol. 47, p. 266-277.

Tucker, S.C. 1975. Floral development in *Saururus cernuus* (Saururaceae). 1. Floral initiation and stamen development. - Amer. J. Bot., vol. 69, no. 9, p. 993-1007.

Tucker, S.C. 1976. Floral development in *Saururus cernuus* (Saururaceae). 2. Carpel initiation and floral vasculature. - Amer. J. Bot., vol. 63, no. 2, p. 289-301.

Tucker, S.C. 1981. Inflorescence and floral development in *Houttuynia cordata* (Saururaceae). - Amer. J. Bot., vol. 68, no. 8, P. 1017-1032.

Tucker, S.C. 1982. Inflorescence and floral development in the Piperaceae. 2. Floral ontogeny of *Piper*. - Amer. J. Bot., vol. 69, no. 9, p. 1384-1401.

Tucker, S.C. 1985. Initiation and development of inflorescence and flower in *Anemopsis californica* (Saururaceae). - Amer. J. Bot., vol. 72, no. 1, p. 20-31.

Tucker, S.C. 1986. Inflorescence and floral development in the Piperaceae. 1. *Piperomia*. - Amer. J. Bot., vol. 67, no. 5, p. 686-702.

Tucker, S.C., Gifford, E.M. 1974. Carpel vascularization of *Drimys lanceolata*. - Phytomorphology, vol. 14, no. 2, p. 197-203.

Uhl, N.W. 1972. Inflorescence and flower structure in *Nypa fruticans* (Palmae). - Amer. J. Bot., vol, 59, no. 7, p. 729-743.

Uhl, N.W. 1976. Developmental studies in Ptychospermae (Palmae). 2, The staminate and pistillate flowers. - Amer. J. Bot., vol. 63, no. 1, p. 97-109.

Upchurch, G.R. (Jr.) 1984. Cuticle evolution in Early Cretaceous angiosperms from the Potomac Group of Virginia and Maryland. - Ann. Mo. Bot. Gard., vol. 71, no. 2, p. 522-550.

Vakhrameev, V.A. 1952. Stratigraphy and fossil flora of Western Kazakhstan. Moscow, Leningrad: Akad. nauk SSSR, 340 p. (in Russian).

Vakhrameev, V.A. 1981. Pollen of *Classopollis*: indicator of Jurassic and Cretaceous climates. - Palaeobotanist, vol. 28/29, p. 301-307.

Vakhrameev, V.A., Kotova, I.Z. 1977. Ancient angiosperms and their associated plants from the Lower Cretaceous of Transbaikalia. - Paleont. Zh. (Moscow), no. 4, p. 101-109 (in Russian).

Vakhrameev, V.A., Krassilov, V.A. 1979. Reproductive structures of angiosperms from the Albian of Kazakhstan. Paleont. Zh. (Moscow), no. 1, p. 121-128 (in Russian).

Van Campo, M. 1971. Précisions nouvelles sur les structures comparées des pollens de gymnospermes et d'angiospermes. - C. r. Hebd. séances Acad. sci., t. 272D, p. 2071-2074.

Van Leeuven, P., Punt, W., Hoen, P.P., 1988. Polygonaceae. - The Northwest Europaean Pollen Flora, vol. 43, p. 81-151.

Vane-Wright, R.I.S., Shulz, S., Boppre, M. 1992. The cladistics of *Amauris* batterflies: congruence, consensus and total evidence. - Cladistics, vol. 8. p. 125-138.

Vasanthy, G., Pocock, S.A.J. 1986. Radial through rotated symmetry of striate pollen of the Acanthaceae. - Canad. J. Bot., vol. 64, p. 3050-3058.

Vasil, I.K., Johri, M.M. 1964. The style, stigma and pollen tube, 1. - Phytomorphology, vol. 14, no.3, p. 352-369.

Vaudois, B. 1981. Archégone et sac embryonnaire (développement d'une conception récente). - Ann. Biol., t. 20, fasc. 2, p. 7-9.

Velenovsky, J. 1889. Kvetena Ceskego Cenomanu. - Kgl. Böhm. Ges. Wiss. Abh., Bd. 3, S. 1-75.

Velenovsky, J., Viniklar, L. 1926. Flora cretacea Bohemiae, 1. Praha, 57 p.

Vijayaraghavan, M.R. 1964. Morphology and embryology of a vesselless dicotyledon - *Sarcandra irvingbailleyi* Swamy, and systematic position of the Chloranthaceae. - Phytomorphology, vol. 42. no. 3, p. 429-441.

Vink, W. 1978. The Winteraceae of the Old World. 3. Notes on the ovary of *Takhtajania*. - Blumea, vol. 24, p. 521-525.

Vishnu-Mittre, 1953. A male flower of the Pentoxyleae with remarks on the structure of the female cones of the group. - Palaeobotanist, vol. 2, p. 75-84.

Van Konijnenburg-van Cittert, J.H.A. 1971. In situ gymnosperm pollen from the Middle Jurassic of Yorkshire. - Acta Bot. Neerl., vol. 20, p. 1-97.

Waha, M. 1985. Ultrastruktur und systematische Bedeutung des Pollens bei *Bocageopsis, Ephedranthus, Malmea* und *Unonopsis* (Annonaceae). - Plant. Syst. Evol., vol. 150, no. 3/4, p. 165-177.

Waha, M. 1987. Sporoderm development of pollen tetrads in *Asimina triloba* (Annonaceae). - Pollen et Spores, vol. 29, p. 31-44.

Walker, J.W. 1972. Chromosome number, phylogeny, phytogeography of the Annonaceae and their bearing on the (original) basic chromosome numbers of angiosperms. - Taxon, vol. 21, no. 1, p. 57-65.

Walker, J.W. 1974a. Evolution of exine structure in the pollen of primitive angiosperms. - Amer. J. Bot., vol. 61, no. 8, p. 891-902.

Walker, J.W. 1974b. Aperture evolution in the pollen of primitive angiosperms. - Amer. J. Bot., vol. 61, no. 10, p. 1112-1136.

Walker, J.W., Brenner, G.J., Walker, A.G. 1983. Winteraceous pollen in the Lower Cretaceous of Israel: early evidence of a magnolialean angiosperm family. - Science, vol. 220, no. 4603, p. 1273-1275.

Walker, J.W., Walker, A.G. 1984. Ultrastructure of Lower Cretaceous angiosperm pollen and the origin and early evolution of flowering plants. - Ann. Mo. Bot. Gard., vol. 71, no. 2, p. 464-521.

Walkom, A.B. 1919. Mesozoic floras of Queensland. The floras of the Barrum and Styx River Series. - Queensland Geol. Surv., vol. 263, p. 1-77.

Walton, J., Wilson, J.A.R. 1932. On the structure of *Vertebraria*. - Proc. Roy. Soc. Edinburgh., vol. 52, no. 2, p. 200-207.

Ward, J.V. 1986. Early Cretaceous angiosperm pollen from the Cheyenne and Kiowa Formation (Albian) of Kansas, U.S.A. - Palaeontographica B, Bd. 202, S. 1-81.

Ward, L.F. 1905. Status of the Mesozoic floras of the United States. - U.S. Geol. Surv. Monogr., vol. 48, pt. 1, p. 1-119.

Weller, S.G. 1976. Breeding system polymorphism in a heterostylous species. - Evolution, vol. 30, no. 3, p. 442-454.

Watson, J. A revision of the English Wealden Flora, 1. Charales - Ginkgoales. - Bull. Brit. Mus. (Nat. Hist.), Geol., vol. 17, p. 207-254.

Watson, J. 1982. The Cheirolepidiaceae: a short review. - Phyta, D.D. Pant Commemorial vol., p. 265-273.

Watson, J., Sincock, C.A. 1992. Bennettitales of the English Wealden. London: Palaeont. Soc. Publ., no. 388, p. 1-228.

Wettstein, R. 1907. Die Entwicklung der Blüte der angiospermen Pflanzen aus derjenigen der Gymnospermen. - Wissen für alle, H. 45, S. 705-708.

Wettstein, R. 1935. Handbuch der systematischen Botanik. 4. Ausg. Leipzig, Wien, 1152 S.

Wieland, G.R. 1906. American fossil cycads. - Carnegie Inst. Wash. Publ., vol. 34, p. 1-296.

Wiggins, V.D. 1982. *Expressipollis striatus* n. sp. to *Anacolosidites striatus* n. sp. An Upper Cretaceous example of suggested pollen aperture evolution. - Grana, vol. 21, p. 39-49.

Wilder, G.J. 1985. Anatomy of noncostal portion of lamina in the Cyclanthaceae (Monocotyledoneae), 1. Epidermis. - Bot. Gaz., vol. 146, no. 1, p. 82-105.

Wilson, C.L. 1937. The phylogeny of the stamen. - Amer. J. Bot., vol. 24, p. 686-699.

Wing, S.L., Tiffney, B.H. 1987. The reciprocal interaction of angiosperm evolution and tetrapod herbivory. - Rev. Palaeobot. Palynol., vol. 50, p. 179-210.

Wodehouse, R.P. 1935. Pollen grains, their structure, identification and significance in science and medicine. N.Y.: McGraw-Hill, 574 p.

Worsdell, W.C. 1900a. The structure of the female "flower" in Coniferae. A historical study. - Ann. Bot. (Gr. Brit.), vol. 14, p. 39-82.

Wu Xiang-wu, He Yuan-liang, Mei Sheng-wu 1986. Discovery of *Ephedrites* from the Lower Jurassic Xiaomeigou Formation, Qinghai. - Acta Palaeobot. Palynol. Sinica, vol. 1, 19-28.

Yakovlev, M.S. 1973.The main types of embryonic development in higher plants. - In: Embryology of higher plants. Kishinev: Shtiintsa, p. 16-26 (in Russian).

Zavada, M. 1983. Comparative morphology of monocot pollen and evolution trends of apertures and wall structures. - Bot. Rev., vol. 49, no. 4, p. 331-379.

Zavada, M.S. 1984a. Angiosperm origins and evolution based on dispersed fossil pollen ultrastructure. - Ann. Mo. Bot. Gard., vol. 71, no. 2, p. 444-463.

Zavada, M.S. 1984b. The relation between pollen exine sculpturing and self-incompatibility mechanisms. - Plant Syst. Evol., vol. 147, nos. 1/2, p. 63-78.

Zavada, M.S. 1991. The ultrastructure of pollen found in the dispersed sporangia of *Arberiella* (Glossopteridaceae). - Bot. Gaz., vol. 152, p. 248 - 255.

Zavada, M.S., Crepet, W.L. 1986. Pollen grain wall structure of *Caytonanthus arberi* (Caytoniales). - Plant Syst. Evol., vol. 153, nos. 3/4, p. 259-264.

Zavada, M.S., Dilcher, D.L. 1986. Comparative pollen morphology and its relationship to phylogeny of pollen in Hamamelidae. - Ann. Mo. Bot. Gard., vol. 73, no. 2, p. 348-381.

Zavada, M.S., Taylor, T.N. 1986a. The role of selfincompatibility and sexual selection in the gymnosperm-angiosperm transition: a hypothesis. - Amer. Natur., vol. 128, no. 4, p. 538-548.

Zavada, M.S., Taylor, T.N. 1986b. Pollen morphology of Lactoridaceae. - Plant Syst. Evol., vol. 154, p. 31-39.

Zherikhin, V.V. 1979. Evolution and replacements of Cretaceous and Cenozoic faunistic assemblages. Moscow: Nauka, 198 p. (in Russian).

Zhou Zhiyan (Chow Tseyan) 1995. On some Cretaceous pseudofrenelopsids with a brief review of cheirolepidiaceous conifers in China. - Rev. Palaeobot. Palynol., vol. 84, p.419-438.

Zimmermann, W. 1930. Die Phylogenie der Pflanzen. Jena: G. Fischer, 452 S.

Zimmermann, W. 1952. Main results of the “telome theory”. - Palaeobotanist, vol. 1, p. 456-470.

Index

A

B

C

D

E

F

G

N

O

P

In 1997 appears:

Fungi Bulgaricae. Tomus 3. Ordo Spaeropsidales. Pars 1. Anamorphae cum conidorum septorum.

by S. G. Vanev, E. F. Sameva & G. G. Bakalova

ISBN 954-430-234-4 (vol. 3, pt.1) (Academic Publishing House)
ISBN 954-642-026-3 (vol. 3, pt.1) (Pensoft)

Marketed and distributed beyond Bulgaria by Pensoft.

Format 165x235, softcover, appr. 300 pp., 15 tables with figs. Bilingual, main text in Bulgarian, *all keys, distributional and ecological data translated into English* (appr. 50 pages English summary), all figure captions with Latin species names. Publication date: August 1997.

The present volume of *Fungi Bulgaricae* summarizes data on one of the taxonomically most difficult groups, from a region which has never been reviewed so far. In total, 444 anamorphic species belonging to 52 genera and 15 anamorphic genera are considered. Fourteen species belonging to the genus *Discosia* are described as new to the science. The volume includes keys for determination, descriptions of the most important morphological features of the taxa, symptoms of diseases caused by parasitic species on the host plants, lists of the hosts, a distribution in Bulgaria, a general geographic distribution, a systematic guide to the taxa, alphabetical index of both the fungi and the hosts. The keys and figures are mostly original, *all translated into English and hence fully usable to the Western reader*, which in itself is a precedent within *Fungi Bulgaricae.* The species descriptions are made on the basis of materials originating from several herbaria. Richly illustrated, contains also an extensive bibliography from the area considered. The monograph is invaluable not only to mycologists and botanists, but also to specialists in phytopathology and quarantine officers.

Contents include: FORWARD • GENERAL PART: A review on the investigations on the pycnidial fungi - Principles of classification and taxonomic criteria - Morphological features of the pycnidial fungi and their taxonomic value - Economic importance • SPECIAL PART: Systematic guide to the taxa - Keys for determinations of anamorphic genera - Keys for determination of species of the genera *Apiocarpella, Ascochyta, Camarosporium, Darluca, Dichomera, Dilophospora, Diplodia, Diplodina, Discosia, Fuckelia, Heteropatella, Septocyta, Septoria, Stagonospora, Wojnowicia* • APPENDICES: Tables of figures - Bulgarian-Latin dictionary of the botanical terms - A list of abbreviations of the authors' names - Alphabetical index of the host plants - Alphabetical index of the spaeropsidal fungi - English translation of the keys, the distributional, ecological and pathological data • LITERATURE.